SYSTEMS OF FREQUENCY DISTRIBUTIONS FOR WATER AND ENVIRONMENTAL ENGINEERING

A multitude of processes in hydrology and environmental engineering are either random or entail random components that are characterized by random variables. These variables are described by frequency distributions. This book provides an overview of different systems of frequency distributions, their properties, and their applications to the fields of water resources and environmental engineering. A variety of systems are covered, including the Pearson system, the Burr system, and systems commonly applied in economics, such as the D'Addario, Dagum, Stoppa, and Esteban systems. The latter chapters focus on the Singh system and the frequency distributions deduced from Bessel functions, maximum entropy theory, and the transformations of random variables. The final chapter introduces the genetic theory of frequency distributions. Using real-world data, this book provides a valuable reference for researchers, graduate students, and professionals interested in frequency analysis.

VIJAY P. SINGH is a University Distinguished Professor, a Regents Professor, and Caroline and William N. Lehrer Distinguished Chair in Water Engineering at Texas A&M University. He has published more than 1,220 journal articles, 30 books, 70 edited reference books, 113 book chapters, and 314 conference papers in the areas of hydrology, groundwater, hydraulics, irrigation, pollutant transport, copulas, entropy, and water resources. He has received 95 national and international awards, including the Arid Lands Hydraulic Engineering Award; the Richard R. Torrens Award; the Norman Medal; the EWRI Lifetime Achievement Award given by the American Society of Civil Engineers (ASCE); the Ray K. Linsley Award and Founder's Award, given by the American Institute of Hydrology; the Crystal Drop Award and the Ven Te Chow Award, given by the International Water Resources Association; and three honorary doctorates. He is a Distinguished Member of ASCE, and a fellow of EWRI, AWRA, IWRS, ISAE, IASWC, and IE. He has served as President of the American Institute of Hydrology (AIH), serves on the editorial boards of more than 25 journals and 3 book series, and is President-Elect of AAWRE-ASCE.

LAN ZHANG currently works as Post-Doc Research Scholar in the Department of Agricultural and Biological Engineering at Texas A&M University. She received her BS in mechanical engineering from Dalian Polytechnic University, her MS in water resources sciences from Beijing Normal University, and her PhD in civil and environmental engineering from Louisiana State University. She has published more than 40 articles in the areas of hydrology, copulas, water quality, entropy, and water resources. She has been working on the copula and its applications in hydrology and water resource engineering for more than 10 years.

SYSTEMS OF FREQUENCY DISTRIBUTIONS FOR WATER AND ENVIRONMENTAL ENGINEERING

VIJAY P. SINGH
Texas A&M University

LAN ZHANG
Texas A&M University

CAMBRIDGE
UNIVERSITY PRESS

University Printing House, Cambridge CB2 8BS, United Kingdom

One Liberty Plaza, 20th Floor, New York, NY 10006, USA

477 Williamstown Road, Port Melbourne, VIC 3207, Australia

314–321, 3rd Floor, Plot 3, Splendor Forum, Jasola District Centre, New Delhi – 110025, India

79 Anson Road, #06–04/06, Singapore 079906

Cambridge University Press is part of the University of Cambridge.

It furthers the University's mission by disseminating knowledge in the pursuit of education, learning, and research at the highest international levels of excellence.

www.cambridge.org
Information on this title: www.cambridge.org/9781108494649
DOI: 10.1017/9781108859530

First published 2020

A catalogue record for this publication is available from the British Library.

ISBN 978-1-108-49464-9 Hardback

VPS: wife Anita, son Vinay, daughter Arti, daughter-in-law Sonali, son-in-law Vamsi, and grandchildren Ronin, Kayden, and Davin

LZ: husband Bret Rath and son Caelan

Contents

Preface

A multitude of processes in hydrometeorology, hydrology, geohydrology, hydraulics, and environmental and water resources engineering are either random or entail random components that are characterized by random variables. These variables are described by frequency distributions that encompass a broad range. In textbooks on statistical methods in hydrology and hydraulics, frequency distributions occupy a prominent place but there is seldom a discussion on where the distributions come from or how these distributions are derived. Statistical literature shows that there are different systems or families of distributions and the distributions used in hydrology and water and environmental engineering originate from one or the other of these systems. Understanding the origination of the distributions helps uncover their underlying hypotheses and may help estimate their parameters and make informed inferences. Currently, there does not appear to be a book covering these systems. This is what constituted the motivation for this book.

The subject matter of the book is divided into 12 chapters. Introducing the theme of the book in Chapter 1, the Pearson system is discussed in Chapter 2. This is the first system that was introduced almost a century and a quarter ago and can be considered as a foundational system, for the differential equation proposed for the system laid the seeds for some other systems. The Pearson system comprises 12 distributions some of which are frequently employed in environmental and water engineering. Each of these distributions is derived and estimation of their parameters is discussed.

Chapter 3 discusses the Burr system, which consists of a set of 12 distributions that exhibit different characteristics and some of these distributions are commonly used in environmental and water engineering. This system employs a hypothesis that relates the probability density function to the cumulative distribution function and its complement. Each distribution of the system is derived in the chapter and a method of parameter estimation is presented. An analogy is drawn between this system and the Pearson system.

Chapter 4 presents the D'Addario system, which is comprised of six distributions that result from the integration of a probability-generating function and a transformation function. Examples of the distributions include Pareto type I, Pareto type II, lognormal type I, lognormal type II, Amoroso, and Davis distributions. The Amoroso distribution leads to 11 special cases. These distributions are derived in the chapter.

The subject matter of Chapter 5 is the Dagum system, which consists of a set of 11 frequency distributions, some of which are commonly employed in water engineering. The system is based on a hypothesis for the elasticity of the cumulative distribution function. A set of logical-empirical postulates, including parsimony, interpretation of parameters, efficiency of parameter estimation, model flexibility, goodness of fit, ease of computation, and algebraic manipulation that should be used for deriving distributions are also discussed.

The Stoppa system is discussed in Chapter 6. This system employs the elasticity of the cumulative distribution function and a differential equation. Its special cases consist of generalized power distribution, generalized exponential distribution, generalized Pareto distribution, and different Stoppa distributions. The generalized distributions consist of several distributions as special cases. The Stoppa system is a generalized system of distributions and is closely related to the Dagum system. Also, several Burr distributions can be derived from the Stoppa or the Dagum system.

Chapter 7 deals with the Esteban system, which uses a slightly different definition of distribution elasticity. The system comprises generalized gamma distribution, generalized beta distribution of first kind, and generalized beta distribution of second kind. These distributions include as special or limiting cases a wide spectrum of frequency distributions used in hydrologic, hydraulic, environmental, and water resources engineering.

The subject matter of Chapter 8 is the Singh system, which may be considered as the generalized Burr and Stoppa system. The system employs certain hypotheses on the relation between probability density function (PDF) and cumulative distribution function (CDF), based on empirical data. A large number of distributions can be derived from this system. The chapter discusses the derivation of CDFs of these distributions.

Chapter 9 deals with frequency distributions that are derived using Bessel functions and cumulants. Beginning with a discussion of Bessel function distributions, including moments of distributions, Bessel function line, inverse Gaussian distribution, and other distributions, the chapter goes on to discuss frequency distributions by series approximations, including Chebyshev-Hermite polynomials, cumulants, series approximation of a frequency distribution with Gram-Charlier type A series, Edgeworth series with baseline Gaussian distribution, and Gram-Charlier/Edgeworth series with non-Gaussian distribution.

Chapter 10 employs the principle of maximum entropy. Entropy maximization provides a general framework for deriving any probability distribution subject to appropriate constraints. This chapter discusses this framework and derives a number of distributions that satisfy different constraints.

A wide spectrum of frequency distributions, used in hydrologic, hydraulic, environmental, and water resources engineering, are derived using transformations of some basic frequency distributions. The basic distributions that have been used are normal, logistic, beta, Laplace, and other distributions, and the transformations used are logarithmic, power, and exponential. Chapter 11 derives the distributions obtained by transformation and transformations applied to basic distributions.

The concluding Chapter 12 deals with the genetic theory of frequency distributions. Starting with the basic concept of elementary errors, the chapter discusses Charlier type A curve and Charlier type B curve. Then, it delves into the extensions that lead to different frequency distributions.

This book is meant for graduate students and faculty members who are in the fields of hydrology, hydraulics, geohydrology, water quality engineering, hydrometeorology, environmental engineering, and water resources engineering. It can also be used as a reference book for courses on statistical methods in environmental and water engineering. It is hoped that the book will help people understand frequency distributions and their application.

Acknowledgments

The subject matter of this book is based on the literature on mathematical statistics, econometrics, and environmental and water engineering. There are scores of researchers who have made seminal contributions in the area of frequency distributions without which the book would not have been possible. It is a pleasure to acknowledge their contributions as specifically as possible in the body of the text and any omission on our part has been entirely inadvertent and we offer our apologies in advance. We would be greatly benefitted and be obliged if the readers transmitted to us any errors, omissions, or criticisms.

We acknowledge our families for their love, support, sacrifices, patience, encouragement, and endurance. Without them, our lives would be less than meaningful. Vijay Singh acknowledges his wife Anita for her help in countless ways and allowing him to do his academic and professional work. Without her sense of caring and sacrifice each day it would have been impossible for him to accomplish what he does. His son Vinay is a bedrock of stability in the family and offers help and support without seeking any acknowledgment. His daughter Arti symbolizes caring, affection, and warmth. It is difficult to think of a more loving and caring daughter. His daughter-in-law Sonali symbolizes togetherness and unity and infectious smile. He is blessed to have her as his daughter-in-law. His son-in-law Vamsi represents calmness, coolness, and concurrence, which are rare these days. His grandchildren Ronin, Kayden, and Davin bring joy and meaning to his life and represent his hope and future. He is grateful to all of them for what they are and what they bring to him. This book is humbly dedicated to them and is a small acknowledgment of what they mean to him. Lan Zhang acknowledges her husband Brett and son Caelan for their love and support.

1

Introduction

1.1 Random Variables in Environmental and Water Engineering

A multitude of hydrometeorologic, hydrologic, geohydrologic, hydraulic, and environmental processes are either random or entail random components that are characterized by random variables. These variables are described by frequency distributions that encompass a broad range. Now the processes and their random components are briefly discussed.

1.1.1 Rainfall

Rainfall is perhaps the most important component of the hydrologic cycle and constitutes input to hydrologic models. The rainfall process is governed by atmospheric processes that result from the interactions between atmosphere, hydrosphere, and the Earth system. The Earth system encompasses the land surface, pedosphere, lithosphere, cryosphere, and anthropogenic influences. The rainfall process is highly heterogeneous in space and time and is characterized, from a hydrologic viewpoint, by intensity, duration, depth of rainfall at a point or over an area; time interval between rainfall events; number of rainfall events in a given time, say, a month, season, or year; extreme rainfall in a year; and areal coverage. These characteristics are random in nature and are therefore described by frequency distributions.

Rainfall depth is often described by an exponential or Weibull distribution. Rainfall duration may be described by an exponential distribution or the distributions skewed to the right. It may be noted that rainfall intensity is given by rainfall depth divided by duration so it may also be described by the same frequency distribution as is for rainfall depth. The number of rainfall events is described by the Poisson or gamma distribution. Extreme rainfall is described by the lognormal, Pearson type III, extreme value type I, or generalized extreme value distribution.

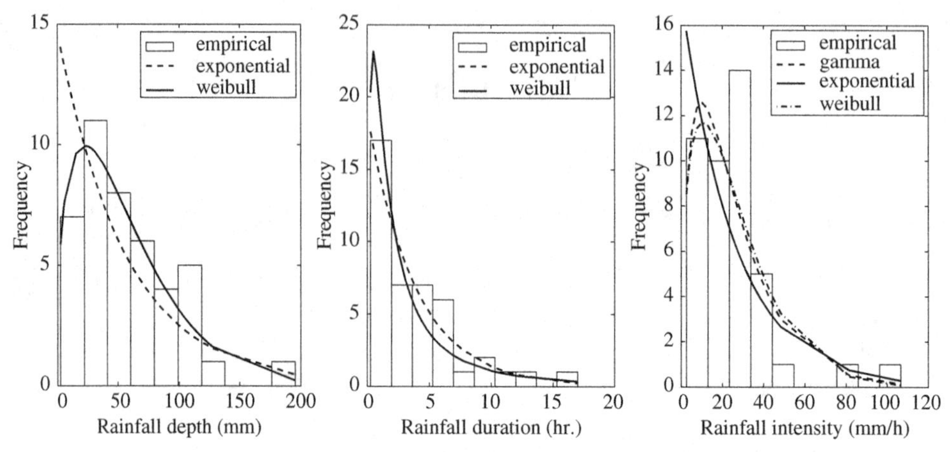

Figure 1.1 Histograms and probability density function plots for rainfall variables, including rainfall depth, duration, and intensity.

Using rainfall depth, rainfall duration and rainfall intensity sample data, Figure 1.1 plots the histogram and the probability density functions estimated from different distribution candidates. It is seen that one may choose different probability distribution functions to model the same set of rainfall variables.

1.1.2 Temperature

Although temperature is a direct component of the hydrologic cycle, it has a decided impact on evapotranspiration, soil moisture depletion, groundwater recharge, vegetation and crop growth, snow and ice melt, freezing and thawing, bacterial and viral growth and spread, and human and ecosystem health. Although temperature field is relatively stable, its maximum and minimum values in a given time, say, month or year; number of days with maximum temperature or maximum temperature; and the starting day of the maximum or minimum temperature are random variables. The maximum temperature is often assumed to follow the generalized extreme value distribution; the minimum temperature is described by extreme value type III or Weibull distribution. Using the monthly temperature as an example, Figure 1.2 graphs the frequency histogram, lognormal, and normal density functions. The frequency histogram indicates two modes. The lognormal and normal distribution might be considered as the two possible candidates without taking the mixture distribution into consideration. In addition, using the number of warm days (i.e., the days warmer than normal during any given month) as an example, Figure 1.3 graphs the frequency histogram and fitted generalized extreme value distribution.

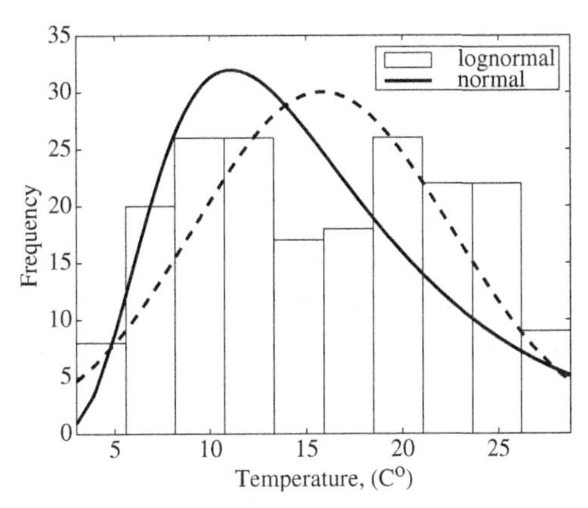

Figure 1.2 Histogram and probability density function plots for monthly temperature.

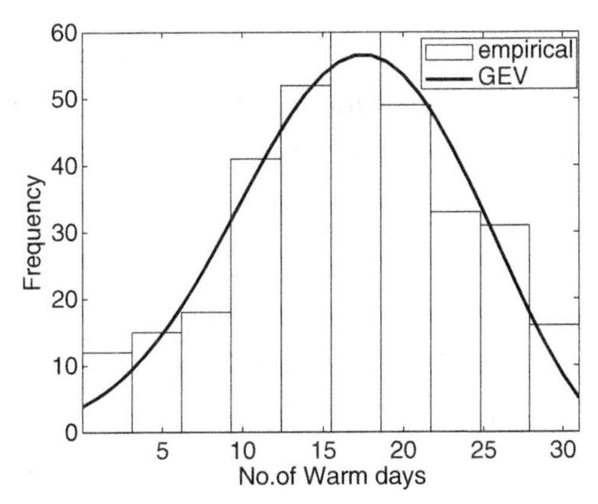

Figure 1.3 Histogram and probability density function plot for number of warm days.

1.1.3 Frost, Fog, and Sunshine Hours

Although sunshine hours at a particular location are relatively stable, they are not the same from year to year. The number of sunshine hours during a given period at a given place can be a random variable and might be described by the normal distribution.

Frost occurs during cold days and has important implications for agriculture. For example, mustard, when it is flowering, is severely impacted by frost persisting for a week or more. The number of frost days (annually) may be considered as random variable and might be described by the normal distribution.

Fog occurs where there is lack of sunshine and has important implications for agriculture, transportation systems, and outdoor recreational systems. The annual number of foggy days may be considered as a random variable and might be described by the Poisson distribution.

1.1.4 Wind

Wind is a regular part of weather. However, extreme winds are becoming too frequent these days. Winds are described by velocity, duration, and direction. Weibull and Rayleigh distribution have been widely applied to model wind velocity (Seguro and Lambert, 2000; Jowder, 2006; Bilir et al., 2015; Pishgar-Komleh et al., 2015; among others). Rayleigh distribution has been found to model wind direction (McWilliams et al., 1979).

1.1.5 Snowfall

In cold regions, snowfall is an important component of the hydrologic cycle. It is a highly heterogeneous process, varying in space and time. The depth of snowfall, duration of snowfall, number of snowfall events in a season, time interval between two snowfall events, and extreme snowfall event in a year are random variables and can be described by frequency distributions. For example, gamma distribution can be used to model extreme snowfall.

1.1.6 Runoff

When it rains, runoff is generated. On urban and small watersheds, overland flow or surface runoff is dominant. Corresponding to rainfall events, runoff events can be defined by their amount, peak value, time to peak, and duration. Although for a given runoff event, these characteristics can be determined deterministically using simple hydrologic models. However, when each characteristic is considered for a duration of time, say a year or several years, then the sample of values of the characteristic can be considered as a random sample and that runoff characteristic might be described by the frequency distributions that are commonly applied to rainfall variables discussed in the preceding text.

1.1.7 Flood

Floods are a common feature in almost every country and cause damage worth billions of dollars, disrupt life and transportation, and cause untold misery almost each year in one part of the world or the other. In hydrology, floods are described by extreme streamflow hydrographs that are characterized by peak, time to peak, volume, duration, and interarrival time. Floods are a random phenomenon when

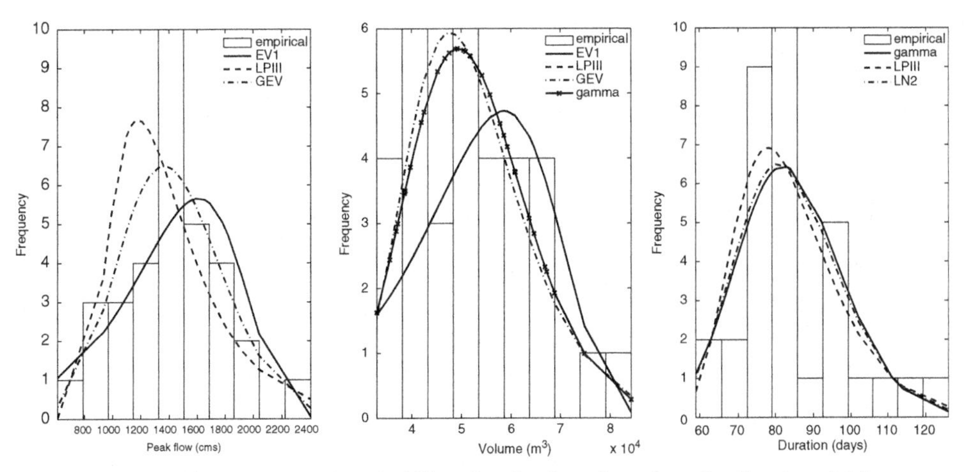

Figure 1.4 Histograms and probability density function plots for flood variables.

maximum from each year or maximum floods above a selected threshold are considered. In that case, the peak value is the instantaneous yearly maximum streamflow. Because of their ubiquitous application to hydraulic design, frequency distributions of floods have been most extensively investigated. Yearly maximum streamflow is often described by the Pearson type III (PIII) distribution in China, log-Pearson type III (LPIII) distribution in the United States, generalized extreme value distribution (GEV) in Europe, extreme value type I (EVI) distribution in certain Asian countries, and 3-parameter log-normal distribution. The interarrival time is described by the exponential distribution, and streamflow volume and duration by the gamma distribution. Figure 1.4 plots the histogram and potential probability distribution candidates for flood variables, including peak flow, flood volume, and flood duration. It is seen that for this dataset, one may choose the GEV distribution to model peak flow. In the case of flood volume, LPIII, EV1, and gamma distributions yield very similar performances. These three distributions (i.e., gamma, LPIII, and LN2) yield very similar performances for flood duration.

1.1.8 Drought

Drought is a creeping phenomenon and impacts all walks of life. Each year, one or the other country is severely impacted by drought, and losses caused as a result are in billions of dollars. There are different types of droughts, such as hydrometeorologic, hydrologic, agricultural, groundwater, and socioeconomic. Hydrometeorologic drought is characterized by precipitation deficit, hydrologic drought by streamflow deficit, agricultural drought by soil moisture deficit, groundwater drought by lowering of water table, and socioeconomic drought by social and financial distress.

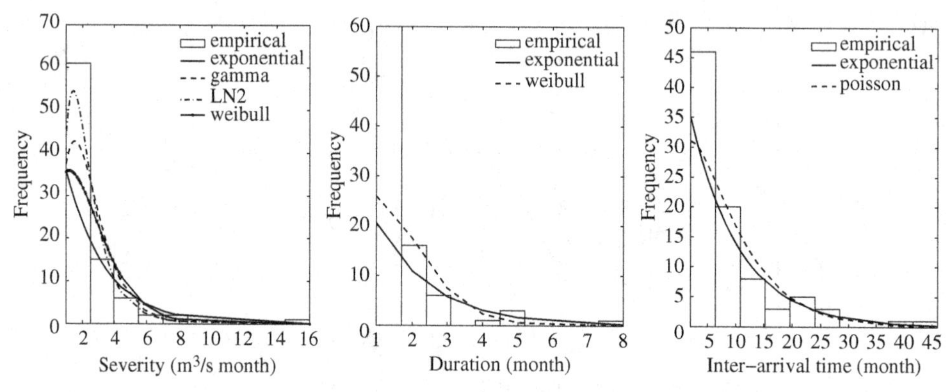

Figure 1.5 Histograms and probability density function plots for drought variables.

Unlike flood, there is no universal definition of drought. For example, drought severity has a relative meaning that may vary from place to place. Nevertheless, regardless of type, drought can be characterized by severity, duration, spatial extent, and interarrival time. All these characteristics are random variables and can be described by frequency distributions. Examples are distributions skewed to the right. Figure 1.5 plots the frequency histogram and probability density functions for drought variables, including drought severity, drought duration, and interarrival time. It is seen that drought severity may be modeled with one of the four distributions, including exponential, gamma, lognormal, and Weibull distributions; Weibull and exponential distributions yield similar performances to model drought duration; and both exponential and Poisson distributions may applied to model drought interarrival time.

1.1.9 Hydrogeology

Hydrogeologic processes are considered as deterministic, for they follow the laws of geophysics. However, there are certain characteristics that may exhibit randomness and may therefore be described by frequency distributions. For example, the distribution of hydraulic conductivity of a geologic formation, such as an aquifer, over space can be considered as a random variable and is often described by the lognormal distribution. Likewise, the average thickness of aquifers can be considered as a random variable that may follow the normal distribution.

1.1.10 Water Quality

Water quality is of fundamental importance in environmental and water engineering and is characterized by a number of parameters. Although these parameters are determined deterministically, they constitute random samples when assembled

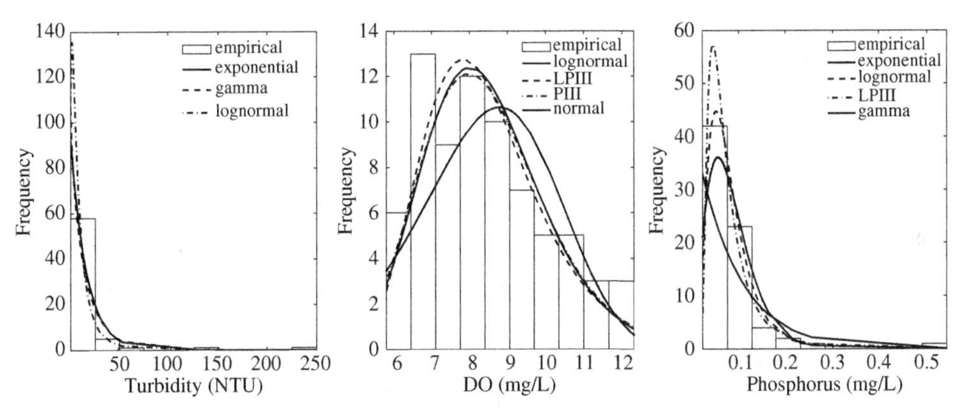

Figure 1.6 Histograms and probability density plots for water quality variables.

over a period, say a few years, or over space, say along a river reach. The sample values can then be described by frequency distributions. In many cases, monthly or yearly water quality variables are skewed to the right. Figure 1.6 plots the histogram and probability density function candidates for monthly water quality variables, including turbidity, dissolved oxygen (DO), and phosphorus. It is seen that the frequency distributions of turbidity and phosphorus are clearly skewed to the right. The right skewed exponential, gamma, LPIII, and lognormal distributions may be applied as probability density functions for turbidity and phosphorus. With the unique characteristics DO, lognormal, LPIII, and PIII, distributions yield similar performances and may be applied as probability density functions for the DO variable. Comparison of the histogram and normal density function indicates that the frequency distribution of DO is still right skewed.

1.2 Systems of Frequency Distributions

It is clear from the preceding discussion that a wide spectrum of frequency distributions are used in hydrologic, hydrogeologic, hydraulic, environmental, and water resources engineering. These frequency distributions employ different hypotheses and have different origins or are derived from different generating systems or families. For example, lognormal, gamma, Pearson and log-Pearson type III, extreme value type I, generalized extreme value, Burr XII, log-logistic, and Pareto distributions are used for flood frequency analysis but do not originate from the same systems or families. Likewise, exponential and Weibull distributions are used for rainfall frequency analysis; 2- and 3-parameter log-logistic distributions are used for analysis of streamflow data, precipitation data, and hazard data; Pareto distributions are used for modeling large exceedances; and Weibull distribution are used for frequencies of low flows and reliability analysis belong to different generating systems. There are different systems from which these frequency distributions are derived and there are different systems of classifications of distributions that are briefly outlined here.

1.2.1 Stoppa System

Stoppa (1990) classified frequency distributions into three groups. The first group comprises three distributions, including Champernowne (1952) distribution, Fisk (1961) distribution, and power distribution (Mandelbrot, 1960). These distributions are derived from probabilistic arguments. The second group of distributions includes gamma distribution (Salem and Mount, 1974), beta distribution (Thurow, 1970), generalized beta distribution (McDonald, 1984), and 3-parameter lognormal distribution (Chieppa and Amato, 1981). These distributions are based on the goodness-of-fit to the observed empirical data that is the common practice in hydrology. However, it is not clear what the hypotheses are used for deriving these distributions. The third group consists of the Pearson system (1895), Burr system (1942), Dagum system, Stoppa system (1990), D'Addario system (1949), and Pareto system (1897), which are derived from differential equations.

1.2.2 Dagum System

Dagum (1990, 2006) recognized three generating systems for classifying frequency distributions: (1) Pearson system (1894, 1895), (2) D'Addario (1949) system, and (3) Dagum system (1980a, 1980b, 1983). Dagum (1977) presented a set of logical-empirical postulates for deriving a particular distribution, which include parsimony, interpretation of parameters, efficiency of parameter estimation, model flexibility, and goodness of fit. Ease of computation and algebraic manipulation may also be added to this set.

1.2.3 Johnson System

Johnson (1949) classified distributions into five groups: (1) Pearson system, (2) expansions, (3) transformed distributions, (4) Bessel function distributions, and (5) miscellaneous. Some distributions may belong to more than one category, while some distributions may not belong to any category. Theoretical arguments have been advanced to justify a particular system of distributions.

1.2.4 General Classification

Singh (2018) suggested nine systems of frequency distributions based on different approaches that seem to have been employed: (1) differential equation involving probability density functions, (2) differential equation involving cumulative distribution function, (3) distribution elasticity, (4) generating functions, (5) genetic theory, (6) Bessel functions, (7) expansions, (8) transformations, and (9) entropy maximization. The frequency distributions resulting from these systems include virtually all the distributions that are known and used in hydrometeorology, hydrology, geohydrology, hydraulics, and environmental engineering.

1.3 Need for Systems of Frequency Distributions

A multitude of frequency distributions have been applied in environmental and water engineering. However, it is not always clear in engineering how different distributions have been derived and what their underlying hypotheses are. It will be interesting to understand these hypotheses that may help select an appropriate distribution for a given set of data. Also, the distribution parameters can be estimated in a more meaningful manner. If different systems of deriving the distributions can be established, then it may be plausible to establish their connections. It may then be possible to develop a universal system from which all distributions or at least a great number of them can be derived. Thus, a study of these systems is of both theoretical and practical interest.

1.4 Organization of the Book

Introducing the theme of the book in the introductory chapter, the subject matter of the book is divided into 10 frequency distribution systems and each system is dealt with in a separate chapter. Each system leads to a number of distributions each of which is derived individually. The derivation sheds light on the hypotheses underlying a given frequency distribution.

References

Bilir, L., Imir, M., Devrim, Y., and Albostan, A. (2015). Seasonal and yearly wind speed distribution and wind power density analysis based on Weibull distribution function. *International Journal of Hydrogen Energy* 40, pp. 15301–15310.

Burr, I. W. (1942). Cumulative frequency functions. *Annals of Mathematical Statistics* 13, pp. 215–232.

Champernowne, D. (1952). The graduation of income distribution. *Econometrica* 20, pp. 591–625.

Chieppa, M., and Amato, P. (1981). A new estimation procedure for the three-parameter lognormal distribution. In *Statistical Distributions Scientific Work*, edited by C. Tsallis et al., Vol. 5, pp. 133–140.

D'Addario, R. (1949). Richerche sulla curva dei redditi. *Giornale Degli Economisti e Annali di Economia* 8, pp. 91–114.

Dagum, C. (1977). A new model of personal income distribution: Specification and estimation. *Economic Appliquee* 30, no. 3, pp. 413–436.
Dagum, C. (1980a). The generation and distribution of income, the Lorenz curve and Ginni ratio. *Economic Appliquee* 33, pp. 327–367.
Dagum, C. (1980b). Generating systems and properties of income distribution models. *Metron* 38, pp. 3–26.
Dagum, C. (1983). Income distribution models. In *Encyclopedia of Statistical Sciences*, Vol. 4, edited by S. Kotz, N. L. Johnson, and C. Read, pp. 27–34. New York: Wiley.
Dagum, C. (1990). Generation and properties of income distribution functions. In *Income and Wealth Distribution, Inequality and Poverty*, edited by C. Dagum and M. Zenga, pp. 1–17. Berlin: Springer.
Fisk, P. (1961). The graduation of income distributions. *Econometrica* 29, pp. 171–185.
Johnson, N. L. (1949). Systems of frequency curves generated by methods of translation. *Biometrika* 36, pp. 147–176.
Jowder, F. (2006). Weibull and Rayleigh distribution functions of wind speeds in Kingdom of Bahrain. *Wind Energy* 30, pp. 439–445.
Mandelbrot, B. (1960). The Pareto-Levy law and the distribution of income. *International Economic Review* 1, pp. 79–106.
McDonald, J. B. (1984). Some generalized functions for the size distribution of income. *Econometrics* 52, no. 3, pp. 647–663.
McWilliams, B., Newmann, M. M., and Sprevak, D. (1979). The probability distribution of wind velocity and direction. *Wind Engineering* 3, no. 4, pp. 269–273.
Pareto, V. (1897). *Cours d'Economie Politique*. New edition by G. H. Bousquet and G. Busino (1964). Geneva, Switzerland: Librairie Droz.
Pearson, K. (1894). Contributions to the mathematical theory of evolution. *Philosophical Transactions of the Royal Society* 184, pp. 71–110.
Pearson, K. (1895). Contributions to the mathematical theory of evolution. II. Skew variation in homogeneous material. *Transactions of the Royal Society London* A186, pp. 343–415.
Pishgar-Komleth, S. H., Keyhani, A., and Sefeedpari, P. (2015). Wind speed and power density analysis based on Weibull and Rayleigh distributions (a case study: Firouzkooh county of Iran). *Renewable and Sustainable Energy Reviews* 42, pp. 313–322. doi: 10.1016/j.rser.2014.10.028.
Salem, A., and Mount, T. (1974). A convenient descriptive model of income distribution: The gamma density. *Econometrica* 42, pp. 1115–1127.
Seguro, J. V., and Lambert, T. W. (2000). Modern estimation of the parameters of the Weibull wind speed distribution for wind energy analysis. *Journal of Wind Engineering and Industrial Aerodynamics* 85, no. 1, pp. 75–84.
Singh, V. P. (2018). Systems of frequency distributions for water and environmental engineering. *Physica A* 506, pp. 50–74.
Stoppa, G. (1990). A new generating system of income distribution models. *Quarderni di Statistica e Mathematica Applicata alle Science Economico-Sociali* 12, pp. 47–55.
Thurow, L. C. (1970). Analyzing the American income distribution. *American Economic Review* 48, pp. 261–269.

2

Pearson System of Frequency Distributions

2.1 Introduction

A wide spectrum of frequency distributions is used in hydrologic, hydraulic, environmental, and water resources engineering. Of these distributions, normal, exponential, gamma, and Pearson type III distributions, which are members of the Pearson system, are used frequently. For example, gamma, Pearson type III, and log-Pearson type III are used for flood frequency analysis; exponential distribution is used for rainfall frequency analysis; and normal distribution is used as an error distribution, in drought modeling, in least squares analysis, and in a wide variety of situations. It is therefore important to understand the Pearson system and its underlying hypothesis and derivatives.

2.2 Differential Equation of Pearson System

Pearson (1894, 1895, 1916) proposed a differential equation that satisfies certain statistical conditions. By solving the differential equation, he derived a system of 12 frequency distributions that together are popularly known as the Pearson system or Pearson family. Amongst these distributions, Pearson type III and its logarithmic version are two of the most commonly used distributions in hydrology and water resources engineering. For the Pearson system, the differential equation is expressed with the probability density function (PDF) as the dependent variable. Some of the distributions arising from the differential equation do not have closed form of the cumulative probability distribution function (CDF).

The differential equation proposed by Pearson (1895) is of the form:

$$\frac{1}{f(x)}\frac{df(x)}{dx} = \frac{a - x}{c_1 + c_2 x + c_3 x^2} \tag{2.1}$$

where $f(x)$ is the PDF of random variable X whose specific value is x; and a, c_1, c_2, and c_3 are constants that determine the type of solution and the consequent form of the distribution. Equation (2.1) can be expressed simply as product of two functions, $g(x)$ and $f(x)$, as

$$\frac{df(x)}{dx} = g(x)f(x), g(x) = \frac{a-x}{c_1 + c_2x + c_3x^2} \tag{2.2}$$

Equation (2.1) is based on the hypothesis that the causes contributing to the frequency distribution are not independent and are not likely to produce equal deviations. From Equation (2.1) it can be seen that if $f(x) = 0$, then $\frac{df(x)}{dx} = 0$. If $x = a$, we have: (a) $\frac{df(x)}{dx} = 0, f(x) = \max\ (f(x))$, i.e., a is the mode for $f(x)$; and (b) $\frac{df(x)}{dx} = 0, f(x) = \min\ (f(x))$, and function $f(x)$ is of U-form.

More specifically, if $c_2 = 0$ and $c_3 = 0$, then Equation (2.1) can be rewritten as

$$\frac{1}{f(x)}\frac{df(x)}{dx} = \frac{a-x}{c_1} \tag{2.3}$$

Solution of Equation (2.3) can be written as

$$f(x) = A\exp\left(-\frac{(a-x)^2}{2c_1}\right), \int_{-\infty}^{\infty} f(x)\, dx = 1 \tag{2.4}$$

where A is a constant chosen such that the area under $f(x)$ is unity. Equation (2.4) shows that c_1 must be positive and if $A = \sqrt{2\pi c_1}$, then the corresponding distribution is normal with the expected value as a and standard deviation as $\sqrt{c_1}$. This shows that Equation (2.1) is a generalization of the differential Equation (2.3), which yields the normal distribution. The normal distribution is a limiting case of all the distributions of the Pearson system.

Solution of Equation (2.1) is dependent on the roots of its denominator:

$$c_1 + c_2x + c_3x^2 = 0 \tag{2.5}$$

Let the roots of Equation (2.5) be a_1, a_2; then the form of $f(x)$ may be solved in general as:

$$f(x) = C(x - a_1)^{m_1}(x - a_2)^{m_2} \tag{2.6}$$

where C is the constant of integration, and m_1 and m_2 are the corresponding exponents related to the roots.

The roots of Equation (2.5) can be: (i) real and of opposite signs, (ii) real and of the same sign, and (iii) complex. These three type of roots lead to three main cases designated as type I, type VI, and type IV, which can be deduced from the discriminant $\alpha = c_2^2 - 4c_1c_3$ or the criterion $(k : k = c_2^2/4c_1c_3)$ (Elderton and Johnson, 1969).

For the three main types, type I (real roots of opposite sign) has limited range in both directions and skewness, which is the beta distribution of the first kind; type VI (real roots of same sign) has unlimited range in one direction and skewness, which is the beta distribution of the second kind (F distribution); and type IV (conjugate complex roots) has unlimited range in both directions and skewness. The remaining nine frequency distributions (curves) may be considered as the transition distributions (curves) for the Pearson family. Among these nine transitional distributions (curves), types II, III, VIII, IX, and X distributions may be considered as special or limiting cases of type I distribution. Types V and XI distributions may be considered as the limiting and special cases of type VI distribution. In addition, normal distribution is a special case of type V distribution (Edgeworth, 1898). It is interesting to note that type I is the beta distribution of the first kind (Thurow, 1970; McDonald, 1984), type III is the gamma distribution (Bartels, 1977; McDonald, 1984), type V is the inverse gamma distribution, type VII is the generalization of Student's t distribution (Kloek and van Dijk, 1977), type X is the exponential distribution, and type XI is the Pareto distribution. Thus, the Pearson system is intended to account for virtually all forms of asymmetry or skewness and all possible supports or domains of observed distribution functions.

One of the major characteristics of the Pearson system is that its first m moments (if existed) can be expressed in terms of the four parameters contained in Equation (2.1). To that end, let second, third, and fourth central moments be denoted, respectively, as μ_2, μ_3, and μ_4, where μ_2 is the second central moment or variance-a measure of dispersion, μ_3 is the third central moment-a measure of skewness, and μ_4 is the fourth central moment-a measure of kurtosis. Following Elderton and Johnson (1969) and Craig (1936), the Pearson system [Equation 2.1] may be solved through the first $(j + 1)$ moments (i.e., the first four moments) as:

$$a\mu_j + jc_1\mu_{j-1} + (j + 1)c_2\mu_j + (j + 2)c_3\mu_{j+2} = \mu_{j+1}, j = 0, 1, 2, 3 \quad (2.7)$$

In Equation (2.7), μ_j represents the j-th central moments with $\mu_0 = 1$. For the normalized variable X (i.e., $x = \frac{y-\mu_y}{\sigma_y}$), we have $\mu_1 = 0$ and $\mu_2 = 1$.

Furthermore, let

$$\alpha_3^2 = \frac{\mu_3^2}{\mu_2^3} = \beta_1; \alpha_4 = \frac{\mu_4}{\mu_2^2}; d = \frac{2\alpha_4 - 3\alpha_3^2 - 6}{\alpha_4 + 3} \quad (2.8)$$

We may express the parameters $\{a, c_1, c_2, c_3\}$ of Equation (2.1) through the first four moments as:

$$a = -\frac{\alpha_3}{2(1 + 2d)}; c_1 = \frac{2 + d}{2(1 + 2d)}; c_2 = \frac{\alpha_3}{2(1 + 2d)}; c_3 = \frac{d}{2(1 + 2d)} \quad (2.9a)$$

for normalized variable X

or

$$a = -\frac{\sigma\alpha_3}{2(1+2d)}, c_1 = \frac{\sigma^2(2+d)}{2(1+2d)}, c_2 = \frac{\sigma\alpha_3}{2(1+2d)}, c_3 = \frac{d}{2(1+2d)} \quad (2.9b)$$

for $X = Y - \mu$

Equation (2.9) is valid unless $d = -1/2$. The moment ratios provide a complete taxonomy of the Pearson distributions and can be used for selecting an appropriate PDF that is fitted by the method of moments. Equation (2.9a) further shows that the two parameters (α_3^2, d) or $(\alpha_3^2, \alpha_4) = (\beta_1, \beta_2)$ are fundamental and all Pearson distributions can be represented in the (α_3^2, d) or (β_1, β_2) diagram (Pearson, 1916; Craig, 1936). Craig (1936) showed that for the normalized variable X, the following conditions hold for the Pearson system:

$$\alpha_3^2 \le \alpha_4; d \in \left(-2 + \frac{4\alpha_4 - 3\alpha_3^2}{\alpha_4 + 3}, 2 - \frac{\alpha_3^2 + 4}{\alpha_4 + 4}\right) \Rightarrow d \in (-2, 2) \quad (2.10)$$

As discussed earlier, the denominator permits $f(x)$ to accommodate a large variety of shapes of observed distributions. The criterion k defined earlier can be rewritten as: $\frac{c_2^2}{4c_1c_3} = \frac{\alpha_3^2}{d(2+d)} = \frac{\alpha_3^2(\alpha_4+3)^2}{4\left(4\alpha_4-3\alpha_3^2\right)\left(2\alpha_4-3\alpha_3^2-6\right)}$ as a function of skewness squared and kurtosis.

2.3 Generalization of Pearson System

Toranzos (1952) hypothesized that a frequency distribution of a random variable can be expressed as a product of two functions, say $g(x)$ and $h(x)$:

$$f(x) = g(x)h(x)\exp\left(-\frac{1}{2}a^2(x-b)^2\right), a, b \text{ constants} \quad (2.11)$$

which can be derived from the differential equation:

$$\frac{1}{f(x)}\frac{df(x)}{dx} = \frac{Q_{m+1}(x)}{P_m(x)} \quad (2.12)$$

where $Q_{m+1}(x)$ is a polynomial of degree $m + 1$, and $P_m(x)$ is another polynomial of degree m. Equation (2.12) generalizes the Pearson system given by Equation (2.1) and can be expressed by the long division of polynomials as

$$\frac{1}{f(x)}\frac{df(x)}{dx} = a_0 + bx + \frac{Q_{m-1}(x)}{P_m(x)} \quad (2.13)$$

where a_0 and b are constants. Clearly, Equation (2.13) with $a_0 = 0$, $b = 0$, and $m = 2$ yields the Pearson system [Equation 2.1]. Integration of Equation (2.13) leads to

$$f(x) = C\exp\left(a_0 x + \frac{b}{2}x^2 + G(x)\right), G(x) = \int \frac{Q_{m-1}(x)}{P_m(x)}dx \tag{2.14}$$

For integration, function $G(x)$ can be decomposed into integrals of forms

$$\int \frac{dx}{(x-c)^{w_0}}, \int \frac{Ax+B}{(r+sx+x^2)^{w_1}}dx \tag{2.15}$$

If $b < 0$ and the roots of denominator are simple, then it can be shown that $f(x)$ will be a Gaussian function multiplied by Pearson's functions. Exponential factors occur when multiple roots exist.

For illustration, let $m = 1$; then Equation (2.12) can be expressed as

$$\frac{1}{f(x)}\frac{df(x)}{dx} = \frac{A_0 + A_1 x + A_2 x^2}{k_0 + k_1 x}; \; A_i, i = 1, 2, 3 \text{ constants and } k_0 \text{ and } k_1 \text{ are constants} \tag{2.16}$$

Changing the origin of x, Equation (2.16) can be written as

$$\frac{1}{f(x)}\frac{df(x)}{dx} = \frac{a_0 + a_1 x + a_2 x^2}{x}; \; a_i, i = 0, 1, 2 \text{ constants} \tag{2.17}$$

Integration of Equation (2.17) results in

$$f(x) = Cx^{a_0}\exp\left(\left(\frac{a_1}{a_2} + x\right)^2\right) \tag{2.18}$$

where C is constant of integration. Equation (2.18) yields bell-shaped curves if $a_2 < 0$ and $a_0 > 0$. Taking $a_0 = c$, $a_1/a_2 = b$, and $a_2 = -a^2$, Equation (2.18) can be expressed as

$$f(x) = Cx^c\exp\left(-\frac{a^2}{2}(x-b)^2\right) \tag{2.19}$$

Equation (2.19) resembles a normal distribution. The value of $f(x)$ will be 0 at $x_1 = 0$ and $x_2 = \infty$ and will have one maximum. Equation (2.18) generates interesting distributions. For example, for $a_0 < 0$ and $a_2 > 0$, $f(x)$ becomes U-shaped with minimum at $x = b$ and infinite at $x = 0$ and $x = \infty$. For $a_0 < 0$ and $a_2 < 0$, $f(x)$ has a zero at $x = \infty$, a minimum near the origin, and a maximum near $x = b$, which becomes infinite at the origin and the curve is U-shaped. For $a_0 > 0$ and $a_2 > 0$, $f(x)$ has a zero at $x = 0$, a maximum near the origin, and a minimum near $x = b$ and becomes infinite at $x = \infty$. Parameters of Equation (2.19) can be estimated by the method of moments.

2.4 Pearson Distributions

Previously, we have discussed the effects of the discriminant of the quadratic function on the distributions that may be derived from the Pearson system with the general density function written as Equation (2.6). In this section, we will first discuss the nonnegative and negative discriminants for deriving different distributions based on the denominator of Equation (2.1), then we will discuss all 12 Pearson distributions.

2.4.1 Nonnegative Discriminant

In this case, we have $\alpha = c_2^2 - 4c_1c_3 \geq 0$ (equivalently $k = \frac{c_2^2}{4c_1c_3} \leq 0$ *or* $k \geq 1$). Two real roots (a_1, a_2) may be written as: $a_1 = \frac{-c_2+\sqrt{c_2^2-4c_1c_3}}{2c_3}, a_2 = \frac{-c_2-\sqrt{c_2^2-4c_1c_3}}{2c_3}$, the quadratic function may be rewritten as: $c_1 + c_2x + c_3x^2 = c_3(x - a_1)(x - a_2)$. The solution of the differential equation [Equation 2.1] can be expressed as:

$$\begin{aligned} f(x) &= \exp\int \frac{a-x}{c_3(x-a_1)(x-a_2)}dx = \exp\left\{\int \frac{1}{c_3(a_1-a_2)}\left[\frac{a-a_1}{x-a_1} + \frac{a_2-a}{x-a_2}\right]dx\right\} \\ &= \exp\left\{\frac{1}{c_3(a_1-a_2)}[(a-a_1)\ln(x-a_1) + (a_2-a)\ln(x-a_2) + C]\right\} \end{aligned} \tag{2.20}$$

From Equation (2.20), we can show that the PDF $f(x)$ may be written in general as:

$$f(x) \propto (x-a_1)^{\frac{a_1-a}{c_3(a_1-a_2)}}(x-a_2)^{\frac{a_2-a}{c_3(a_1-a_2)}} \tag{2.21}$$

2.4.2 Negative Discriminant

In this case, we have $\alpha = c_2^2 - 4c_1c_3 < 0$ (equivalently $k = \frac{c_2^2}{4c_1c_3} \in (0, 1)$) such that the quadratic function yields complex roots. Now let us set $z = x + \frac{c_2}{2c_3}$, and substitute $x = z - \frac{c_2}{2c_3}$ into the quadratic function $c_1 + c_2x + c_3x^2$ and we have:

$$\begin{aligned} c_1 + c_2x + c_3x^2 &= c_1 + c_2\left(z - \frac{c_2}{2c_3}\right) + c_3\left(z - \frac{c_2}{2c_3}\right)^2 \\ &= c_3z^2 + \frac{4c_1c_3 - c_2^2}{4c_3} = c_3\left(z^2 + \frac{4c_1c_3 - c_2^2}{4c_3^2}\right) \end{aligned} \tag{2.22}$$

In Equation (2.22), we have: $\frac{4c_1c_3-c_2^2}{4c_3^2} > 0$. Now, the solution of the differential equation [i.e., Equation 2.1] may then be solved using the new variable z (i.e., the monotone linear transform of the variable x) as:

$$f(z) = \left\{ \int \frac{a - \left(z - \frac{c_2}{2c_3}\right)}{c_3\left(z^2 + \frac{4c_1c_3-c_2^2}{4c_3^2}\right)} dz \right\}$$

$$= \exp\left\{ -\frac{1}{c_3}\left[\frac{1}{2}\ln\left(z^2 + \frac{4c_1c_3 - c_2^2}{4c_3^2}\right) - \left(\frac{a + \frac{c_2}{2c_3}}{\sqrt{\frac{4c_1c_3 - c_2^2}{4c_3^2}}}\right)\tan^{-1}\left(\frac{z}{\sqrt{\frac{4c_1c_3 - c_2^2}{4c_3^2}}}\right)\right] + C \right\} \tag{2.23}$$

2.4.3 Pearson Type 0 Distribution

Pearson type 0 distribution is the Gaussian distribution with the condition of Equation (2.1) as: $c_1 > 0$, and $c_2 = c_3 = 0$. The governing differential equation [Equation 2.1] then becomes

$$\frac{1}{f(x)}\frac{df(x)}{dx} = \frac{a - x}{c_1} \tag{2.24}$$

Solution of Equation (2.24) may be easily obtained as:

$$f(x) = C\exp\left(-\frac{(a - x)^2}{2c_1}\right) = C\exp\left(-\frac{(x - a)^2}{2c_1}\right); C = \frac{1}{\sqrt{2\pi c_1}} \tag{2.25}$$

Finally, the PDF so derived is the normal distribution, i.e., $N(a, c_1)$, and is sketched in Figure 2.3.

2.4.4 Pearson Type I Distribution

Pearson type I distribution is a main type. There exist two real roots of opposite sign ($a_1 < 0 < a_2$, $c_3 \neq 0$) for the quadratic function. Using the PDF [Equation 2.21] derived from the nonnegative discriminant under the condition of $a_1 \le x \le a_2$, we can rewrite Equation (2.21) as:

$$f(x) \propto (x-a_1)^{-\frac{a_1-a}{c_3(a_1-a_2)}}(x-a_2)^{\frac{a_2-a}{c_3(a_1-a_2)}} = (a_1-x)^{-\frac{a_1-a}{c_3(a_1-a_2)}}(a_2-x)^{\frac{a_2-a}{c_3(a_1-a_2)}}$$
$$\Rightarrow f(x) \propto \left(1-\frac{x}{a_1}\right)^{-\frac{a_1-a}{c_3(a_1-a_2)}}\left(1-\frac{x}{a_2}\right)^{\frac{a_2-a}{c_3(a_1-a_2)}} \tag{2.26}$$

Setting $x = a_1 + x'(a_2 - a_1)$, $x' \in [0, 1]$ we have:

$$f(x') \propto \left[\frac{a_1-a_2}{a_1}x'\right]^{-\frac{a_1-a}{c_3(a_1-a_2)}}\left[\frac{a_2-a_1}{a_2}(1-x')\right]^{\frac{a_2-a}{c_3(a_1-a_2)}} \tag{2.27}$$

Rearranging Equation (2.27), we have:

$$f(x') = C(x')^{m_1}(1-x')^{m_2} \tag{2.28}$$

where $m_1 = -\frac{a_1-a}{c_3(a_1-a_2)}, m_2 = \frac{a_2-a}{c_3(a_1-a_2)}, C = \frac{\Gamma(m_1+m_2+2)}{\Gamma(m_1+1)\Gamma(m_2+1)}$

Comparing Equation (2.28) with the beta distribution, it is seen that the transformed variable $x' = \frac{x-a_1}{a_2-a_1}$ follows the beta distribution $B(m_1 + 1, m_2 + 1)$ under the condition of $m_1, m_2 > -1$. The distribution is sketched in Figure 2.3.

In general, we may conclude the Pearson type I distribution as follows:

(1) $c_2^2 - 4c_1c_3 > 0, c_3 \neq 0$. There are two roots for the denominator of Equation (2.1) with opposite sign; criterion $k = \frac{c_2^2}{4c_1c_3} < 0$.
(2) $\alpha_3 \neq 0$, i.e., skewed.
(3) The domain is limited in both directions, i.e., $x \in (a_1, a_2) : a_1 < 0 < a_2$.
(4) The function may be bell shaped $(m_1, m_2 > 0)$, U-shaped $(m_1 < 0, m_2 < 0)$, and J-shaped (m_1, m_2 taking on opposite signs).
(5) From Craig (1936), the constraints of variable d in Equation (2.8) are: $d \in (-1, 0)$ and $d \neq -\frac{1}{2}$.

2.4.5 Pearson Type II Distribution

Pearson type II distribution is a transition type. It is a special case of type I distribution with $a = c_2 = 0$. According to Equation (2.9), $a = c_2 = 0$ is equivalent to $\alpha_3 = 0$. The discriminant of the quadratic function is then reduced to: $\alpha = -4c_1c_3$. To yield real roots, we have $c_3 \neq 0$ and c_1, c_3 are opposite in signs. The roots of the reduced quadratic function $c_1 + c_3x^2$ are given as

$$a_1 = -a_2 = -\sqrt{-\frac{c_1}{c_3}} < 0 \tag{2.29}$$

Substituting Equation (2.29) and $a = 0$ into Equation (2.28) we have:

$$m = m_1 = m_2 = -\frac{1}{2c_3} = -\frac{5\alpha_4 - 9}{2\alpha_4 - 6} \tag{2.30}$$

Substituting Equations (2.29)–(2.30) back into Equation (2.21) or (2.26), we have:

$$f(x) = C(x - a_1)^{m_1}(x + a_1)^{m_2} = C\left(x^2 - a_1^2\right)^{-\frac{1}{2c_3}} \text{ or } f(x) = C\left(1 - \frac{x^2}{a_1^2}\right)^{-\frac{1}{2c_3}} \tag{2.31}$$

The distribution is graphed in Figure 2.3.

In general, we may conclude the Pearson type II distribution as:

(1) $c_2 = 0$, c_1, c_3 have opposite signs. There are two real roots with the same magnitude but opposite sign with the criterion $k = \frac{c_2^2}{4c_1c_3} = 0$.
(2) Domain is limited in both directions, i.e., $-\left(-\frac{c_1}{c_3}\right)^{0.5} < x < \left(-\frac{c_1}{c_3}\right)^{0.5}$.
(3) The PDF is symmetric about the y-axis. The symmetry yields $\alpha_3 = 0$.
(4) It is the special case of type I distribution, $d \in (-1, 0), d \neq -\frac{1}{2}$.
(5) With the constraints on variable d, we have $\alpha_4 < 3$.
(6) The Pearson type II distribution is bell shaped if $d \in (-0.5, 0)$ or equivalently $\alpha_4 \in (1.8, 3)$. The Pearson type II distribution is U-shaped if $d \in \left(-1, -\frac{1}{2}\right)$ or equivalently $\alpha_4 < 1.8$ (Craig, 1936; Elderton and Johnson, 1969).

2.4.6 Pearson Type III Distribution

Pearson type III distribution is a transition type. It is also a special case of Pearson type I distribution. In this case $c_3 = 0$, and the quadratic function in Equation (2.1) reduces to a linear function $[c_1 + c_2x]$ with only one root $a_1 = -\frac{c_1}{c_2}, c_1, c_2 \neq 0$. Solving Equation (2.1), we have:

$$\ln f = \int \frac{a - x}{c_1 + c_2x} dx = \frac{ac_2 + c_1}{c_2^2} \ln(c_1 + c_2x) - \frac{x}{c_2} + C'. \tag{2.32}$$

$$f(x) = C(c_1 + c_2x)^{\frac{ac_2 + c_1}{c_2^2}} \exp\left(-\frac{x}{c_2}\right) = C(x_1 + c_2x)^{\frac{c_1}{c_2^2} - 1} \exp\left(-\frac{x}{c_2}\right) \tag{2.33}$$

Setting $c_1 + c_2x = x'$, Equation (2.33) is rewritten as:

$$f(x') = Cx'^{\left(\frac{c_1}{c_2^2}-1\right)} \exp\left(-\frac{x'-c_1}{c_2^2}\right) = C\exp\left(\frac{c_1}{c_2^2}\right)x'^{\left(\frac{c_1}{c_2^2}-1\right)} \exp\left(-\frac{x'}{c_2^2}\right)$$
$$= C_0x'^{\left(\frac{c_1}{c_2^2}-1\right)} \exp\left(-\frac{x'}{c_2^2}\right), x' > 0,\ C_0 = \frac{1}{c_2^{\frac{2c_1}{c_2^2}}\Gamma\left(\frac{c_1}{c_2^2}\right)}, c_1 > 0, \frac{c_1}{c_2^2} > 1 \tag{2.34}$$

Equation (2.34) shows that x' follows a gamma distribution [i.e., gamma $\left(\frac{c_1}{c_2^2}, c_2^2\right)$] with the gamma density function written in the form as: $f(x) \propto x^{\alpha-1}\exp(-\beta x)$. Equivalently, $c_1 + c_2x \sim Gamma\left(\frac{c_1}{c_2^2}, \frac{1}{c_2^2}\right)$, if $c_1 + c_2x > 0$, i.e., $x > a_1$.

In general, we may conclude the Pearson type III distribution, graphed in Table 2.3, as:

(1) $c_3 = 0$. There is only one real root with the criterion $k = \frac{c_2^2}{4c_1c_3} = \infty$.
(2) The domain is limited in one direction, i.e., $x > -c_1/c_2$.
(3) Applying Equation (2.8) (Craig, 1936), $d = 0$ for Pearson type III distribution.
(4) The density function is skewed. There exists a relation between skewness (α_3) and kurtosis (α_4) as: $2\alpha_4 = 3\alpha_3^2 + 6; \alpha_3 > 0$.
(5) The Pearson type III distribution can be bell shaped but it may also be J-shaped (Elderton and Johnson, 1969).

2.4.7 Pearson Type IV Distribution

Pearson type IV distribution is the main type with the discriminant of quadratic function [Equation 2.1] less than zero. Similar to type I distribution, the type IV distribution is skewed. The general solution of the differential equation [Equation 2.1] is given as Equation (2.23).

Let $m_1 = \frac{1}{2c_3}$, $D^2 = \frac{4c_1c_3-c_2^2}{4c_3^2}$ and $m_2 = -\frac{2ac_3+c_2}{2c_3^2D}$. Then, Equation (2.23) may be rewritten as:

$$f(z) \propto \left(z^2 + D^2\right)^{-m_1} \exp\left(-m_2\tan^{-1}\left(\frac{z}{D}\right)\right) \tag{2.35}$$

Equation (2.35) may be further rewritten as:

$$f(z) = C_0\left(1 + \left(\frac{z}{D}\right)^2\right)^{-m_1} \exp\left(-m_2\tan^{-1}\left(\frac{z}{D}\right)\right) \tag{2.36}$$

In Equation (2.36), C_0 is the constant.

As shown in Equation (2.36), there is no constraint on the values that the variable may take on, such that $z \in \mathbb{R}$, which is equivalent to $x \in \mathbb{R}$. m_1, m_2 are the shape parameters where $m_1 > \frac{1}{2}$ [i.e., $c_3 \in (0,1)$]. D is the scale parameter with $D > 0$.

Equation (2.36) gives the PDF for the transformed variable $z : z = x + \frac{c_2}{2c_3}$. Finally, we may write the PDF for the original variable x as:

$$f(x) = \frac{\left|\frac{\Gamma\left(m_1 + \frac{m_2}{2}i\right)}{\Gamma(m_1)}\right|^2}{DB\left(m_1 - \frac{1}{2}, \frac{1}{2}\right)} \left[1 + \left(\frac{x}{D}\right)^2\right]^{-m_1} \exp\left(-m_2 \tan^{-1}\left(\frac{x}{D}\right)\right) \tag{2.37}$$

In general, we may conclude for the Pearson type IV distribution as:

(1) $c_3 \in (0,1)$. There are two complex roots with the criterion $k = \frac{c_2^2}{4c_1c_3} \in (0,1)$.
(2) The domain is unlimited in both directions, i.e., $x \in \mathbb{R}$.
(3) The probability distribution is skewed, i.e., $\alpha_3 \neq 0$.
(4) The Pearson type IV distribution is bell shaped.

2.4.8 Pearson Type V Distribution

Pearson type V distribution is a transitional type. For this type, the roots of the quadratic function are real and equal with $c_1 = \frac{c_2^2}{4c_3}$. The differential equation [Equation 2.1] can then be solved as:

$$\ln f = \int \frac{a - x}{c_3\left(x + \frac{c_2}{2c_3}\right)^2} dx = -\frac{1}{c_3} \ln\left(x + \frac{c_2}{2c_3}\right) - \frac{1}{c_3}\left(a + \frac{c_2}{2c_3}\right)\left(x + \frac{c_2}{2c_3}\right)^{-1} + C \tag{2.38}$$

$$f(x) = C_0\left(x + \frac{c_2}{2c_3}\right)^{-1/c_3} \exp\left(-\frac{1}{c_3}\frac{a + \frac{c_2}{2c_3}}{x + \frac{c_2}{2c_3}}\right); \quad C_0 = \frac{\left(\frac{c_2}{c_3}\left(\frac{1}{2c_3} - 1\right)\right)^{\frac{1}{c_3} - 1}}{\Gamma\left(\frac{1}{c_3} - 1\right)} \tag{2.39}$$

In general, we may conclude for the Pearson type V distribution as:

(1) $\alpha_3 \neq 0$. There are equal real roots, i.e., the criterion $k = \frac{c_2^2}{4c_1c_3} = 1$.
(2) $d > 0$ such that the first m moments exist.
(3) The domain is unlimited in one direction as: $x \in \left(-\frac{c_2}{2c_3}, \infty\right)$.
(4) The Pearson type V distribution is always bell shaped.
(5) The Pearson type V distribution may also be called the inverse gamma distribution with shape parameter: $\frac{1}{c_3} - 1$ and scale parameter: $\frac{1}{c_3}\left(a + \frac{c_2}{2c_3}\right)$.

2.4.9 Pearson Type VI Distribution

Pearson type VI distribution is the last main type of the Pearson family. In this case, there are two real roots with the same sign (but opposite to the sign of α_3) for the quadratic function. The other conditions for Pearson type VI distribution are $\alpha_3 \neq 0$ and $d > 0$. The function derived in Equation (2.21) may still be applied for Pearson type VI distribution. Let $x - a_2 = x'$, we have $x - a_1 = x - a_2 + a_2 - a_1 = x' + A$, $A = a_2 - a_1$. Equation (2.21) may then be written as:

$$f(x') \propto (x' + A)^{-m_1} x'^{-m_2} \sim (x' + A)^{\frac{a_1 - a}{c_3 A}} x'^{-\frac{a_2 - a}{c_3 A}} \sim \left(\frac{x'}{A}\right)^{-\frac{a_2 - a}{c_3 A}} \left(1 + \frac{x'}{A}\right)^{\frac{a_1 - a}{c_3 A}} \tag{2.40}$$

From Equation (2.40), it is seen that x'/A follows the beta prime distribution (also called beta distribution of second kind or inverted beta distribution).

In general, we may conclude for the Pearson type VI distribution as:

(1) $\alpha_3 \neq 0$, $d > 0$, criterion $k = \frac{c_2^2}{4c_1c_3} > 1$.
(2) The domain unlimited in one direction, $x \in (-A, +\infty)$.
(3) Pearson type VI distribution is usually bell shaped but it may also be J-shaped.

2.4.10 Pearson Type VII Distribution

Pearson type VII distribution is a transition type. It may also be considered as a special case of Pearson type IV distribution. In this case, $c_2 = 0$, $a = 0$. c_1 and c_3 are of the same sign. The PDF may then be written as:

$$f(x) = C(x - a_1 i)^{-\frac{1}{2c_3}} (x + a_1 i)^{-\frac{1}{2c_3}} = C_0 \left(1 + \frac{x^2}{a_1^2}\right)^{-\frac{1}{2c_3}} \tag{2.41}$$

In Equation (2.41), $a_1 = \sqrt{\frac{c_1}{c_3}} = \sqrt{\frac{2+d}{d}} > 0$, $C_0 = \frac{1}{a_1\sqrt{2\pi}} \frac{\Gamma\left(\frac{1}{2c_3}\right)}{\Gamma\left(\frac{1}{2c_3} - \frac{1}{2}\right)}$.

In general, we can conclude for the Pearson type VII distribution as:

(1) $\alpha_3 = 0$, $a = 0$, $c_2 = 0$. The criterion $k = \frac{c_2^2}{4c_1c_3} = 0$.
(2) The domain is unlimited in both directions, i.e., $x \in (-\infty, +\infty)$.
(3) $\alpha_4 > 3$.
(4) Pearson type VII distribution is bell shaped and symmetric.
(5) The Student t distribution is a special case of the Pearson type VII distribution.

2.4.11 Pearson Type VIII Distribution

Pearson type VIII distribution is a transition type and is a special case of Pearson type I distribution (i.e., $k = \frac{c_2^2}{4c_1c_3} < 0$). According to Craig (1936), other conditions of Pearson type VIII distribution are:

$$d < -\frac{1}{2}; (2+3d)\alpha_3^2 = 4(1+2d)^2(2+d) \tag{2.42}$$

The preceding two conditions are equivalent to the conditions in Elderton and Johnson (1969):

$$\begin{cases} 5\alpha_4 - 6\alpha_3^2 - 9 < 0 \\ \left(10\alpha_4 - 12\alpha_3^2 - 18\right)^2\left(4\alpha_4 - 3\alpha_3^2\right) - \left(8\alpha_4 - 9\alpha_3^2 - 12\right)\alpha_3^2(\alpha_4+3)^2 = 0 \end{cases} \tag{2.43}$$

(i) Limiting case of Pearson type I distribution: $\alpha_3 = 0$.

In this case, the Pearson type VIII distribution is the uniform distribution (also called rectangular distribution) as:

$$f(x) = \frac{1}{a_2 - a_1}, a_2 > a_1 \tag{2.44}$$

In Equation (2.44), a_1, a_2 are the roots of the quadratic function. This limiting case of Pearson type I distribution is also the limiting case for Pearson type IX and XII distributions.

(ii) General case: $\alpha_3 \neq 0$

In this case, from the condition on criterion $k = \frac{c_2^2}{4c_1c_3} < 0$, there are two distinctive roots a_1, a_2. With the conditions listed for Equation (2.43), we have $m_1 < 0$, $m_2 = 0$ in Equation (2.6) or equivalently $a_2 = a$ in Equation (2.21). Equation (2.21) is then reduced to:

$$f(x) \propto (x-a_1)^{-\frac{a_1-a_2}{c_3(a_1-a_2)}} = (x-a_1)^{-\frac{1}{c_3}} \tag{2.45}$$

Substituting d and c_3 in Equations (2.8) and (2.9a) into Equation (2.45), we obtain

$$f(x) \propto (x-a_1)^{-\frac{2\left(5\alpha_4-6\alpha_3^2-9\right)}{2\alpha_4-3\alpha_3^2-6}} = (x-a_1)^{\frac{2\left(5\alpha_4-6\alpha_3^2-9\right)}{3\alpha_3^2-2\alpha_4+6}} \tag{2.46}$$

To fulfill the condition of $m_1 < 0$ with the condition given for Equation (2.43), we obtain $3\alpha_3^2 - 2\alpha_4 + 6 > 0$. Let $m = \frac{5\alpha_4 - 6\alpha_3^2 - 9}{3\alpha_3^2 - 2\alpha_4 + 6} < 0$. Then we obtain the PDF for Pearson type VIII distribution as:

$$f(x) = C_0(x - a_1)^{2m} = C\left(1 - \frac{x}{a_1}\right)^{2m}; C = \frac{(-a_1)^{2m}(1 + 2m)}{(a_2 - a_1)^{1+2m}} \tag{2.47}$$

2.4.12 *Pearson Type IX Distribution*

Similar to the Pearson type VIII distribution, Pearson type IX distribution is also a transition type and is a special case of Pearson type I distribution (i.e., $k = \frac{c_2^2}{4c_1c_3} < 0$). According to Craig (1936), the conditions of Pearson type IX distribution are:

$$-\frac{1}{2} < d < 0; (2 + 3d)\alpha_3^2 = 4(1 + 2d)^2(2 + d) \tag{2.48}$$

The conditions in Equation (2.48) are equivalent to the conditions in Elderton and Johnson (1969) which is:

$$\begin{cases} 5\alpha_4 - 6\alpha_3^2 - 9 > 0 \\ \left(10\alpha_4 - 12\alpha_3^2 - 18\right)^2\left(4\alpha_4 - 3\alpha_3^2\right) - \left(8\alpha_4 - 9\alpha_3^2 - 12\right)\alpha_3^2(\alpha_4 + 3)^2 = 0 \end{cases} \tag{2.49}$$

which as the same as type VIII distribution, and there are two distinctive roots a_1, a_2. With the conditions given for Equations (2.48)–(2.49), we have $m_1 = 0$, $m_2 > 0$ in Equation (2.6) or equivalently $a_1 = a$ in Equation (2.21). Equation (2.21) is then reduced to:

$$f(x) \propto (a_2 - x)^{\frac{a_2 - a_1}{c_3(a_1 - a_2)}} = (a_2 - x)^{-\frac{1}{c_3}} \tag{2.50}$$

Substituting d and c_3 in Equations (2.8) and (2.9a) into Equation (2.50), we again obtain

$$f(x) \propto (a_2 - x)^{-\frac{2\left(5\alpha_4 - 6\alpha_3^2 - 9\right)}{2\alpha_4 - 3\alpha_3^2 - 6}} = (a_2 - x)^{\frac{2\left(5\alpha_4 - 6\alpha_3^2 - 9\right)}{3\alpha_3^2 - 2\alpha_4 + 6}} \tag{2.51}$$

To fulfill the condition of $m_2 > 0$ with the condition given for Equation (2.49), we obtain $3\alpha_3^2 - 2\alpha_4 + 6 > 0$. Let $m = \frac{5\alpha_4 - 6\alpha_3^2 - 9}{3\alpha_3^2 - 2\alpha_4 + 6} > 0$, we obtain the PDF for Pearson type IX distribution as:

$$f(x) = C\left(1 - \frac{x}{a_2}\right)^{2m}; c = \frac{a_2^{2m}(1 + 2m)}{(a_2 - a_1)^{1+2m}} \tag{2.52}$$

2.4.13 Pearson Type X Distribution

Pearson type X distribution is the limiting case for Pearson type III, type IX, and type XI distributions. With $\alpha_3^2 = 4, \alpha_4 = 9 \Rightarrow d = 0,$ the Pearson type III distribution [Equation 2.34] may be rewritten as:

$$f(x) = \exp(-x),\ x > 0 \tag{2.53}$$

2.4.14 Pearson Type XI Distribution

Pearson type XI distribution is a special case of Pearson type VI distribution with $\alpha_3 \neq 0, k > 1, 0 < d < \frac{2}{5}, (2 + 3d)\alpha_3^2 = 4(1 + 2d)^2(2 + d)$. Pearson type XI distribution is skewed with J-shape. The PDF can be written using Equation (2.40) as:

$$f(x) \propto \left(1 + \frac{x}{A}\right)^{\frac{a_1 - a}{c_3 A}}, \frac{a_1 - a}{c_3 A} < 0 \tag{2.54}$$

Comparing Equation (2.54) with the PDF of the Pareto distribution, it is seen that Pearson type XI distribution is a Pareto distribution. It is special case of Pearson VI distribution.

2.4.15 Pearson Type XII Distribution

The Pearson type XII distribution is a transition type with $d \to -\frac{1}{2}$, or equivalently $1 + 2d = 5\alpha_4 - 6\alpha_3^2 - 9 = 0$. In this case, the differential equation [Equation 2.1] may be rewritten as:

$$\frac{1}{f}\frac{df}{dx} = \frac{2\alpha_3}{x^2 - 2\alpha_3 x - 3} \tag{2.55}$$

The roots of the quadratic function are real as: $x_1 = \alpha_3 - \left(\alpha_3^2 + 3\right)^{0.5}, x_2 = \alpha_3 + \left(\alpha_3^2 + 3\right)^{0.5}$.

Similar to Pearson type I distribution, the PDF of the Pearson type XII distribution may be written as

$$f(x) = C\left(\frac{a_2 - x}{x - a_1}\right)^{\frac{\alpha_3}{\left(\alpha_3^2 + 3\right)^{0.5}}}\ a_2 > 0 > a_1, \alpha_3 \geq 0; C = \frac{1}{(a_2 - a_1)B(1 - m_2, 1 + m_2)} \tag{2.56}$$

The uniform distribution is a limiting case for Pearson XII distribution, i.e., $\alpha_3 = 0$. In general, the Pearson type XII distribution is a special case of Pearson type I distribution ($\alpha_3 > 0$). The Pearson type XII distribution has twisted J-shape.

2.5 Graphical Representation of Shapes Based on the Relation of α_3^2 versus d and α_3^2 versus α_4

Pearson (1916) presented the diagram of $\alpha_3^2 - \alpha_4$. Pearson's diagram was constructed directly based on the skewness squared (third moment) and kurtosis (fourth moment). Craig (1936) developed an alternative diagram with the use of $\alpha_3^2 - d$. In this alternative diagram, d is a function of skewness squared (α_3^2) and kurtosis (α_4). To better illustrate and connect the two diagrams, Table 2.1 lists the equivalent conditions for both diagrams. As seen in Table 2.1 as well as the discussions in earlier sections, Pearson (1895, 1916) directly used the skewness-squared (α_3^2) and kurtosis (α_4) to evaluate the shapes and conditions of the 12 Pearson distributions, while Craig (1936) applied the indirect measure d a function of α_3^2 and α_4 to evaluate the shapes and conditions using d versus α_3^2. Figures 2.1 and 2.2 graph the relations with the use of both methods.

2.5.1 *Graphical Representation of Pearson Distributions*

Previously, we have derived all 12 Pearson distributions, including three main types (i.e., types I, IV, and VI) and nine transitional types, as well as the evaluation of shapes and conditions for all 12 distributions. Here, we will simply represent different distributions based on their unique conditions. Table 2.2 lists the pertinent information of each distributions, and Figure 2.3 graphs all 12 distributions with different shapes, including the most popular normal distribution (graphed with Type X – exponential distribution). Here we will use three main type of distributions as well as Pearson Type III distribution to further illustrate the procedure to compute the parameters of the Pearson distributions.

For all 12 distributions, the kurtosis and coefficients of the differential equation [Equation 2.1] can be computed exactly in the same manner with the use of Equation (2.6) and Equation (2.9a).

2.5.2 *Type I(U): $d = -0.8; \alpha_3^2 = 3$*

Using the type I(U) as an example, the three type I shapes (i.e., U, J, and B) can be computed using the same formula as follows:

Applying Equation (2.6) we obtain the kurtosis: $\alpha_4 = \frac{3d+3\alpha_3^2+6}{2-d} = \frac{3(-0.8)+3(3)+6}{2-(-0.8)} = 4.5$

Applying Equation (2.9a), we obtain the coefficients of the differential equation [Equation 2.1] as:

$$a = -\frac{\alpha_3}{2(1+2d)} = -\frac{3^{0.5}}{2(1+2(-0.8))} = 1.44; c_2 = -a = -1.44$$

Table 2.1 $d - \alpha_3^2$ (Craig) and the equivalent $\alpha_3^2 - \alpha_4$ (Pearson) criteria for Pearson distributions

	Craig		Pearson	
Type	$d = \frac{2\alpha_4 - 3\alpha_3^2 - 6}{\alpha_4 + 3}$	$d - \alpha_3^2$	$k = \frac{c_2^2}{4c_1c_3}$	$\alpha_3^2 - \alpha_4$
I-U	$d \in \left(-1, \frac{2}{3}\right)$	$\alpha_3^2 > 0$	$k < 0$	$8\alpha_4 - 9\alpha_3^2 - 12 < 0$
I-J	$d \in \left(-\frac{2}{3}, -\frac{1}{2}\right) \cup \left(-\frac{1}{2}, 0\right)$	$\alpha_3^2(2+3d) < 4(1+2d)^2(d+2)$	$\alpha_4 - \alpha_3^2 - 1 > 0$ $2\alpha_4 - 3\alpha_3^2 - 6 < 0$ $5\alpha_4 - 6\alpha_3^2 - 9 \neq 0$	$5\alpha_4 - 6\alpha_3^2 - 9 < 0 \cap 8\alpha_4 - 9\alpha_3^2 - 12 > 0$ or $\begin{cases} 5\alpha_4 - 6\alpha_3^2 - 9 > 0 \\ \alpha_3^2(\alpha_4+3)^2(8\alpha_4 - 9\alpha_3^2 - 12) > \\ 4(5\alpha_4 - 6\alpha_3^2 - 9)^2(4\alpha_4 - 3\alpha_3^2) \end{cases}$
I-B	$d \in \left(-\frac{1}{2}, 0\right)$	$\alpha_3^2(2+3d) > 4(1+2d)^2(d+2)$		$\begin{cases} 5\alpha_4 - 6\alpha_3^2 - 9 > 0 \\ \alpha_3^2(\alpha_4+3)^2(8\alpha_4 - 9\alpha_3^2 - 12) < \\ 4(5\alpha_4 - 6\alpha_3^2 - 9)^2(4\alpha_4 - 3\alpha_3^2) \end{cases}$
II-U	$d \in \left(-1, -\frac{1}{2}\right)$	$\alpha_3^2 = 0$	$k = 0$	$\alpha_3 = 0,\ \alpha_4 \in (1.8, 3)$
II-B	$d \in \left(-\frac{1}{2}, 0\right)$	$\alpha_3^2 = 0$		$\alpha_3 = 0,\ \alpha_4 \in (0, 1.8)$
III-B	$d = 0$	$\alpha_3^2 \in (0, 4)$	$k \to \infty$	$\alpha_3 > 0, \alpha_3^2 \in (0, 4), 2\alpha_4 - 3\alpha_3^2 - 6 = 0$
III-J	$d = 0$	$\alpha_3^2 \geq 4$		$\alpha_3 > 0, \alpha_3^2 > 4, 2\alpha_4 - 3\alpha_3^2 - 6 = 0$
IV-B	$d > 0$	$\alpha_3 \neq 0, \alpha_3^2 < 4d(d+2)$	$k \in (0, 1)$	$2\alpha_4 - 3\alpha_3^2 - 6 > 0, 8\alpha_4 - 15\alpha_3^2 - 36 < 0$
V-B	$d > 0$	$\alpha_3^2 = 4d(d+2)$	$k = 1$	$\begin{cases} 2\alpha_4 - 3\alpha_3^2 - 6 > 0 \\ 4(4\alpha_4 - 3\alpha_3^2) > \\ 4(4\alpha_4 - 3\alpha_3^2)(2\alpha_4 - 3\alpha_3^2 - 6) \end{cases}$
VI-B	$d > 0$	$\alpha_3^2 > 4d(d+2)$ $\alpha_3^2(2+3d) <$ $4(1+2d)^2(2+d)$	$k > 1$	$\begin{cases} 2\alpha_4 - 3\alpha_3^2 - 6 > 0 \\ 4(4\alpha_4 - 3\alpha_3^2) > \\ 4(4\alpha_4 - 3\alpha_3^2)(2\alpha_4 - 3\alpha_3^2 - 6) \\ \alpha_3^2(\alpha_4+3)^2(8\alpha_4 - 9\alpha_3^2 - 12) < \\ 4(5\alpha_4 - 6\alpha_3^2 - 9)^2(4\alpha_4 - 3\alpha_3^2) \end{cases}$

Table 2.1 (*cont.*)

	Craig		Pearson	
Type	$d = \frac{2\alpha_4 - 3\alpha_3^2 - 6}{\alpha_4 + 3}$	$d - \alpha_3^2$	$k = \frac{c_2^2}{4c_1c_3}$	$\alpha_3^2 - \alpha_4$
VI-J	$d > 0$	$\alpha_3^2 > 4d(d+2)$ $\alpha_3^2(2+3d) > 4(1+2d)^2(2+d)$		$\begin{cases} 2\alpha_4 - 3\alpha_3^2 - 6 > 0 \\ 4\left(4\alpha_4 - 3\alpha_3^2\right) > \\ 4\left(4\alpha_4 - 3\alpha_3^2\right)\left(2\alpha_4 - 3\alpha_3^2 - 6\right) \\ \alpha_3^2(\alpha_4+3)^2\left(8\alpha_4 - 9\alpha_3^2 - 12\right) > \\ 4\left(5\alpha_4 - 6\alpha_3^2 - 9\right)^2\left(4\alpha_4 - 3\alpha_3^2\right) \end{cases}$
VII-B	$d > 0$	$\alpha_3 = 0$	$k = 0$	$\alpha_3^2 = 0, \alpha_4 > 3$
VIII-Uniform	$d = -\frac{1}{2}$	$\alpha_3 = 0$		$\alpha_3^2 = 0, \alpha_4 = 1.8$
VIII-J	$d \in \left(-\frac{2}{3}, -\frac{1}{2}\right)$	$\alpha_3 \neq 0$ $\alpha_3^2(2+3d) = 4(1+2d)^2(2+d)$	$k < 0$	$\begin{cases} 5\alpha_4 - 6\alpha_3^2 - 9 < 0, \alpha_3 \neq 0 \\ \alpha_3^2(\alpha_4+3)^2\left(8\alpha_4 - 9\alpha_3^2 - 12\right) = \\ 4\left(5\alpha_4 - 6\alpha_3^2 - 9\right)^2\left(4\alpha_4 - 3\alpha_3^2\right) \end{cases}$
IX-J	$d \in \left(-\frac{1}{2}, 0\right)$	$\alpha_3 \neq 0$ $\alpha_3^2(2+3d) = 4(1+2d)^2(2+d)$	$k < 0$	$\begin{cases} 5\alpha_4 - 6\alpha_3^2 - 9 > 0, \alpha_3 \neq 0 \\ \alpha_3^2(\alpha_4+3)^2\left(8\alpha_4 - 9\alpha_3^2 - 12\right) = \\ 4\left(5\alpha_4 - 6\alpha_3^2 - 9\right)^2\left(4\alpha_4 - 3\alpha_3^2\right) \end{cases}$
X-J	$d = 0$	$\alpha_3^2 = 4$	$k = \infty$	$\alpha_3^2 = 4, \alpha_4 = 9$
XI-J	$d > 0$	$\alpha_3 \neq 0$ $\alpha_3^2(2+3d) = 4(1+2d)^2(2+d)$	$k > 1$	$\begin{cases} 2\alpha_4 - 3\alpha_3^2 - 6 > 0 \\ 4\left(4\alpha_4 - 3\alpha_3^2\right) > \\ 4\left(4\alpha_4 - 3\alpha_3^2\right)\left(2\alpha_4 - 3\alpha_3^2 - 6\right) \\ \alpha_3^2(\alpha_4+3)^2\left(8\alpha_4 - 9\alpha_3^2 - 12\right) = \\ 4\left(5\alpha_4 - 6\alpha_3^2 - 9\right)^2\left(4\alpha_4 - 3\alpha_3^2\right) \end{cases}$
XII-J	$d = -\frac{1}{2}$			$5\alpha_4 - 6\alpha_3^2 - 9 = 0, \alpha_3 \neq 0$

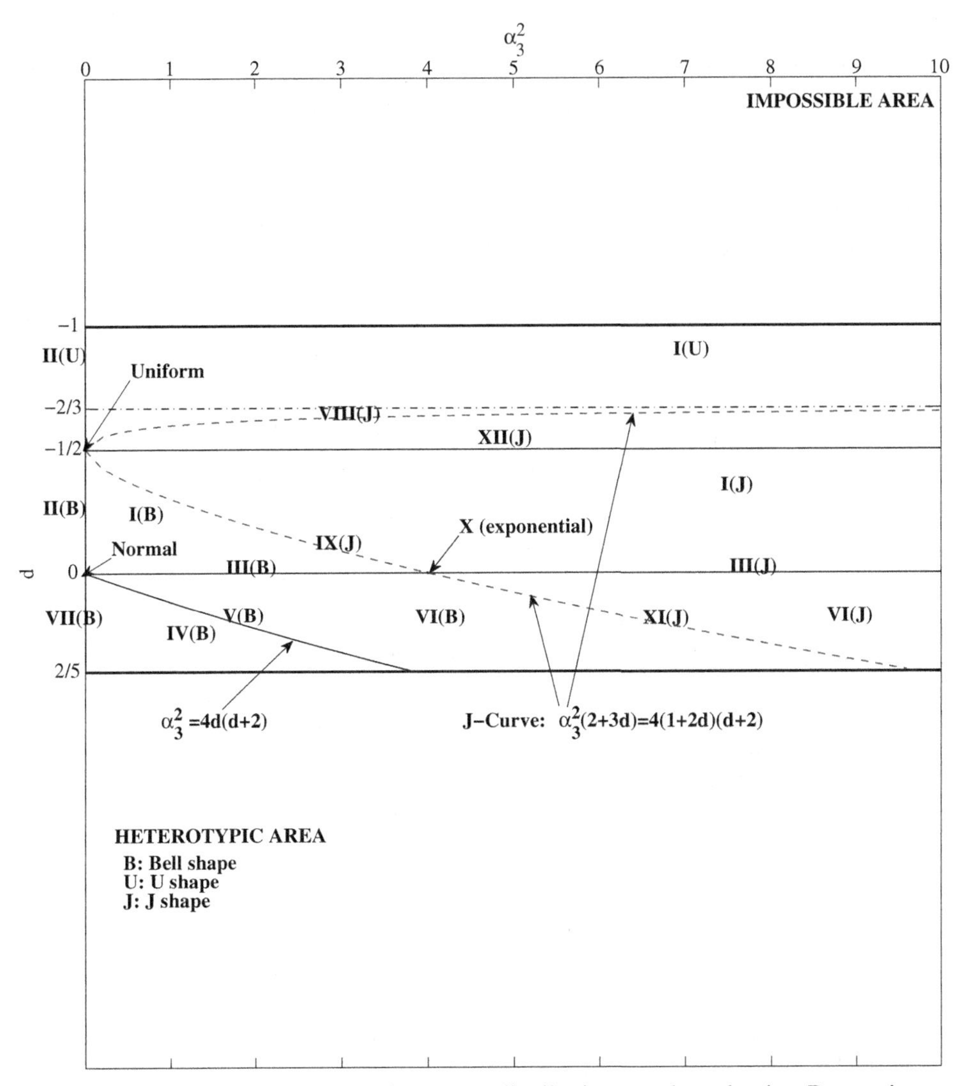

Figure 2.1 Shapes and types of Pearson distributions evaluated using Pearson's method (Pearson, 1895, 1916).

$$c_1 = \frac{2+d}{2(1+2d)} = \frac{2-0.8}{2(1+2(-0.8))} = -1;$$
$$c_3 = \frac{d}{2(1+2d)} = -\frac{0.8}{2(1+2(-0.8))} = 0.67$$

Setting $c_1 + c_2x + c_3x^2 = 0$, we can compute the roots as: $a_1 = -0.55$; $a_2 = 2.72$

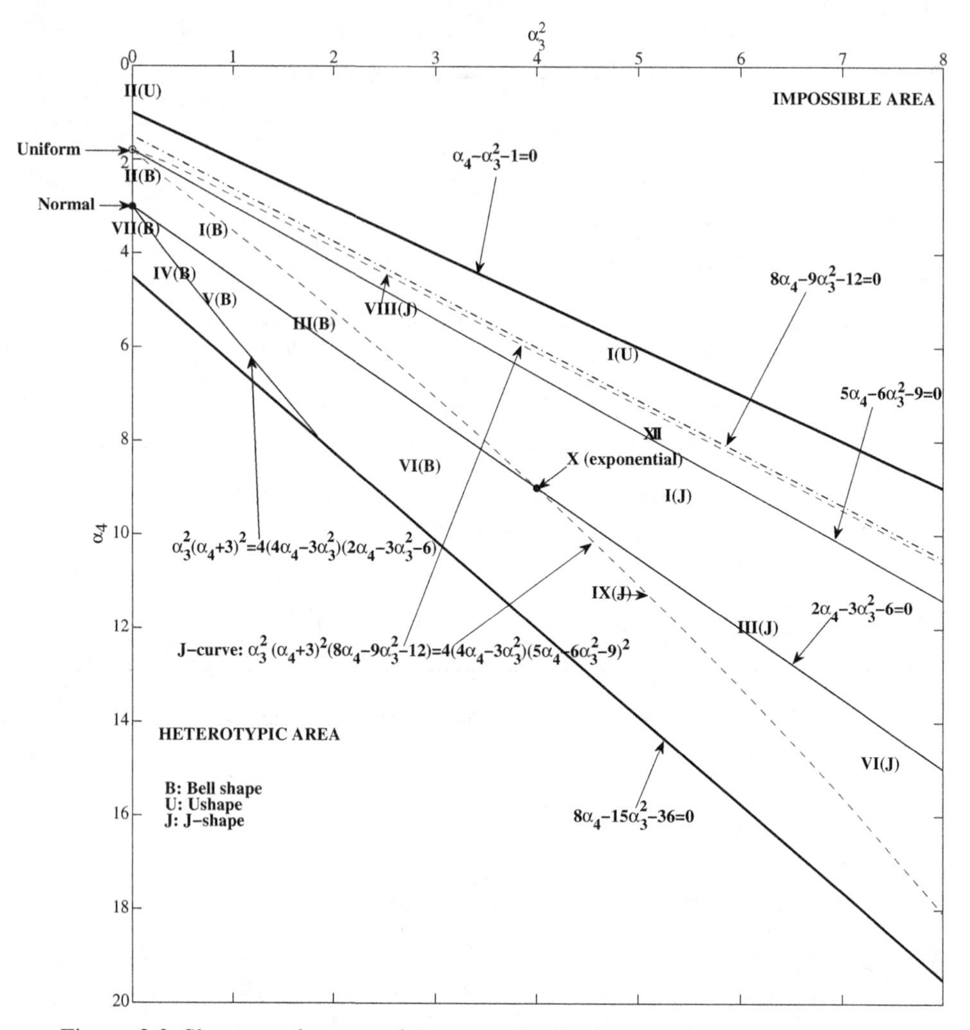

Figure 2.2 Shapes and types of Pearson distributions evaluated using Craig's method (Craig, 1936).

Applying Equation (2.28), we obtain the exponents as:

$$m_1 = -\frac{a_1 - a}{c_3(a_1 - a_2)} = -\frac{-0.55 - 1.44}{0.67(-0.55 - 2.72)} = -0.92;$$

$$m_2 = \frac{a_2 - a}{c_3(a_1 - a_2)} = \frac{2.72 - 1.44}{0.67(-0.55 - 2.72)} = -0.58.$$

and the PDF of the distribution (i.e., the beta distribution) may be written using Equations (2.26)–(2.28) as:

Table 2.2 *Parameters of different Pearson type distributions*

	Criteria			Coefficients of Differential Equation (2.1)				Exponents	
Type	d	α_3^2	α_4	a	c_1	c_2	c_3	m_1	m_2
I(U)	–0.80	3.00	4.5	1.44	–1.00	–1.44	0.67	–0.92	–0.58
I(J)	–0.40	6.00	8.25	–6.12	4.00	6.12	–1.00	–0.76	1.76
I(B)	–0.10	1.00	4.14	–0.63	1.19	0.63	–0.06	1.22	14.78
II(U)	–0.80	0	1.29	0.00	–1.00	0.00	0.67	–0.75	–0.75
II(B)	–0.20	0	2.45	0.00	1.50	0.00	–0.17	3.00	3.00
III(B)[1]	0	1.50	5.25	–0.61	1.00	0.61	0.00	[2.67, 0.38]	
III(J)	0	5.00	10.5	–1.12	1.00	1.12	0.00	[0.80, 1.25]	
IV(B)[2]	0.2	1.00	5.33	–0.36	0.79	0.36	0.07	[7.14, –13.76,0.87]	
V(B)[3]	0.32	3.00	9.52	–0.53	0.71	0.53	0.10	[9.19, 21.98]	
VI(B)	0.20	1.96	6.93	–0.50	0.79	0.50	0.07	11.45	–25.24
VI(J)	0.20	8.50	17.83	–1.04	0.79	1.04	0.07	13.74	0.26
VII(B)	0.20	0.00	3.67	0.00	0.79	0.00	0.07	7.00	7.00
VIII-uniform	–0.60	1.80	1.8	0	–3.50	0	1.50	0	0.00
VIII(J)	–0.63	4.00	6.17	3.73	–2.55	–3.73	1.18	–0.85	0.00
IX(J)	1.73	3.00	7.61	–1.05	1.16	1.05	–0.05	0.00	18.94
X(J)[4]	0	4	9	–1	1	1	0	1	
XI(J)	0.15	6.00	13.26	–0.94	0.82	0.94	0.06	0.00	–16.91
XII(J)	–0.5	5	13.26	N/A	N/A	N/A	N/A		

[1] Parameters for gamma distribution $\left(\frac{c_1}{c_2^2}, c_2^2\right)$.

[2] Parameters for Pearson type IV distribution $\left(\frac{1}{2c_3}, -\frac{2ac_3+c_2}{2c_3^2 D}, D = \frac{\sqrt{4c_1c_3-c_2^2}}{2c_3}\right)$.

[3] Parameters for Pearson type V distribution $\left(\frac{1}{c_3} - 1, \frac{1}{c_3}\left(a + \frac{c_2}{2c_3}\right)\right)$.

[4] Parameter for exponential distribution $f(x) = \exp(-x)$, parameter = 1.

$$f(x) \propto \left(1 - \frac{x}{-0.55}\right)^{-0.92} \left(1 - \frac{x}{2.72}\right)^{-0.58} \sim Cx'^{-0.92}(1 - x')^{-0.58},$$

$$x' = \frac{x - a_1}{a_2 - a_1} = \frac{x + 0.55}{3.27} \in (0, 1), C = \frac{\Gamma(-0.92 - 0.58 + 2)}{\Gamma(-0.92 + 1)\Gamma(-0.58 + 1)} = 0.07$$

2.5.3 Type III(B): $d = 0; \alpha_3^2 = 1.5$

Pearson type III distribution is popularly applied in hydrological frequency analysis (e.g., flood). This is the same as the type I distribution. We first obtain:

$$\alpha_4 = \frac{3d + 3\alpha_3^2 + 6}{2 - d} = \frac{3(0) + 3(1.5) + 6}{2 - 0} = 5.25$$

$$a = -\frac{\alpha_3}{2(1 + 2d)} = -\frac{1.5^{0.5}}{2} = -0.61; c_2 = -a = 0.61$$

$$c_1 = \frac{2 + d}{2(1 + 2d)} = \frac{2 - 0}{2(1 + 2(0))} = 1; c_3 = \frac{d}{2(1 + 2d)} = -\frac{0}{2(1 + 2(0))} = 0$$

With the coefficients computed, the shape and scale parameters for the Pearson III distribution (i.e., gamma distribution) are given as:

Shape parameter: $\frac{c_1}{c_2^2} = \frac{1}{0.61^2} = 2.67$ and scale parameter: $c_2^2 = 0.38$.

and the corresponding PDF may be written using Equations (32)–(33) as:

$$f(x) \propto (x_1 + 0.61x)^{\frac{1}{0.38} - 1} \exp\left(-\frac{x}{0.61}\right) \sim Cx'^{1.67} \exp\left(-\frac{x'}{0.38}\right), x' = 1 + 0.61x;$$

where $C = \frac{1}{0.61^{\left(\frac{2}{0.38}\right)} \Gamma\left(\frac{1}{0.38}\right)} = 9.2092$.

2.5.4 Type IV(B): $d = 0.2, \alpha_3^2 = 1$

First we obtain:

$$\alpha_4 = \frac{3d + 3\alpha_3^2 + 6}{2 - d} = \frac{3(0.2) + 3(1) + 6}{2 - 0.2} = 5.33$$

$$a = -\frac{\alpha_3}{2(1 + 2d)} = -\frac{1^{0.5}}{2(1 + 2(0.2))} = -0.36; c_2 = -a = 0.36$$

$$c_1 = \frac{2 + d}{2(1 + 2d)} = \frac{2 + 0.2}{2(1 + 2(0.2))} = 0.79; c_3 = \frac{d}{2(1 + 2d)} = \frac{0.2}{2(1 + 2(0.2))} = 0.07.$$

$0.79 + 0.36x + 0.07x^2 = 0$ has the complex roots as: $a_{1,2} = -2.57 \pm 2.16i$.

Setting $m_1 = \frac{1}{2c_3} = \frac{1}{2(0.07)} = 7.14; D = \frac{\sqrt{4c_1c_3 - c_2^2}}{2c_3} = \frac{\sqrt{4(0.79)(0.07) - 0.36^2}}{2(0.07)} = 0.87$,
$m_2 = -\frac{1}{c_3 D}\left(a + \frac{c_2}{2c_3}\right) = -\frac{1}{0.07(0.87)}\left(-0.36 + \frac{0.36}{2(0.07)}\right) = -13.76$

Let $Z = x + \frac{c_2}{2c_3} = x + 2.57$, we can write the density function using Equation (2.36) as:

$$f(Z) = f(x + 2.57) \propto \left(1 + \left(\frac{x + 2.57}{0.87}\right)^2\right)^{-7.14} \exp\left(13.76 \tan^{-1}\left(\frac{x + 2.57}{0.87}\right)\right)$$

2.5.5 *Type VI:* d = 0.2, α_3^2 = 1.96

First, we obtain

$$\alpha_4 = \frac{3d + 3\alpha_3^2 + 6}{2 - d} = \frac{3(0.2) + 3(1.96) + 6}{2 - 0.2} = 6.93$$

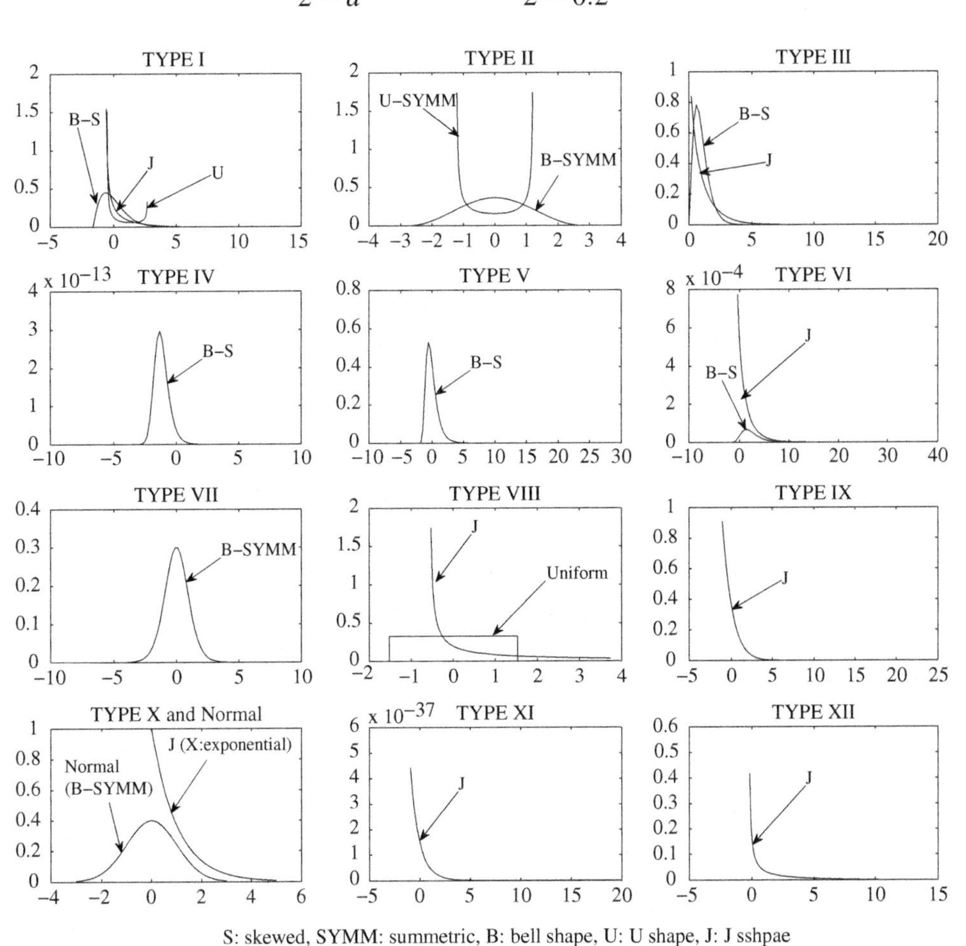

Figure 2.3 Representation of all 12 Pearson distributions under different conditions.

$$a = -\frac{\alpha_3}{2(1+2d)} = -\frac{1.96^{0.5}}{2(1+2(0.2))} = -0.5; c_2 = -a = 0.5$$

$$c_1 = \frac{2+d}{2(1+2d)} = \frac{2+0.2}{2(1+2(0.2))} = 0.79; c_3 = \frac{d}{2(1+2d)} = \frac{0.2}{2(1+2(0.2))} = 0.07.$$

The roots of the equation $0.79 + 0.36x + 0.07x^2 = 0$ are: $a_1 = -4.78$, $a_2 = -2.36$

$$m_1 = -\frac{a_1 - a}{c_3(a_1 - a_2)} = -\frac{-4.78 + 0.5}{0.07(-2.36 + 4.78)} = 25.7;$$

$$m_2 = \frac{a_2 - a}{c_3(a_1 - a_2)} = -\frac{-2.36 + 0.5}{0.07(-2.36 + 4.78)} = -11.78$$

Using Equation (2.40) and letting $x' = x - a_2$ we have the PDF as:

$$f(x') \propto (x' + 2.42)^{-25.7} x'^{11.78} \sim C\left(1 + \frac{x'}{2.42}\right)^{-25.7} \left(\frac{x'}{A}\right)^{11.78}, C = \frac{1}{B(12.78, 13)}.$$

2.6 Application

In this section, annual rainfall, peak flow, and maximum daily precipitation data are used to illustrate the application of the Pearson system. The chosen probability distribution candidates are normal, exponential, gamma, Pearson III, and Log-Pearson III distributions. The normal distribution is the limiting case of Pearson type I, III, IV, V, VI, and VII distributions, and gamma distribution is the limiting case of Pearson type III distribution. For parameter estimation, the maximum likelihood estimation method is applied to normal, exponential, and gamma distributions. The indirect method of moment (IDMOM) is applied to the log-Pearson III distribution. The log-likelihood functions for normal, exponential, and gamma distributions are expressed as:

Normal Distribution

$$f(x; \mu, \sigma) = \frac{1}{\sqrt{2\pi}\sigma} \exp\left(-\frac{(x-\mu)^2}{2\sigma^2}\right) \tag{2.57a}$$

$$\ln L = \sum_{i=1}^{n} \ln f(x_i) = -\frac{n}{2} \ln(2\pi) - n \ln \sigma - \frac{1}{2\sigma^2} \sum_{i=1}^{n} (x_i - \mu)^2 \tag{2.57b}$$

Maximizing the log-likelihood function, the parameters of normal distribution are expressed as:

$$\mu = \bar{x}; \sigma^2 = \frac{\sum_{i=1}^{n} (x_i - \mu)^2}{n} \approx \frac{\sum_{i=1}^{n} (x_i - \bar{x})^2}{n} = S^2 \tag{2.57c}$$

In Equations (2.57b) and (2.57c), n represents the sample size.

Gamma Distribution

$$f(x; \alpha, \beta) = \frac{\beta^\alpha}{\Gamma(\alpha)} x^{\alpha-1} \exp(-\beta x) \tag{2.58a}$$

$$\ln L = n\alpha \ln\beta - n \ln\Gamma(\alpha) + (\alpha - 1)\sum_{i=1}^{n} \ln x_i - \beta \sum_{i=1}^{n} x_i \tag{2.58b}$$

Maximizing the log-likelihood function, the parameters of gamma distribution may be estimated by solving the following equations:

$$\frac{\alpha}{\beta} = \bar{x}; \ \ln\beta - \psi(\alpha) = \frac{\sum_{i=1}^{n} \ln x}{n} = \overline{\ln x} \tag{2.58c}$$

In Equations (2.58b) and (2.58c), n represents the sample size.

Pearson III Distribution

From Section 2.4.6, the Pearson III distribution may be rewritten as:

$$f(x; \alpha, \beta, r) = \frac{1}{\beta\Gamma(\alpha)} \left(\frac{x - r}{\beta}\right)^{\alpha-1} \exp\left(-\frac{x - r}{\beta}\right); \alpha, \beta > 0; x > r \tag{2.59a}$$

Its log-likelihood function is expressed as:

$$\ln L = -n\alpha \ln\beta - n \ln\Gamma(\alpha) + (\alpha - 1)\sum_{i=1}^{n} \ln(x_i - r) - \frac{\sum_{i=1}^{n} (x_i - r)}{\beta} \tag{2.59b}$$

Maximizing the log-likelihood function, the parameters of the Pearson-III distribution may be estimated by solving the following equations:

$$\ln\beta + \psi(\alpha) = \frac{\sum_{i=1}^{n} \ln(x_i - r)}{n}; \alpha\beta = \frac{\sum_{i=1}^{n} (x_i - r)}{n}; (\alpha - 1)\beta = \frac{\sum_{i=1}^{n} \frac{1}{x_i - r}}{n} \tag{2.59c}$$

Log-Pearson III Distribution

Rather than applying the maximum likelihood estimation method, the IDMOM is applied for parameter estimation. The PDF of the log-Pearson III distribution is written as:

$$f(x;\alpha,\beta,r) = \frac{1}{x\beta\Gamma(\alpha)}\left(\frac{\ln x - r}{\beta}\right)^{\alpha-1}\exp\left(-\frac{\ln x - r}{\beta}\right) \tag{2.60a}$$

Equation (2.60a) shows that $y = \ln x$ follows the Pearson III distribution. Thus, the IDMOM method is essentially the method of moments in the logarithm domain. The first three moments (i.e., mean, variance, and skewness) for the variable in the logarithm domain are expressed as:

$$my = \frac{\sum_{i=1}^{n} y_i}{n}; s^2y = \frac{\sum_{i=1}^{n}(y_i - my)^2}{n-1}; ky = \frac{\frac{1}{n}\sum_{i=1}^{n}(y_i - \bar{y})^3}{(s^2y)^{\frac{3}{2}}}; y = \ln x \tag{2.60b}$$

and the parameters of the log-Pearson III distribution may be expressed explicitly as:

$$\alpha = \left(\frac{2}{ky}\right)^2; \beta = \left(\frac{s^2y}{\beta}\right)^{\frac{1}{2}}; r = my - \left(s^2y \times \beta\right)^{\frac{1}{2}} \tag{2.60c}$$

Using the rainfall data provided in the appendix, Table 2.3 lists the sample statistics, including sample mean, variance, skewness, and kurtosis. It is seen that the dataset is right skewed with the sample skewness of 1.01. Based on the sample statistics and the basic common feature of the annual data, both normal and gamma distributions are applied to fit the annual rainfall data. The parameters estimated are listed in Table 2.4. Figure 2.4 compares the empirical frequency with the frequency computed from the fitted parametric distributions. It is seen from the figure that both normal and gamma distributions yield similar performances. However, the gamma distribution catches the peak frequency better.

Table 2.3 *Sample statistics of annual rainfall (mm), peak flow (cms), and maximum daily precipitation (mm)*

Sample Statistics	Mean	Variance	Skewness	Kurtosis
Annual rainfall (mm)	957.51	3.06E + 04	1.01	6.3
Peak flow (cms)	283.27	1.17E + 04	1.64	7.08
Maximum daily precipitation (mm)	56.24	346.47	0.99	4.08

Table 2.4 *Parameters estimated for annual rainfall (mm) and peak flow (cms) and maximum daily precipitation (mm)*

Variable	Distribution (Estimation Method)	Parameters Estimated		
Annual rain	Normal (MLE)	957.51	174.87	
(mm)	Gamma (MLE)	32.09	0.034	
Peak flow	Pearson (MLE)	121.86	1.45	104.78
(cms)	Log-Pearson (IDMOM)[1]	0.051	47.25	3.19
Maximum	Exponential (MLE)	56.24		
Daily precipitation	Gamma (MLE)	10.14	0.18	
(mm)	Log-Pearson (IDMOM)[1]	0.049	59.58	1.54

[1] IDMOM

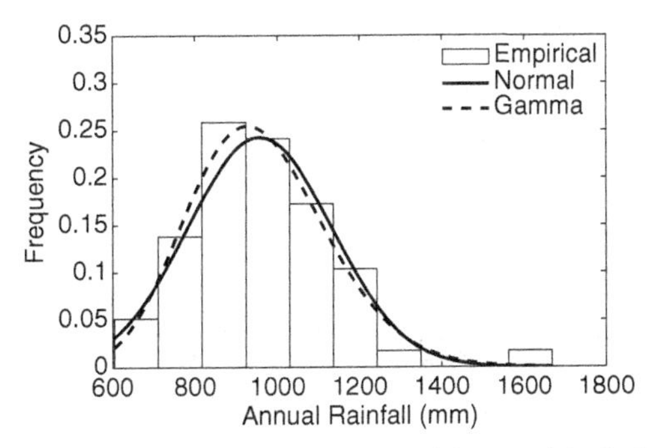

Figure 2.4 Comparison of fitted frequency with empirical frequency: annual rainfall amount.

Using peak flow data provided in the appendix, Tables 2.3–2.4 list the sample statistics and parameters estimated for Pearson III and log-Pearson III distributions. As annual rainfall data, MLE was applied to estimate the parameters of the Pearson III distribution, while the IDMOM was applied to estimate the parameters of the log-Pearson III distribution. Figure 2.5 compares the empirical frequency with the frequency computed from the fitted parametric distributions. It is seen from the figure that the log-Pearson III distribution performed better than did the Pearson III distribution.

Using the maximum daily precipitation data provided in the appendix, Tables 2.3 and 2.4 list the sample statistics and the parameters estimated for the exponential, gamma and log-Pearson III distributions. Figure 2.6 compares the empirical frequencies with the frequencies computed from the fitted parametric distributions. It is seen from the figure that the gamma and log-Pearson distributions yield similar values of frequencies. The gamma distribution performs slightly better than does the log-Pearson distribution and the exponential distribution may not be applied for study maximum daily precipitation.

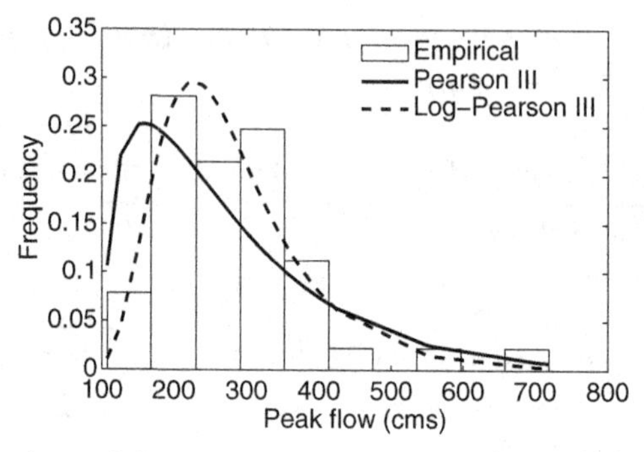

Figure 2.5 Comparison of fitted frequency with empirical frequency: peak flow.

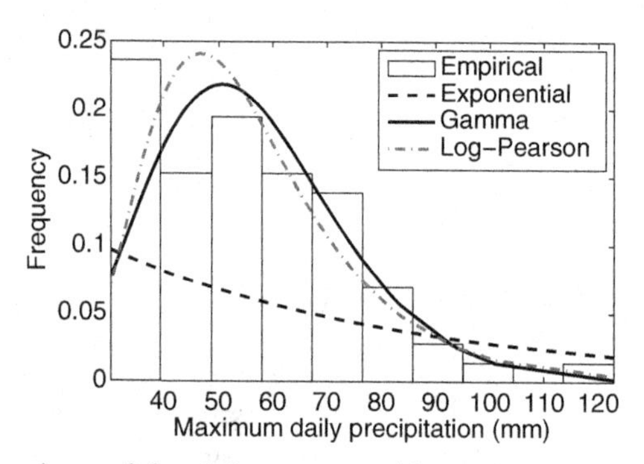

Figure 2.6 Comparison of fitted frequency with empirical frequency: maximum daily precipitation.

Applications to annual rainfall, peak flow, and maximum daily precipitation data show that the distributions in the Pearson system may be applied to analyze the frequency distribution after proper empirical study of observations.

2.7 Conclusion

The Pearson system can be regarded as perhaps the most basic system, for other systems were inspired by it. It leads to several frequency distributions that are frequently used in water and environmental engineering. This system also paves the way for a variety of hydrological inferences, such as distribution selection, moment diagram, and parameter estimation, to name but a few.

References

Bartels, C. P. A. (1977). *Economic Aspects of Regional Welfare*. Leiden: Martinus Nijhoff.
Craig, C. C. (1936). A new exposition and chart for Pearson system of frequency curves. *The Annuals of Mathematical Statistics* 7, no. 1, pp. 16–28.

Edgeworth, F. Y. (1898). On the representation of statistics by mathematical formulae. *Journal of the Royal Society* 1, pp. 670–700.
Elderton, W. P., and Johnson, N. L. (1969). *System of Frequency Curves*. Cambridge: Cambridge University Press.
Kloek, T., and van Dijk, N. K. (1977). Further results on efficient estimation of income distribution parameters. *Journal of Econometrics* 8, pp. 61–74.
McDonald, J. B. (1984). Some generalized functions for the size distribution of income. *Econometrics* 52, no. 3, pp. 647–663.
Pearson, K. (1894). Contributions to the mathematical theory of evolution. *Philosophical Transactions of the Royal Society* 184, pp. 71–110.
Pearson, K. (1895). Contributions to the mathematical theory of evolution. II. Skew variation in homogeneous material. *Transactions of the Royal Society London* A186, pp. 343–415.
Pearson, K. (1916) IX. Mathematical contributions to the theory of evolution.-XIX. Second supplement to a memoir on skew variation. *Philosophical Transactions of the Royal Society of London. Series A. Containing Papers of a Mathematical or Physical Character* 126 (538–548), 201–204.
Thurow, L. C. (1970). Analyzing the American income distribution. *American Economic Review* 48, pp. 261–269.
Toranzos, F. I. (1952). An asymmetric bell-shaped frequency curve. *Annals of Mathematical Statistics* 23, no. 3, pp. 467–469.

3
Burr System of Frequency Distributions

3.1 Introduction

Burr (1942) proposed a system of 12 frequency distributions, frequently referred to Burr I distribution through Burr XII distribution. Amongst these distributions, the Burr XII is one of the most commonly used distributions in hydrologic, hydraulic, environmental, and water resources engineering and hydrometeorology. These distributions are mostly 1- to 2-parameter distributions wherein the parameters are shape parameters that can be estimated using moment ratios. However, using a linear transformation, Burr (1942) also showed that the number of parameters of these distributions could be augmented by two, and these additional parameters would correspond to the origin and location and would relate to the mean and standard deviation. Burr (1942) also developed the theory of cumulative moments for estimating parameters of the system of distributions. This chapter discusses these distributions and the method of parameter estimation.

3.2 Characteristics of Probability Distribution Functions

For a random variable X, let the cumulative distribution function (CDF) be denoted as $F(x)$ and the probability density function (PDF) as $f(x)$. The random variable can vary from $-\infty$ to $+\infty$. Then, $F(x)$ is a nondecreasing function from 0 at $x \to -\infty$ to 1 at $x \to \infty$. The random variable can be continuous or discrete. Accordingly, $F(x)$ will be continuous or a step ladder unction. Here we will only deal with the continuous case but the discrete can be handled in a straightforward manner.

It is assumed that $F(x)$ is in contact with its asymptotes. To be specific, $F(x)x^k$ and $[1 - F(x)]x^k$ are bounded as x tends to $-\infty$ and $+\infty$, respectively. A particular expression for $F(x)$ over a range will be bounded in one of both directions and $F(x)$ will be 0 below the finite lower limit and one above the upper limit. By definition, the probability of $X = x$ between a and b is

$$P(a \leq x \leq b) = F(b) - F(a) \tag{3.1}$$

For the continuous random variable, we can write

$$F(x) = \int_{-\infty}^{x} f(x)dx, f(x) = \frac{dF(x)}{dx} \tag{3.2}$$

$$P(a \leq x \leq b) = \int_{a}^{b} f(x)\, dx \tag{3.3}$$

Burr (1942) argued that the direct fitting of $F(x)$ to data is more advantageous as Equations (3.1) to (3.3) suggest. Also, determining $f(x)$ from $F(x)$ is much simpler than determining $F(x)$ from $f(x)$, at least in some cases. Now the problem remains one of finding suitable CDFs and then determining their parameters.

3.3 Burr Hypothesis

Keeping the characteristics of F(x) enumerated in the preceding text in mind, Burr (1942) hypothesized that the relation between PDF and the CDF can be expressed in the form a differential equation as

$$\frac{dF(x)}{dx} = F(x)(1 - F(x))g(x, F(x)) \tag{3.4}$$

where $F(x)$ is the probability distribution function (CDF), $g(x, F(x))$ is some function that is positive for $0 \leq F(x) \leq 1$, and the range of x is defined as desired. The form of Equation (3.4) is similar to Pearson differential equation if $g(x, F(x)) = (c_1 + c_2x + c_3x^2)^{-1}$ with $F(x)$, $1 - F(x)$ of Equation (3.4) replaced by $f(x)$ and $a - x$, respectively.

Equation (3.4) shows that $F(x)$ is a nondecreasing function, whereas for some forms of the $g(x, F(x))$ function, we obtain $\frac{dF(x)}{dx} = 0, \forall F(x) = 0$ and $F(x) = 1$. It may be remarked that Verhulst (1845) employed a differential equation that has gone somewhat unnoticed nearly a century earlier. This equation can be regarded as a special case of the Burr differential equation and now constitutes one of the paradigms in the chaos theory (Ausloos and Diricks, 2006).

For simplicity, let $g(x, F(x)) = g(x)$. Then, solution of Equation (3.4) becomes

$$\frac{dF(x)}{dx} = F(x)(1 - F(x))g(x) \tag{3.5}$$

Equation (3.5) can be written as

$$\frac{dF(x)}{F(x)(1 - F(x))} = g(x)dx \tag{3.6}$$

For simplicity of integration, Equation (3.6) can be cast as

$$\frac{dF(x)}{F(x)} + \frac{dF(x)}{1-F(x)} = g(x)dx \tag{3.7}$$

Integrating Equation (3.7), one gets

$$\ln\left(\frac{1-F(x)}{F(x)}\right) = -\int g(x)\,dx \tag{3.8}$$

Equation (3.8) yields

$$F(x) = \frac{1}{1+\exp\left(-\int g(x)\,dx\right)} \tag{3.9}$$

The function $g(x)$ should be chosen such that $F(x)$ increases from 0 to 1 on the specified interval of x. For example, $g(x) = c, g(x) = c\sec^2 x, g(x) = c\cosh x$, $g(x) = \frac{c}{x}, g(x) = \frac{1}{x(c-x)}$ with c > 0 fulfilling the condition of $F(x)$ increasing from 0 to 1. In what follows, we illustrate this property using several examples.

Example 3.1 $g(x) = c, c > 0$

First, from Equation (3.5) we see that $g(x) = c, c > 0$ fulfills (i) $f(x) > 0$ and (ii) $F(x)$ is a nondecreasing function. Substituting $g(x) = c$ into Equation (3.9) we have:

$$\begin{aligned} F(x) &= \frac{1}{1+\exp\left(-\int_{x_0}^{x} c dx\right)} = \frac{1}{1+\exp\left(-cx+cx_0\right)} \\ &= \frac{1}{1+\left(\exp\left(cx_0\right)\right)\exp\left(-cx\right)} = \frac{1}{1+C_0\exp\left(-cx\right)} \end{aligned} \tag{3.10}$$

In Equation (3.10), x_0 denotes the lower limit of the interval for x, and C_0 is the constant. With $c > 0$, $\exp(-cx)$ is a decreasing function if $x \in \mathbb{R}$. When $x \to \infty$, $\exp(-cx) \to 0 \Rightarrow F(x \to \infty) \to 1$. Additionally, when $x \to -\infty$, $\exp(-cx) \to +\infty \Rightarrow F(x \to -\infty) \to 0$. This example shows that $g(x) = c$ fulfills the Burr hypothesis for $x \in \mathbb{R}$.

Example 3.2 $g(x) = c\ \sec^2 x, c > 0$

Using Equation (3.5), we can easily show that $g(x) = c\sec^2 x, c > 0$ fulfills (i) $f(x) > 0$ and (ii) $F(x)$ is a nondecreasing function. Substituting $g(x) = c\sec^2 x$ into Equation (3.9), we have:

$$F(x) = \frac{1}{1+\exp\left(-\int_{x_0}^{x} c\sec^2 x dx\right)} = \frac{1}{1+C_0\exp\left(-c\tan x\right)} \tag{3.11}$$

In Equation (3.11), $\tan x$ is a monotone increasing function from $\left(-\frac{k\pi}{2}, \frac{k\pi}{2}\right), k = 1$, 2, ... When $x \to -\frac{k\pi}{2}$, $\exp\left(-c \tan x\right) \to \infty \Rightarrow F\left(x \to -\frac{k\pi}{2}\right) \to 0$. When $x \to \frac{k\pi}{2}$, $\exp\left(-c\tan x\right) \to 0 \Rightarrow F\left(x \to \frac{k\pi}{2}\right) = 1$. This example shows that $g(x) = c\ \sec^2 x$ fulfills the Burr hypothesis on the interval of $\left\{x : x \in \left(-\frac{k\pi}{2}, \frac{k\pi}{2}\right), k = 1, 2, \ldots\right\}$.

Example 3.3 $g(x) = c \cosh x$
For this g function, we have the hyperbolic cosine function given as:

$$\cosh x = \frac{e^x + e^{-x}}{2} > 0 \tag{3.12}$$

Equation (3.12) indicates that Equation (3.5) is greater than 0 and $F(x)$ is nondecreasing function. The integral of the hyperbolic cosine function is:

$$\int \cosh x dx = \sinh x + C_0 = \frac{1 - e^{-2x}}{2e^{-x}} + C_0 \tag{3.13}$$

It is worth noting that the hyperbolic sine function is a monotone increasing function from $(-\infty, +\infty)$ for $x \in (-\infty, +\infty)$. Now substituting Equation (3.13) into Equation (3.9), we have:

$$F(x) = \frac{1}{1 + \exp\left(-sinh\ x - C_0\right)} = \frac{1}{1 + C \exp\left(-c \sinh x\right)}. \tag{3.14}$$

In Equation (3.14), it is shown for $c > 0$ that when $x \to -\infty \Rightarrow \sinh x \to -\infty \Rightarrow 1 + C \exp(-c \sinh x) \to \infty \Rightarrow F(x \to -\infty) \to 0$ and when $x \to \infty \Rightarrow \sinh x \to \infty \Rightarrow 1 + C \exp(-c \sinh x) \to 1 \Rightarrow F(x \to \infty) \to 1$. This example shows that $g(x) = \cosh x$ fulfills the Burr hypothesis on $x \in \mathbb{R}$.

Example 3.4 $g(x) = \frac{1}{x(c-x)}$
In the case of $g(x) = \frac{1}{x(c-x)}$, $g(x) > 0$ if $x(c - x) > 0 \Rightarrow x \in (0, c)$ such that (i) $f(x) > 0$ and (ii) $F(x)$ is a nondecreasing function. The integration of $g(x)$ function is expressed as:

$$\int \frac{1}{x(c-x)} dx = \frac{1}{c} \ln\left(\frac{x}{c-x}\right) + C_0 \tag{3.15}$$

Now substituting Equation (3.15) into Equation (3.9), we have:

$$F(x) = \frac{1}{1 + \exp\left(-ln\left(\frac{x}{c-x}\right) + C_0\right)} = \frac{1}{1 + C\left(\frac{c-x}{x}\right)} \tag{3.16}$$

in Equation (3.16), when $x \to 0 \Rightarrow 1 + \frac{C(c-x)}{x} \to \infty \Rightarrow F(x \to 0) \to 0$; and when $x \to c \Rightarrow 1 + \frac{C(c-x)}{c} \to 1 \Rightarrow F(x \to c) \to 1$. This example shows that $g(x) = \frac{1}{x(c-x)}$ fulfills the Burr hypothesis on the interval of $x \in (0, c)$.

Now, we have applied several examples to explain the Burr hypothesis. In what follows, we will discuss different types of Burr frequency distributions.

3.4 Burr System of Frequency Distributions

Burr (1942) provided a set of 12 CDFs, which together constitute what is commonly referred to as Burr family or system. The last member of the family, known as Burr XII distribution, is one of the most popular distributions in hydrometeorology and flood hydrology (Mielke, 1973) and is often considered as a reference distribution because of its unique properties. The Burr XII distribution has been generalized by Singh and Maddala (1976), and the generalized version, called Burr-Singh-Maddala (BSM) distribution (Brouwers, 2015), is used in a variety of areas, including econometrics, actuarial sciences, forestry and ecology, chemistry, and water engineering. Two of its approximations, the Weibull and Hill distributions (Brouwers, 2015) are widely employed in material sciences, medical sciences, physico-chemistry, and hydrometeorology. For purposes of illustrating the derivation of the Burr distributions, let us consider these distributions. It should be noted that parameters appearing in the following distributions are positive real numbers.

3.4.1 Burr I Distribution

The $g(x)$ function for Burr I distribution is given as:

$$g(x) = \frac{1}{x(1-x)}; x \in (0,1) \tag{3.17}$$

Substituting Equation (3.17) into Equation (3.9), we have:

$$\begin{aligned} F(x) &= \frac{1}{\exp\left(-\int_0^x \frac{1}{x(1-x)}dx\right)+1} = \frac{1}{\exp\left(-\int_o^x \left(\frac{1}{x}-\frac{1}{1-x}\right)dx\right)+1} \\ &= \frac{1}{\exp\left(ln\left(\frac{1-x}{x}\right)\right)+1} = \frac{1}{\frac{1-x}{x}+1} = x; x \in (0,1) \end{aligned} \tag{3.18}$$

Taking the derivative of Equation (3.18), we obtain the PDF as: $f(x) = \frac{dF(x)}{dx} = 1$. The Burr I distribution is a uniform distribution with $x \in (0,1)$. The $g(x)$ function for Burr I distribution is a U-shape function that is deduced from $g(x) = ((c-x)x)^{-1}$ with $c = 1$. Figure 3.1 plots $g(x), F(x)$, and $f(x)$ of Burr I distribution.

3.4.2 Burr II Distribution

The $g(x)$ function for Burr II distribution is given as

$$g(x) = \frac{re^{-x}(e^{-x}+1)^{r-1}}{(e^{-x}+1)^r - 1}, r > 0, x \in (-\infty, +\infty) \tag{3.19}$$

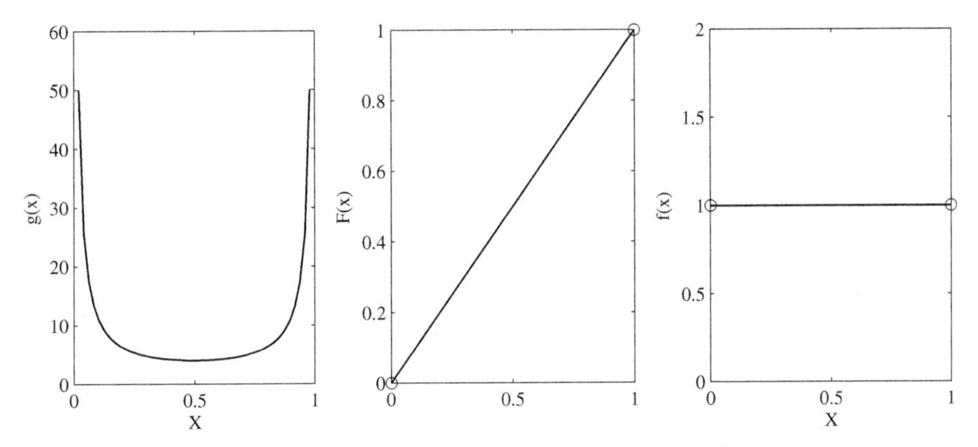

Figure 3.1 Plots of the g(x), CDF and PDF functions for Burr I distribution.

Equation (3.19) shows that $g(x) > 0$ for $r > 0$. In other words, it fulfills the condition that $F(x)$ is a nondecreasing function.

Substituting Equation (3.19) into Equation (3.9), we have:

$$\int_{-\infty}^{x} g(x) = \int_{-\infty}^{x} \frac{re^{-x}(e^{-x}+1)^{r-1}}{(e^{-x}+1)^{r}-1} dx = -\ln\left((e^{-x}+1)^{r}-1\right) \tag{3.20}$$

$$F(x) = \frac{1}{1+\exp\left(-\int_{-\infty}^{x} g(x)dx\right)} = \frac{1}{1+\exp\left(\ln\left((e^{-x}+1)^{r}-1\right)\right)}$$
$$= (\exp(-x)+1)^{-r}; r > 0, x \in (-\infty, +\infty) \tag{3.21}$$

Furthermore, Equation (3.21) shows:

$$x \to -\infty \Rightarrow F(x \to -\infty) \to \infty^{-r} \Rightarrow F(x \to -\infty) \to 0, \text{ and}$$

$$x \to \infty \Rightarrow F(x \to \infty) \to 1^{-r} \Rightarrow F(x \to \infty) \to 1.$$

Now we have shown that the Burr hypothesis is satisfied.

Taking the derivative of Equation (3.21), the PDF of Burr II distribution can be given as:

$$f(x) = \frac{re^{-x}}{(e^{-x}+1)^{r+1}}; r > 0, x \in (-\infty, +\infty) \tag{3.22}$$

It is S-shaped when $r < 1$; it is reverse S-shaped when $r > 1$; and $g(x) = 1$ if $r = 1$. There exists a limit for $g(x)$ as: $\lim_{x\to\infty} g(x) = 1$. The PDF is unimodal and is

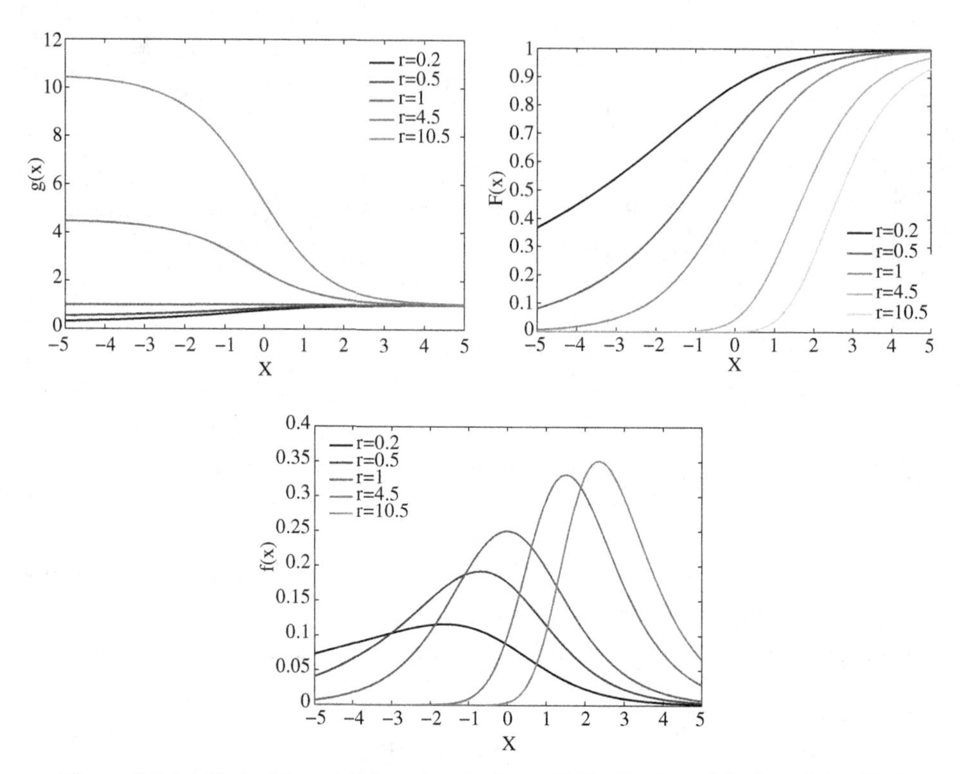

Figure 3.2 (a) Plot of the g(x) function for Burr II distribution; (b) plot of the CDF F(x) of Burr II distribution; and (c) plot of the PDF f(x) of Burr II distribution.

generally bell shaped and skewed, however, it is symmetric about the y-axis when $r = 1$. Figures 3.2a–3.2c plot $g(x), F(x)$, and $f(x)$ for Burr II distribution.

3.4.3 Burr III Distribution

The $g(x)$ function for Burr III distribution is written as:

$$g(x) = -\frac{kr}{x((x^{-k}+1)^{-r}-1)(x^k+1)} = \frac{kr}{x(1-(x^{-k}+1)^{-r})(x^k+1)}; \tag{3.23}$$

$$k > 0, r > 0, x \in (0, +\infty)$$

Equation (3.23) shows that $g(x) > 0$. It indicates that (i) $f(x) > 0$ and (ii) $F(x)$ is a nondecreasing function if $F(x) \in [0, 1]$.

Substituting Equation (3.23) into Equation (3.9), we have:

$$\int_0^x g(x)\,dx = \int_0^x \frac{kr}{x(1-(x^{-k}+1)^{-r})(x^k+1)}dx = -\ln\left((1+x^{-k})^r - 1\right) \tag{3.24}$$

The Burr III distribution can be given as:

$$F(x) = \frac{1}{1 + \exp\left(-\int_0^x g(x)dx\right)} = \frac{1}{1 + \exp\left(ln\left((1 + x^{-k})^r - 1\right)\right)}$$
$$= \frac{1}{(1 + x^{-k})^r} = \left(1 + x^{-k}\right)^{-r}, k > 0, r > 0, x \in (0, +\infty) \tag{3.25}$$

Furthermore, Equation (3.25) shows that:

$$x \to 0 \Rightarrow F(x \to 0) \to (1 + \infty)^{-r} \Rightarrow F(x \to 0) \to 0; \text{ and}$$

$$x \to \infty \Rightarrow F(x \to \infty) \to (1 + 0)^{-r} \Rightarrow F(x \to \infty) \to 1.$$

Now with the discussion of the corresponding $g(x)$ function, the Burr hypothesis is satisfied.

Taking the derivative of Equation (3.25), the PDF of Burr III distribution can be given as:

$$f(x) = rkx^{-k-1}(1 + x^{-k})^{-r-1}. \tag{3.26}$$

The $g(x)$ function of Burr III distribution is L-shaped for any given parameter set $\{k, r\} : g(x) > 0$. The limit of $g(x)$ is $\lim_{x \to \infty} g(x) = 0$. The PDF of Burr III distribution is unimodal. It is L-shaped if any of k or r is less than 1 and is bell shaped and skewed when both of the parameters are greater than 1. Figures 3.3a–3.3c plot $g(x), F(x), f(x)$ for the Burr III distribution.

3.4.4 Burr IV Distribution

The $g(x)$ function for Burr IV distribution is given as:

$$g(x) = \frac{r\left(\frac{c-x}{x}\right)^{\frac{1}{c}-1}}{x^2\left(1 - \left(\left(\frac{c-x}{x}\right)^{\frac{1}{c}} + 1\right)^{-r}\right)\left(1 + \left(\frac{c-x}{x}\right)^{\frac{1}{c}}\right)}; c, r > 0, x \in (0, c) \tag{3.27}$$

Equation (3.27) shows that $g(x) > 0$ on the interval $x \in (0, c)$ with c, $r > 0$. It indicates that: (i) $f(x) > 0$ and (ii) $F(x)$ is a nondecreasing function in the given interval.

Substituting Equation (3.27) into Equation (3.9), we have:

$$\int_0^x g(x)dx = \int_0^x \frac{r\left(\frac{c-x}{x}\right)^{\frac{1}{c}-1}}{x^2\left(1 - \left(\left(\frac{c-x}{x}\right)^{\frac{1}{c}} + 1\right)^{-r}\right)\left(1 + \left(\frac{c-x}{x}\right)^{\frac{1}{c}}\right)} dx$$
$$= -\ln\left(\left(\left(\frac{c}{x} - 1\right)^{\frac{1}{c}} + 1\right)^r - 1\right) \tag{3.28}$$

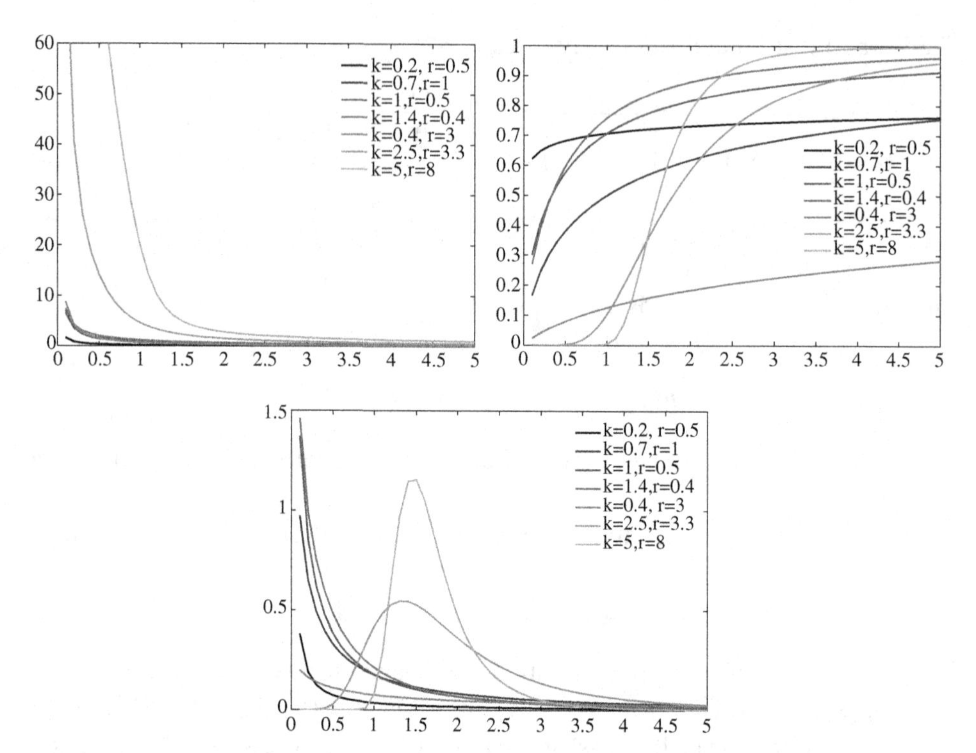

Figure 3.3 (a) Plot of the g(x) function for Burr III distribution; (b) plot of the CDF F(x) of Burr III distribution; and (c) plot of the PDF f(x) of Burr III distribution.

The Burr IV distribution can now be expressed as:

$$F(x) = \frac{1}{1+\exp\left(-\int g(x)dx\right)} = \frac{1}{1+\exp\left(ln\left(\left(\left(\frac{c}{x}-1\right)^{\frac{1}{c}}+1\right)^{r}-1\right)\right)}$$
$$= \frac{1}{\left(\left(\frac{c}{x}-1\right)^{\frac{1}{c}}+1\right)^{r}} = \left(\left(\frac{c}{x}-1\right)^{\frac{1}{c}}+1\right)^{-r} ;c,r>0,x\in(0,c) \tag{3.29}$$

Taking the derivative of Equation (3.29), the PDF of Burr IV distribution can be given as:

$$f(x) = \frac{r\left(\frac{c-x}{x}\right)^{\frac{1}{c}-1}}{x^{2}\left(\left(\frac{c-x}{x}\right)^{\frac{1}{c}}+1\right)^{r+1}} ;c,r,>0,x\in(0,c) \tag{3.30}$$

In case of Burr IV distribution, the $g(x)$ function is U-shaped with $g(x)>0$. Its PDF is generally U-shaped as well. Figures 3.4a–3.4c plot $g(x), F(x),$ and $f(x)$ for Burr IV distribution.

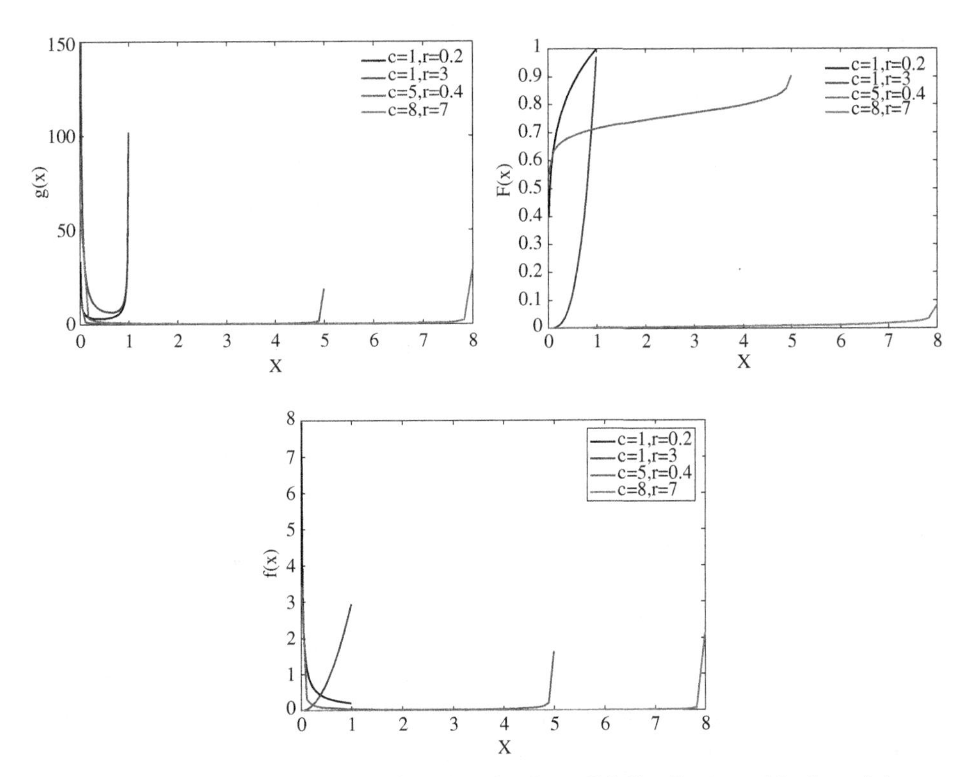

Figure 3.4 (a) Plot of the g(x) function for Burr IV distribution; (b) plot of the CDF F(x) of Burr IV distribution; and (c) plot of the PDF f(x) of Burr IV distribution.

3.4.5 Burr V Distribution

The $g(x)$ function for Burr V distribution is given as

$$g(x) = -\frac{kr\left(\left(ke^{-\tan x}+1\right)^{-r}-1\right)\left(\tan^2(x)+1\right)}{k+e^{\tan x}};k,r>0,x\in\left(-\frac{\pi}{2},\frac{\pi}{2}\right) \quad (3.31)$$

Equation (3.31) shows that $g(x) > 0$. It indicates that: (i) $f(x) > 0$ and (ii) $F(x)$ is a nondecreasing function for $F \in [0, 1]$.

Substituting Equation (3.31) into Equation (3.9), we obtain:

$$\begin{aligned}\int_{-\frac{\pi}{2}}^{x} g(x)\,dx &= \int_{-\frac{\pi}{2}}^{x} -\frac{kr\left(\left(ke^{-\tan x}+1\right)^{-r}-1\right)\left(\tan^2 x+1\right)}{k+e^{\tan x}}\,dx \\ &= -\ln\left(\left(ke^{-\tan x}+1\right)^{r}-1\right)\end{aligned} \quad (3.32)$$

The Burr V distribution can then be expressed as:

$$F(x) = \frac{1}{1 + \exp\left(-\int_{\frac{-\pi}{2}}^{x} g(x)dx\right)} = \frac{1}{1 + \exp\left(ln\left(ke^{-tan\,x} + 1\right)^{r} - 1\right)}$$
$$= \frac{1}{\left(ke^{-\tan(x)} + 1\right)^{r}}; k, r > 0, x \in \left(-\frac{\pi}{2}, \frac{\pi}{2}\right) \tag{3.33}$$

Furthermore, it is seen from Equation (3.33) that $x \to -\frac{\pi}{2} \Rightarrow F\left(x \to -\frac{\pi}{2}\right) \to \frac{1}{\infty} \Rightarrow F\left(x \to -\frac{\pi}{2}\right) \to 0$ and $x \to \infty \Rightarrow F\left(x \to \frac{\pi}{2}\right) \to 1$. With the discussion of $g(x)$, the Burr hypothesis is satisfied.

Taking the derivative of Equation (3.33), the PDF of Burr V distribution can be given as:

$$f(x) = \frac{kr\left(\tan^{2}(x) + 1\right)}{e^{\tan(x)}\left(ke^{-\tan(x)} + 1\right)^{r+1}}; k, r > 0, x \in \left(-\frac{\pi}{2}, \frac{\pi}{2}\right) \tag{3.34}$$

The $g(x)$ function for Burr V distribution is L-shaped with $g(x) > 0$. The PDF of Burr V distribution may be either unimodal or bimodal, depending on the parameter. The Burr V distribution is skewed. Figures 3.5a–3.5c plot $g(x)$, $F(x)$, and $f(x)$ of Burr V distribution with different parameters.

3.4.6 Burr VI Distribution

The $g(x)$ function for Burr VI distribution is given as

$$g(x) = -\frac{ckr\cosh(x)e^{-\sinh x}}{\left(\left(ke^{-c\sinh(x)} + 1\right)^{-r} - 1\right)\left(ke^{-c\sinh(x)} + 1\right)}; c, k, r > 0, x \in (-\infty, \infty) \tag{3.35}$$

Equation (3.35) shows that $g(x) > 0$ with the constants being greater than 0. It indicates that (i) $f(x) = \frac{dF}{dx} > 0$, and (ii) $F(x)$ is a nondecreasing function in the range of (0,1).

Substituting Equation (3.35) into Equation (3.9), we have:

$$\int_{-\infty}^{x} g(x)dx = \int_{-\infty}^{x} -\frac{ckr\cosh(x)e^{-\sinh x}}{\left(\left(ke^{-c\sinh(x)} + 1\right)^{-r} - 1\right)\left(ke^{-c\sinh(x)} + 1\right)}dx$$
$$= -\ln\left(\left(ke^{-c\sinh x} + 1\right)^{r} - 1\right) \tag{3.36}$$

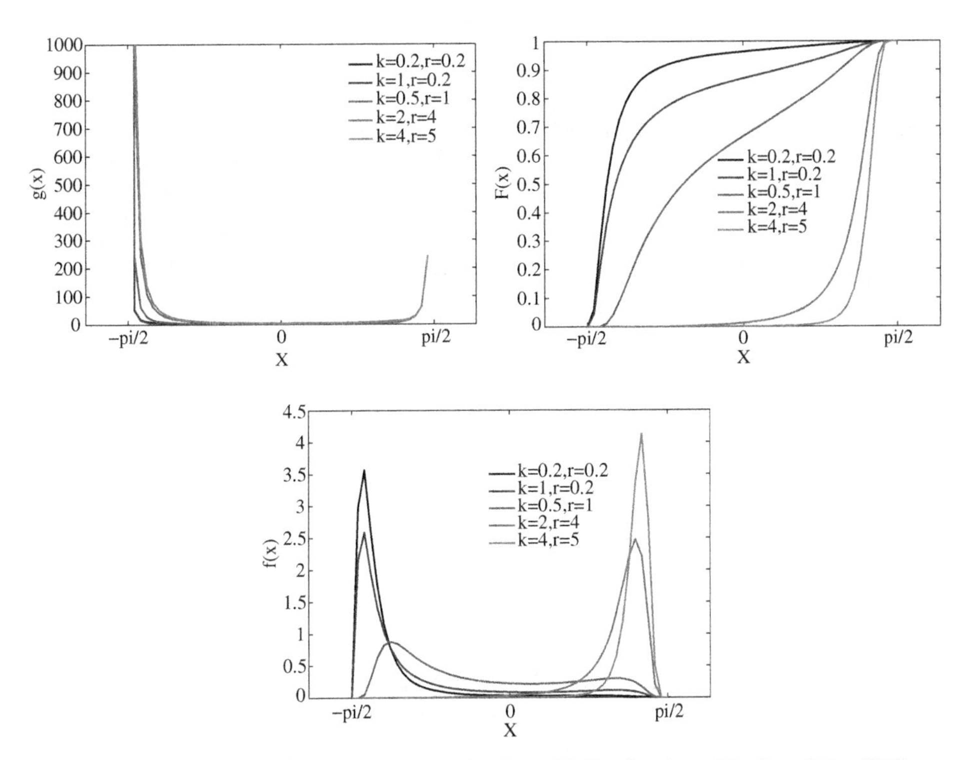

Figure 3.5 (a) Plot of the g(x) function for Burr V distribution; (b) plot of the CDF F(x) of Burr V distribution; and (c) plot of the PDF f(x) of Burr V distribution.

The CDF of Burr VI distribution can then be expressed as:

$$\begin{aligned} \mathrm{F(x)} &= \frac{1}{1+\exp\left(-\int_{-\infty}^{x} g(x)dx\right)} = \frac{1}{1+\exp\left(\ln\left(\left(ke^{-c\sinh x}+1\right)^{r}-1\right)\right)} \\ &= \frac{1}{\left(ke^{-c\sinh x}+1\right)^{r}} = \left(ke^{-c\sinh x}+1\right)^{-r}; c,k,r>0, x\in(-\infty,\infty) \end{aligned} \tag{3.37}$$

From Equation (3.37), it is seen that:

$\mathrm{x} \to -\infty \Rightarrow \mathrm{e}^{-c\sinh x} \to \infty \Rightarrow \mathrm{F(x} \to -\infty) \to \infty^{-\mathrm{r}} \Rightarrow F(x \to -\infty) \to 0$; and

$\mathrm{x} \to +\infty \Rightarrow \mathrm{e}^{-c\sinh x} \to 0 \Rightarrow F(x \to +\infty) \to 1^{-r} \Rightarrow F(x \to +\infty) \to 1$.

Taking the derivative of Equation (3.37), the PDF of Burr VI distribution can be given as:

$$f(x) = \frac{ckr\cosh(x)}{e^{c\sinh(x)}\left(ke^{-c\sinh(x)}+1\right)^{r+1}}; c,k,r>0, x\in(-\infty,\infty) \tag{3.38}$$

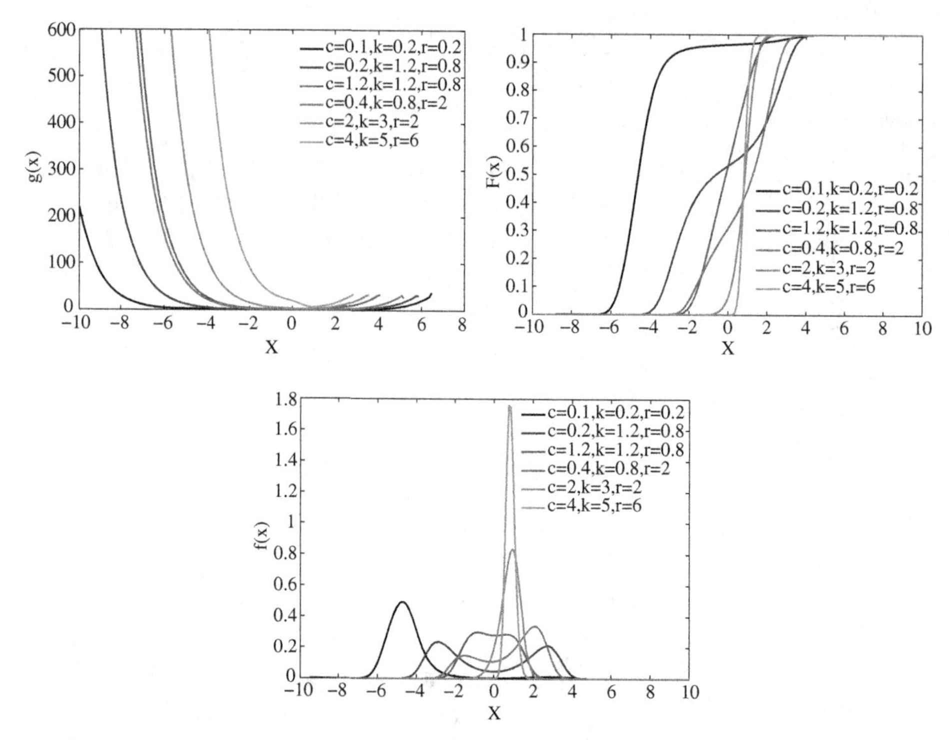

Figure 3.6 (a) Plot of the g(x) function for Burr VI distribution; (b) plot of the CDF F(x) of Burr VI distribution; and (c) plot of the PDF f(x) of Burr VI distribution.

The $g(x)$ function for Burr VI distribution is L-shaped with $g(x) > 0$. The PDF of Burr VI distribution is unimodal if all the parameters are greater than 1 and tends to be bimodal if at least one of the parameters is less than 1. Figures 3.6a–3.6c plot $g(x), F(x)$, and $f(x)$ for Burr VI distribution with different parameter sets.

3.4.7 Burr VII Distribution

The $g(x)$ function for Burr VII distribution is given as

$$g(x) = \frac{r(\tanh(x) - 1)}{\left(\frac{\tanh(x)+1}{2}\right)^r - 1}; r > 0, x \in (-\infty, \infty) \tag{3.39}$$

Substituting $\tanh x = \frac{e^{2x}-1}{e^{2x}+1}$ into Equation (3.39) we have:

$$\frac{\tanh x + 1}{2} = \frac{e^{2x}}{e^{2x}+1} < 1 \Rightarrow \left(\frac{\tanh x + 1}{2}\right)^r - 1 < 0; \tanh x - 1 = -\frac{2}{e^{2x}+1} < 0 \tag{3.40}$$

Equation (3.40) clearly shows that $g(x)$ defined by Equation (3.39) is greater than 0. It indicates that (i) $f(x) = \frac{dF}{dx} > 0$ and (ii) $F(x)$ is a nondecreasing function in the range of [0,1].

Substituting Equation (3.39) into Equation (3.9) we have:

$$\int_{-\infty}^{x} g(x)\,dx = \int_{-\infty}^{x} \frac{r(\tanh x - 1)}{\left(\frac{\tanh x+1}{2}\right)^r - 1}\,dx = 2\text{arctanh}\left(2\left(\frac{\tanh x + 1}{2}\right)^r - 1\right) \quad (3.41)$$

Applying $\text{arctanh}\, x = \frac{1}{2}\ln\left(\frac{1+x}{1-x}\right)$, Equation (3.41) can be rewritten as:

$$\int_{-\infty}^{x} g(x)\,dx = \ln\left(\frac{2^{-r}(\tanh x + 1)^r}{1 - 2^{-r}(\tanh x + 1)^r}\right) \quad (3.42)$$

The CDF of the Burr VII distribution can now be given as:

$$\begin{aligned} \mathrm{F(x)} &= \frac{1}{1 + \exp\left(-\int_{-\infty}^{x} g(x)dx\right)} = \frac{1}{1 + \frac{(1-2^{-r}(\tanh x+1)^r)^r}{2^{-r}(\tanh x+1)}} \\ &= 2^{-r}(\tanh x + 1)^r; x \in (-\infty, +\infty), \mathrm{r} > 0 \end{aligned} \quad (3.43)$$

In Equation (3.43), we can show that:

$$\mathrm{x} \to -\infty \Rightarrow \frac{\tanh x + 1}{2} \to 0 \Rightarrow F(x \to -\infty) \to 0, \text{ and}$$

$$\mathrm{x} \to \infty \Rightarrow \frac{\tanh x + 1}{2} = \frac{1}{1 + e^{-2x}} \to 1 \Rightarrow F(x \to \infty) \to 1$$

Now, with the discussion of $g(x)$ function, Burr hypothesis is satisfied for the Burr VII distribution.

Taking the derivative of Equation (3.43), the PDF of Burr VII distribution is given as:

$$f(x) = \frac{r(\tanh(x) + 1)^{r-1}\text{sech}^2(x)}{2^r} \quad (3.44)$$

Similar to Burr II distribution, $g(x) > 0$ and it is S-shaped if $r < 1$, it is (reverse S-shaped) if $r > 1$, and $g(x) = 2$ if $r = 1$. There exists a limit for the $g(x)$ function as: $\lim_{x\to\infty} g(x) = 1$. The PDF of the Burr VII distribution is unimodal bell shaped and generally skewed. It is symmetric about the y-axis if $r = 1$. Figures 3.7a–3.7c plot $g(x), F(x), \text{and} f(x)$ of Burr VII distribution with different parameters.

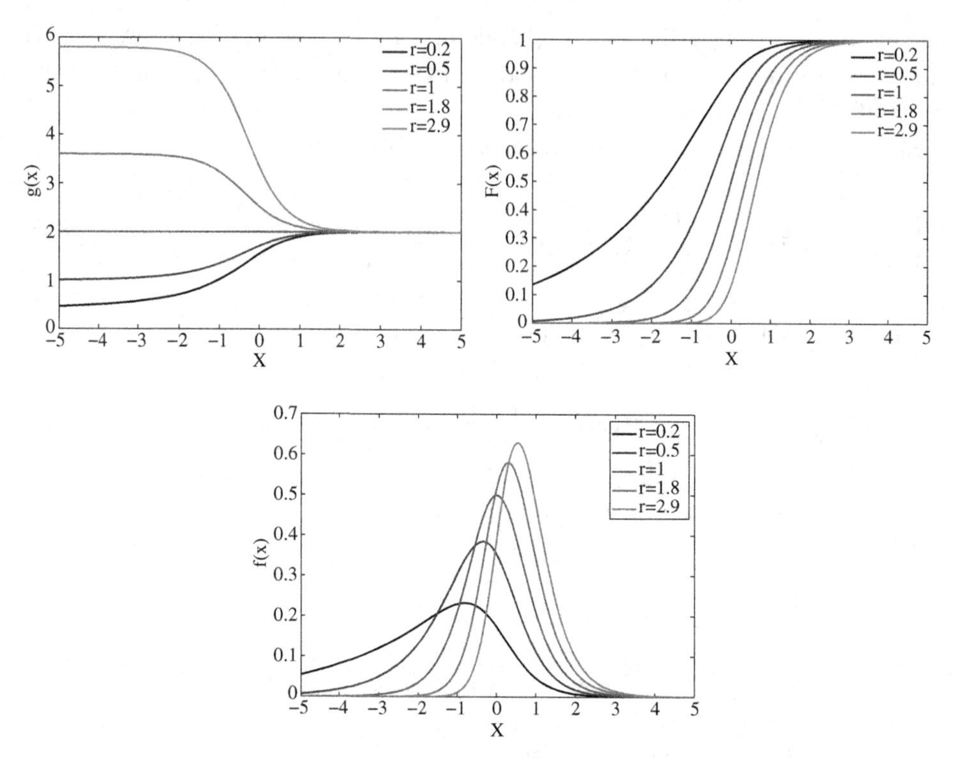

Figure 3.7 (a) Plot of the g(x) function for Burr VII distribution; (b) plot of the CDF F(x) of Burr VII distribution; and (c) plot of the PDF f(x) of Burr VII distribution.

3.4.8 Burr VIII Distribution

The $g(x)$ function for Burr VIII distribution is given as

$$g(x) = \frac{re^x}{\tan^{-1}(e^x)(e^{2x}+1)\left(1-\left(\frac{2\tan^{-1}(e^x)}{\pi}\right)^r\right)}; r > 0, x \in (-\infty, \infty) \tag{3.45}$$

In Equation (3.45), $\tan^{-1}(e^x) \in \left(0, \frac{\pi}{2}\right) \Rightarrow g(x) > 0$. It indicates that: (i) $f(x) = \frac{dF}{dx} > 0$, and (ii) $F(x)$ a nondecreasing function.

Substituting Equation (3.45) into Equation (3.9), we have:

$$\int_{-\infty}^{x} g(x)dx = \int_{-\infty}^{x} \frac{re^x}{\tan^{-1}(e^x)(e^{2x}+1)\left(1-\left(2\tan^{-1}\frac{e^x}{\pi}\right)^r\right)}dx$$

$$= \ln\left(\frac{\left(\frac{2\tan^{-1}(e^x)}{\pi}\right)^r}{1-\left(\frac{2\tan^{-1}(e^x)}{\pi}\right)^r}\right) \tag{3.46}$$

The CDF of Burr VIII distribution can now be expressed as:

$$F(x) = \frac{1}{1 + \exp\left(-\int_{-\infty}^{x} g(x)dx\right)} = \frac{1}{1 + \exp\left(-\ln\left(\frac{\left(\frac{2\tan^{-1}(e^x)}{\pi}\right)^r}{1 - \left(\frac{2\tan^{-1}(e^x)}{\pi}\right)^r}\right)\right)}$$

$$= \frac{1}{1 + \frac{1 - \left(\frac{2\tan^{-1}(e^x)}{\pi}\right)^r}{\left(\frac{2\tan^{-1}(e^x)}{\pi}\right)^r}} = \left(\frac{2\tan^{-1}(e^x)}{\pi}\right)^r; r > 0, x \in (-\infty, +\infty) \tag{3.47}$$

From Equation (3.47), we can clearly show that:
$x \to -\infty \Rightarrow e^x \to 0 \Rightarrow \tan^{-1}(e^x) \to 0 \Rightarrow F(x \to -\infty) \to 0$, and

$$x \to +\infty \Rightarrow e^x \to +\infty \Rightarrow \tan^{-1}(e^x) \to \frac{\pi}{2} \Rightarrow F(x \to \infty) \to 1.$$

Taking the derivative of Equation (3.47), the PDF of Burr VIII distribution can be given as:

$$f(x) = \frac{2re^x\left(\frac{2\tan^{-1}(e^x)}{\pi}\right)^{r-1}}{\pi(e^{2x} + 1)}; r > 0, x \in (-\infty, \infty) \tag{3.48}$$

The $g(x)$ function has a value less than 0. $g(x)$ is S-shaped if $r > 1$. $g(x)$ is reverse S-shaped if $r \leq 1$. The limit of $g(x)$ for Burr distribution may be given as: $\lim_{x\to\infty} g(x) = -1$. The PDF of Burr VIII distribution is generally bell shaped and skewed. It is symmetric about the y-axis if $r = 1$. Figures 3.8a–3.8c plot $g(x), F(x)$, and $f(x)$ for Burr VIII distribution with different parameters.

3.4.9 Burr IX Distribution

The $g(x)$ function for Burr IX distribution is given as

$$g(x) = \frac{re^x(e^x + 1)^{r-1}}{(e^x + 1)^r - 1}; r > 0, x \in (-\infty, \infty) \tag{3.49}$$

For $r > 0$, $x \in (-\infty, +\infty)$, it is clear that $g(x) > 0$. It indicates that: (i) $f(x) = \frac{dF}{dx} > 0$ and (ii) $F(x)$ is a nondecreasing function. Furthermore, the $g(x)$ function is shown to be independent of parameter k and has the following interesting properties:

$$g(x) > 0; \lim_{x\to-\infty} g(x) = 1; \lim_{x\to\infty} g(x) = r \tag{3.49a}$$

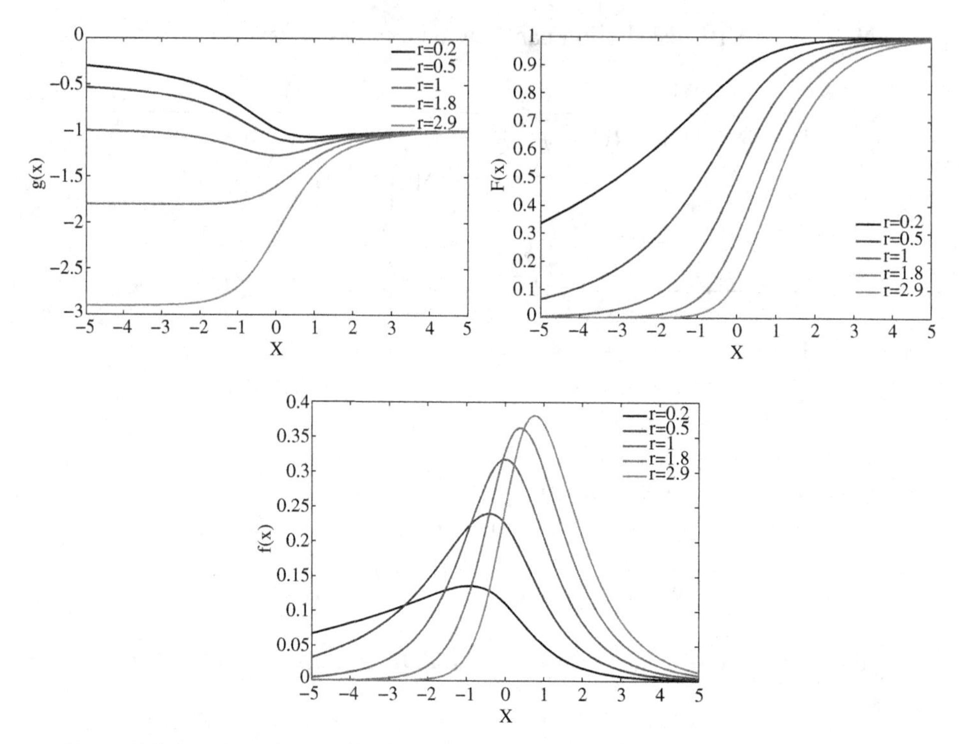

Figure 3.8 (a) Plot of the g(x) function for Burr VIII distribution; (b) plot of the CDF F(x) of Burr VIII distribution; and (c) plot of the PDF f(x) of Burr VIII distribution.

Substituting Equation (3.4) into Equation (3.9), we have:

$$\int_{-\infty}^{x} g(x)\,dx = \int_{-\infty}^{x} \frac{re^{x}(e^{x}+1)^{r-1}}{(e^{x}+1)^{r}-1}\,dx = \ln\left((e^{x}+1)^{r}-1\right) \tag{3.50}$$

The CDF of Burr IX distribution can now be given as:

$$\begin{aligned} F(x) &= \frac{1}{1+\exp\left(-\int_{-\infty}^{x} g(x)dx\right)} = \frac{1}{1+\exp\left(-\ln\left((e^{x}+1)^{r}-1\right)\right)} \\ &= \frac{1}{1+\frac{1}{(e^{x}+1)^{r}-1}} = 1-\frac{1}{(e^{x}+1)^{r}};\, r>0, x\in(-\infty,+\infty) \end{aligned} \tag{3.51}$$

$$F(x) = 1-\frac{1}{(e^{x}+1)^{r}};\, r>0, x\in(-\infty,\infty) \tag{3.52}$$

Recalling the Burr IX distribution given by Burr (1942),

$$F(x) = 1 - \frac{2}{k((1+e^x)^r - 1) + 2}; k > 0, r > 0, x \in (-\infty, \infty) \tag{3.52a}$$

Comparing Equation (3.52a) with Equation (3.52), k may be considered as a free parameter (independent of $g(x)$) with $k > 0$, which may be considered as integration constant. Equation (3.52a) converges to Equation (3.52) if $k = 2$.

Equations (3.52) and (3.52a) indicate that:

$x \to -\infty \Rightarrow e^x + 1 \to 1$ *and* $k((1 + e^x)^r - 1) + 2 \to 2 \Rightarrow F(x \to -\infty) \to 0$; and

$x \to +\infty \Rightarrow (1 + e^x)^r \to +\infty$ *and* $k((1 + e^x)^r - 1) + 2 \to +\infty \Rightarrow F(x \to +\infty) \to 1$.

To this end, we may claim that the Burr hypothesis is satisfied.

Taking the derivative of Equation (3.52a), the general PDF of Burr IX distribution can be given as:

$$f(x) = \frac{2kre^x(e^x+1)^{r-1}}{(k((e^x+1)^r - 1) + 2)^2}; k > 0, r > 0, x \in (-\infty, \infty) \tag{3.53}$$

The PDF of the Burr IX distribution is bell shaped. Both parameters k, r dominate the shape and skewness of the PDF when $r < 1$. However, parameter r dominates the density function if $r > 1$; in other words, the PDF represented by Equation (3.52a) converges to Equation (3.52). In addition, $g(x)$ is S-shaped if $r > 1$. It is reverse S-shaped if $r < 1$. Figures 3.9a–3.9c plot $g(x), F(x)$, and $f(x)$ for Burr IX distribution with different parameters.

3.4.10 Burr X Distribution

The $g(x)$ function for Burr X distribution is given as

$$g(x) = \frac{2rxe^{-x^2}}{\left(1 - (1 - e^{-x^2})^r\right)(1 - e^{-x^2})}; r > 0, x \in (0, \infty) \tag{3.54}$$

With $r > 0$, Equation (3.54) is greater than 0. It indicates: (i) $f(x) = \frac{dF}{dx} > 0$, and (ii) $F(x)$ is a nondecreasing function.

Substituting Equation (3.54) into Equation (3.9), we have

$$\int_0^x g(x)\,dx = \int_0^x \frac{2rxe^{-x^2}}{\left(1 - (1 - e^{-x^2})^r\right)(1 - e^{-x^2})}\,dx = \ln\left(\frac{\left(1 - e^{-x^2}\right)^r}{1 - (1 - e^{-x^2})^r}\right) \tag{3.55}$$

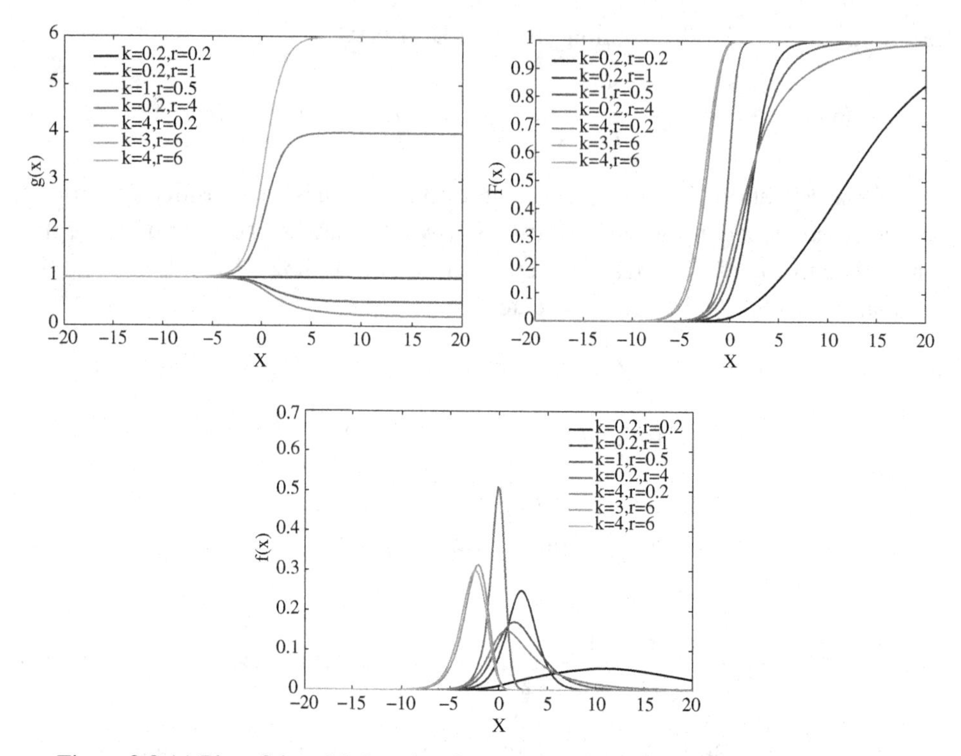

Figure 3.9 (a) Plot of the g(x) function for Burr IX distribution; (b) plot of the CDF F(x) of Burr IX distribution; and (c) plot of the PDF f(x) of Burr IX distribution.

and the CDF of Burr X distribution can be given as:

$$F(x) = \frac{1}{1 + \exp\left(-\int_0^x g(x)dx\right)} = \frac{1}{1 + \dfrac{1-\left(1-e^{-x^2}\right)^r}{\left(1-e^{-x^2}\right)^r}} = \left(1 - e^{-x^2}\right)^r, \tag{3.56}$$

$x \in (0, \infty), r > 0$

Equation (3.56) shows that:

$$x \to 0 \Rightarrow 1 - e^{-x^2} \to 0 \Rightarrow F(x \to 0) \to 0, \text{ and}$$

$$x \to \infty \Rightarrow 1 - e^{-x^2} \to 1 \Rightarrow F(x \to \infty) \to 1.$$

To this end, we have shown that the Burr hypothesis is satisfied.

Taking the derivative of Equation (3.56), the PDF of Burr X distribution can be given as:

$$f(x) = 2rxe^{-x^2}(1 - e^{-x^2})^{r-1};\ r > 0, x \in (0, \infty) \tag{3.57}$$

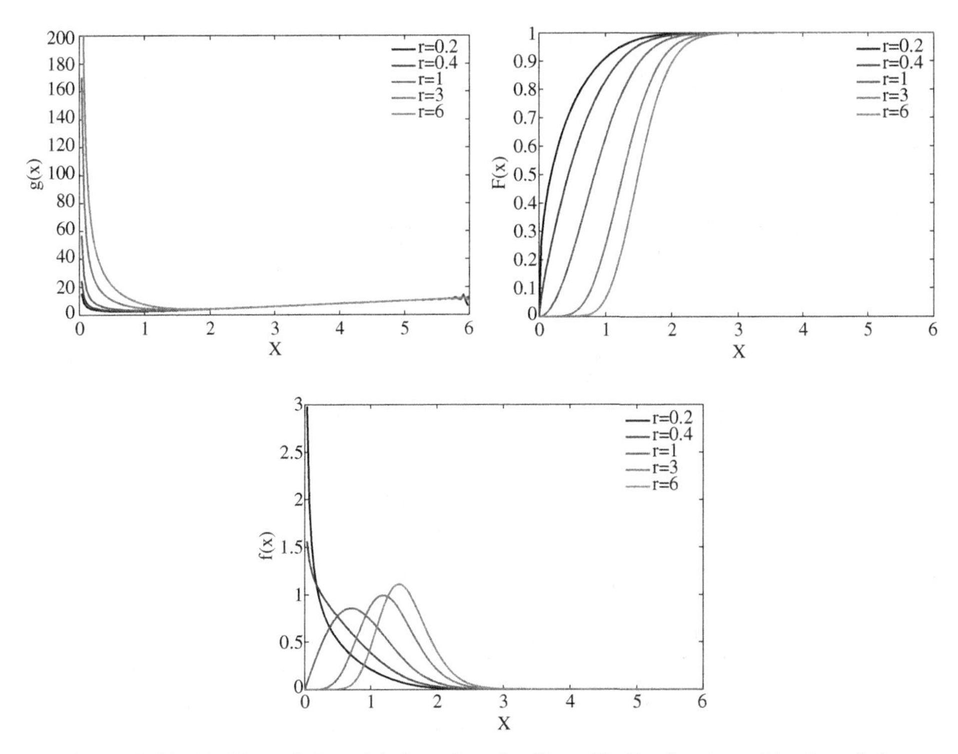

Figure 3.10 (a) Plot of the g(x) function for Burr X distribution; (b) plot of the CDF F(x) of Burr X distribution; and (c) plot of the PDF f(x) of Burr X distribution.

For Burr X distribution, $g(x)$ is L-shaped. Its PDF is L-shaped if $r < 1$. The PDF is bell shaped if $r \geq 1$. Figures 3.10a–3.10c plot $g(x), F(x)$, and $f(x)$ for Burr X distribution with different parameters.

3.4.11 Burr XI Distribution

The $g(x)$ function for Burr XI distribution is given as

$$g(x) = \frac{r(\cos(2\pi x) - 1)}{\left(x - \dfrac{\sin(2\pi x)}{2\pi}\right)\left(\left(x - \dfrac{\sin(2\pi x)}{2\pi}\right)^r - 1\right)}; r > 0, x \in (0, 1) \qquad (3.58)$$

From the prosperity of trigonometric function, $\frac{\cos(2\pi x)-1}{\left(x-\frac{\sin(2\pi x)}{2\pi}\right)^r-1} > 0$. Furthermore, we can show $x - \frac{\sin(2\pi x)}{2\pi} > 0$ as follows: Let $h(x) = x - \frac{\sin(2\pi x)}{2\pi}$, we have: $\frac{dh(x)}{dx} = 1 - \cos(2\pi x) \geq 0$. It indicates that $h(x)$ is nondecreasing function. Furthermore, if $x = 0 \Rightarrow h(0) = 0$, which means $h(x) > 0$ *for* $x \in (0,1)$. To this end, we have shown $g(x) > 0$. It indicates that: (i) $f(x) = \frac{dF}{dx} > 0$, and (ii) $F(x)$ is a nondecreasing function.

Substituting Equation (3.58) into Equation (3.9) we have:

$$\begin{aligned}\int_0^x g(x)\,dx &= \int_0^x \frac{r(\cos(2\pi x) - 1)}{\left(x - \frac{\sin(2\pi x)}{2\pi}\right)\left(\left(x - \frac{\sin(2\pi x)}{2\pi}\right)^r - 1\right)}dx \\ &= \ln\left(\left(x - \frac{\sin(2\pi x)}{2\pi}\right)^r\right) - \ln\left(1 - \left(x - \frac{\sin(2\pi x)}{2\pi}\right)^r\right) \\ &= \ln\left(\frac{\left(x - \frac{\sin(2\pi x)}{2\pi}\right)^r}{1 - \left(x - \frac{\sin(2\pi x)}{2\pi}\right)^r}\right)\end{aligned} \tag{3.59}$$

The CDF for Burr XI distribution can now be written as:

$$\begin{aligned}F(x) &= \frac{1}{1 + \exp\left(-\int_0^x g(x)dx\right)} = \frac{1}{1 + \exp\left(-\ln\left(\frac{\left(x - \frac{\sin(2\pi x)}{2\pi}\right)^r}{1 - \left(x - \frac{\sin(2\pi x)}{2\pi}\right)^r}\right)\right)} \\ &= \frac{1}{1 + \frac{1 - \left(x - \frac{\sin(2\pi x)}{2\pi}\right)^r}{\left(x - \frac{\sin(2\pi x)}{2\pi}\right)^r}} = \left(x - \frac{\sin(2\pi x)}{2\pi}\right)^r; r > 0\ x \in (0,1)\end{aligned} \tag{3.60}$$

From Equation (3.60), we can show that:

$$x \to 0 \Rightarrow F(x \to 0) \to \left(0 - \frac{\sin(0)}{2\pi}\right)^r = 0$$

$$x \to 1 \Rightarrow F(x \to 1) \to \left(1 - \frac{\sin(2\pi)}{2\pi}\right)^r = 1$$

Now, we have shown that the Burr hypothesis is satisfied.

Taking the derivative of Equation (3.60), the PDF of Burr XI distribution can be written as:

$$f(x) = r(1 - \cos(2\pi x))\left(x - \frac{\sin 2\pi x}{2\pi}\right)^{r-1}; r > 0, x \in (0,1) \tag{3.61}$$

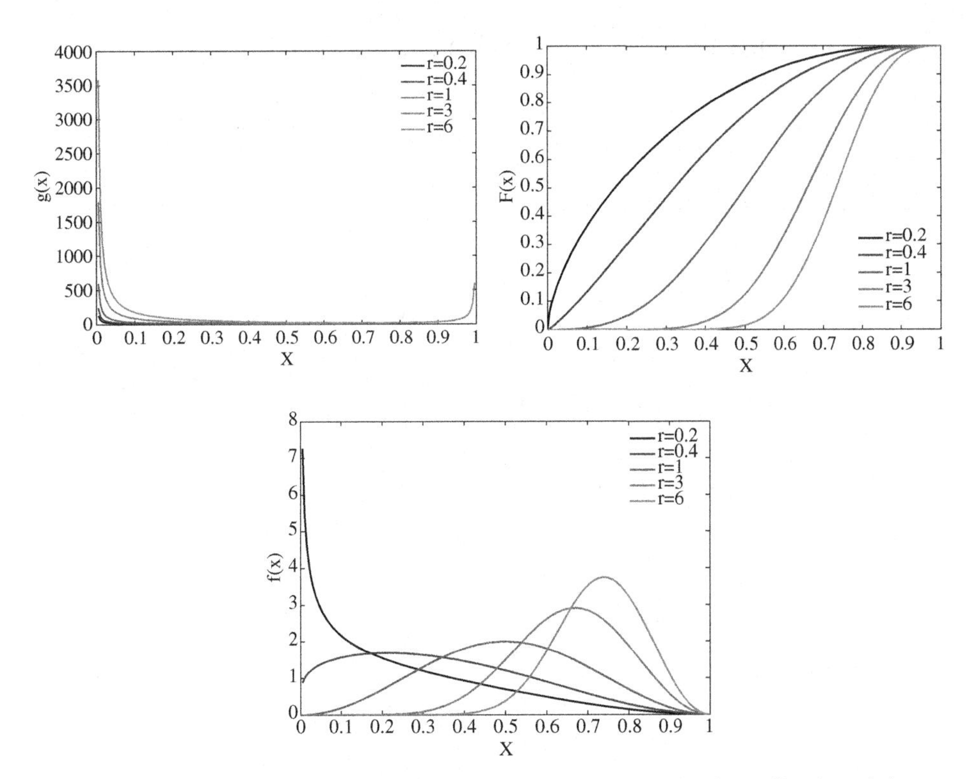

Figure 3.11 (a) Plot of the g(x) function for Burr XI distribution; (b) plot of the CDF F(x) of Burr XI distribution; and (c) plot of the PDF f(x) of Burr XI distribution.

For Burr XI distribution, $g(x) > 0$. $g(x)$ is L-shaped. There is no limit for $g(x)$ when $x \rightarrow 0$. There is an upper limit for $g(x)$ when $x \rightarrow 1$ for any given parameter r. The PDF is generally bell shaped and skewed. However, the PDF may be L-shaped if $r < 1$ and is small enough. Figures 3.11a–3.11c plot $g(x), F(x), \text{and} f(x)$ of Burr XI distribution.

3.4.12 Burr XII Distribution

The $g(x)$ function for Burr XII distribution is given as

$$g(x) = \frac{ckx^{c-1}(1+x^c)^{k-1}}{(1+x^c)^k - 1}; c, k > 0, x \in (0, \infty) \tag{3.62}$$

With c, $k > 0$ and $x \in (0, \infty)$, we have $(1 + x^c)^k - 1 > 0$ such that $g(x) > 0$. Thus, Equation (3.62) indicates that (i) $f(x) = \frac{dF}{dx} > 0$, and (ii) $F(x)$ is a nondecreasing function.

Substituting Equation (3.62) into Equation (3.9), we have:

$$\int_0^x g(x)\,dx = \int_0^x \frac{ckx^{c-1}(1+x^c)^{k-1}}{(1+x^c)^k - 1}\,dx = \ln\left((x^c+1)^k - 1\right) \tag{3.63}$$

The CDF of Burr XII distribution can now be given as:

$$\begin{aligned} F(x) &= \frac{1}{1+\exp\left(-\int_0^x g(x)dx\right)} = \frac{1}{1+\left((x^c+1)^k-1\right)^{-1}} \\ &= \frac{(x^c+1)^k-1}{(x^c+1)^k} = 1-(1+x^c)^{-k}; c,k>0, x\in(0,\infty) \end{aligned} \tag{3.64}$$

From Equation (3.64), we obtain

$$x \to 0 \Rightarrow (1+x^c)^{-k} \to 1 \Rightarrow F(x\to 0) = 1$$

$$x \to +\infty \Rightarrow (1+x^c) \to +\infty \Rightarrow (1+x^c)^{-k} \to 0 \Rightarrow F(x\to +\infty) \to 1$$

Now we have shown the Burr hypothesis is satisfied.

Taking the derivative of Equation (3.64) with respect to x, the PDF of Burr XII distribution can be given as

$$f(x) = \frac{ckx^{c-1}}{(1+x^c)^{k+1}}; c,k>0, x\in(0,\infty) \tag{3.65}$$

For Burr XII distribution, $g(x) > 0$; $\lim_{x\to\infty} g(x) = 0$ for all parameters. $g(x)$ is L-shaped. The PDF is bell shaped if $c > 1$ and is L-shaped $c \leq 1$. It is again seen that parameter c dominates the shape of the PDF. Figures 3.12a–3.12c plot $g(x), F(x)$, and $f(x)$ of Burr XII distributions.

Burr (1942) noted that if $x = \gamma + \delta y$ where γ and δ fix the origin and scale, which can be determined by the mean and standard deviation of X, then the number of parameters for all distributions would increase by two.

3.5 Parameter Estimation by Cumulative Moment Theory

The parameters of the Burr distributions can be determined from $\bar{x}, \sigma$ and two moment ratios r_3 and r_4, which are defined as

$$r_i = \frac{M_i}{\sigma^i} \tag{3.66}$$

where M_i is the i-th central moment and σ is the standard deviation and σ^i is the standard deviation raised to power i. Burr (1942) computed values of r_3 and r_4 for

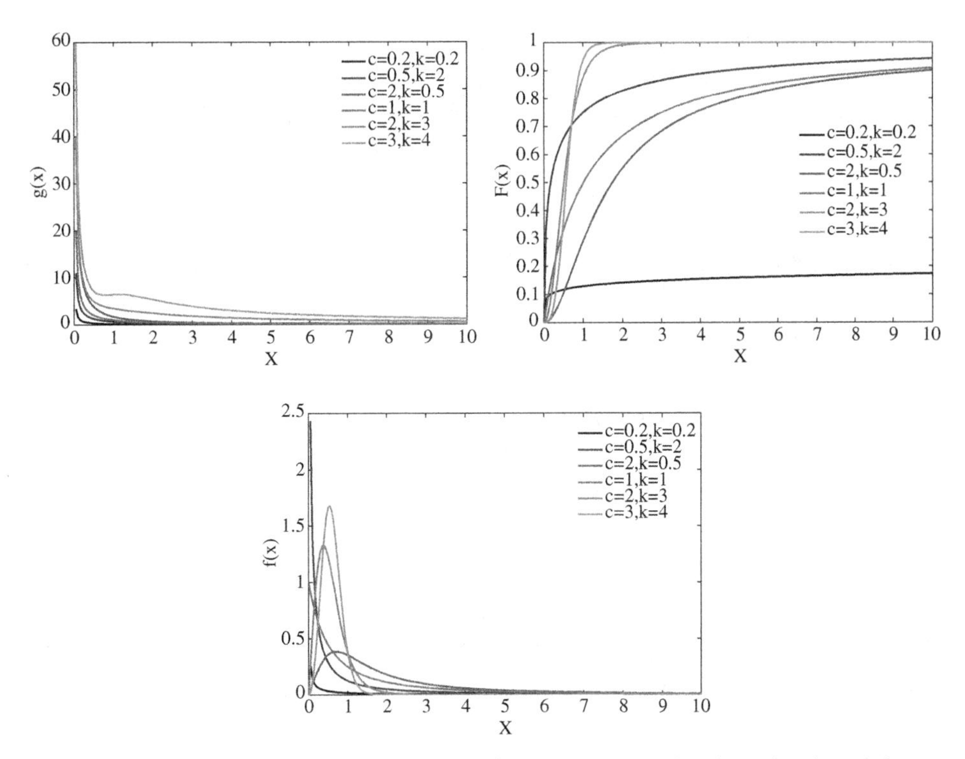

Figure 3.12 (a) Plot of the g(x) function for Burr XII distribution; (b) plot of the CDF F(x) of Burr XII distribution; and (c) plot of the PDF f(x) of Burr XII distribution.

Burr VII through XI distributions, and r_3 varied considerably as the shape parameter of these distributions varied. The parameters of the distributions can be expressed in terms of r_3 and r_4. Burr and Cislak (1968) defined

$$\delta = \frac{2r_4 - 3r_3^2 - 6}{r_4 + 3} \tag{3.67}$$

and discussed the selection of a distribution based on the $r_3{}^2$ and δ diagram and determined distribution parameters using the theory of cumulative moments with mean, standard deviation and these two moment ratios (Burr, 1973). A plot of δ as the ordinate and $r_3{}^2$ as the abscissa covers the Burr system as well as the Pearson system and other distributions. It can be seen from the Pearson system that Equation (3.67) is exactly the same as Equation (2.8). Therefore, Figure 2.1 or 2.2 would exhibit the region of distributions. Burr (1973) found that with the transformation $x = 1/y$, the region of coverage greatly extended and covered virtually all J-shaped and bell-shaped curves. Thus, the first four moments for almost any distribution can be matched with the use of either general distribution or its transformed version in y.

Now the cumulative moment theory (Burr, 1942) is discussed. First, a moment of $F(x)$ is defined. Let $M_j(a)$ denote the j-th cumulative moment about point a, which is defined as (Burr, 1942):

$$M_j(a) = \int_a^{\infty} (x-a)^j (1 - F(x))\, dx - \int_{-\infty}^{a} (x-a)^j F(x)\, dx \tag{3.68}$$

Equation (3.68) can be written as

$$M_j(a) = \int_0^{\infty} (x-a)^j (1 - F(x)) dx - \int_{-\infty}^{0} (x-a)^j F(x) dx - \int_0^{a} (x-a)^j dx \tag{3.69}$$

If $a = 0$, then Equation (3.68) becomes

$$M_j(0) = \int_0^{\infty} x^j (1 - F(x)) dx - \int_{-\infty}^{0} x^j F(x)\, dx \tag{3.70}$$

First, relations between $M_j(a)$ and $M_j(0) = M_j$ are developed. To that end, we expand $(x - a)^j$ using the binomial series:

$$(x-a)^j = \sum_{i=0}^{j} (-a)^i x^{j-i} \tag{3.71}$$

Substituting Equation (3.71) in Equation (3.69) we obtain

$$M_j(a) = \int_0^{\infty} \sum_{i=0}^{j} \binom{j}{i} (-a)^i x^{j-i} (1 - F(x)) dx - \int_{-\infty}^{0} \sum_{i=0}^{j} \binom{j}{i} (-a)^i x^{j-i} F(x) dx - \int_0^{a} (x-a)^j dx \tag{3.72}$$

Equation (3.72), using Equation (3.70), can be written as

$$M_j(a) = \sum_{i=0}^{j} \binom{j}{i} (-a)^i M_{j-i} + \frac{(-a)^{j+1}}{j+1} \tag{3.73}$$

Equation (3.73) can also be cast for M_j by translating the origin to $x = a$, renaming the moments, and replacing $(-a)$ by a as

$$M_j = \sum_{i=0}^{j} \binom{j}{i} a^i M_{j-i}(a) + \frac{a^{j+1}}{j+1} \tag{3.74}$$

It may also be interesting to relate the cumulative moments to ordinary moments. To that end, Equation (3.69) is integrated by parts.

$$M_j(a) = \left.\frac{(x-a)^{j+1}(1-F(x))}{j+1}\right|_a^{\infty} + \int_a^{\infty} \frac{(x-a)^{j+1}}{j+1} f(x)dx - \left.\frac{(x-a)^{j+1}F(x)}{j+1}\right|_{-\infty}^{a} - \int_{-\infty}^{a} \frac{(x-a)^{j+1}}{j+1} f(x)dx \tag{3.75}$$

In Equation (3.75) the first and third terms vanish and the net result is

$$M_j(a) = \frac{1}{j+1}\int_{-\infty}^{\infty} (x-a)^{j+1} f(x)dx = \frac{1}{j+1} N_{j+1}(a) \tag{3.76}$$

in which $N_{j+1}(a)$ is the $(j + 1)$ moment of $f(x)$ about a. If $a = 0$, then

$$M_j = \frac{1}{j+1}\int_{-\infty}^{\infty} x^{j+1} f(x)dx = \frac{1}{j+1} N_{j+1} \tag{3.77}$$

If a is replaced by the centroid, μ, then Equation (3.76) becomes

$$M_j(a) = \frac{1}{j+1}\int_{-\infty}^{\infty} (x-\mu)^{j+1} f(x)dx = \frac{1}{j+1} N_{j+1}(\mu) \tag{3.78}$$

In Equation (3.76) one can expand $(x - a)^{j+1}$ using the binomial series. Then

$$\begin{aligned} M_j(a) &= \frac{1}{j+1}\int_{-\infty}^{\infty} \sum_{i=0}^{j+1} \binom{j+1}{i} (-a)^i x^{j-i+1} f(x)dx \\ &= \frac{1}{j+1}\sum_{i=0}^{j+1} \binom{j+1}{i} (-a)^i N_{j-i+1}(a) \end{aligned} \tag{3.79}$$

Likewise, for the moment about the centroid Equation (3.79) can be written as

$$\begin{aligned} M_j(\mu) &= \frac{1}{j+1}\int_{-\infty}^{\infty} \sum_{i=0}^{j+1} \binom{j+1}{i} (-\mu)^i x^{j-i+1} f(x)dx \\ &= \frac{1}{j+1}\sum_{i=0}^{j+1} \binom{j+1}{i} (-\mu)^i N_{j-i+1} \end{aligned} \tag{3.80}$$

One can also derive $N_j(a)$ in terms of $M_j(a)$. To that end, we write

$$N_j = \int_a^{\infty} x^j f(x)dx + \int_{-\infty}^{a} x^j f(x)dx \tag{3.81}$$

Integrating Equation (3.81) by parts, we obtain

$$N_j = -\left(x^j(1-F(x))\right|_a^{\infty} + j\int_a^{\infty} x^{j-1}(1-F(x))\,dx + x^j F(x)\big|_{\infty}^{a} - j\int_{-\infty}^{a} x^{j-1} F(x)dx \tag{3.82}$$

In Equation (3.82), the first and third terms vanish and Equation (3.82) can be expressed as:

$$N_j = j\int_a^{\infty} x^{j-1}(1 - F(x))dx - j\int_{-\infty}^{a} x^{j-1}F(x)dx = jM_j \tag{3.83}$$

Now we can expand $((x - a) + a))^{j-1}$ as a binomial series:

$$(x - a + a)^{j-1} = \sum_{i=0}^{j-1} \binom{j-1}{i} a^i (x - a)^{j-i-1} \tag{3.84}$$

Equation (3.83), with the use of Equation (3.84), can be written as

$$\begin{aligned} N_j &= j\int_a^{\infty} \sum_{i=0}^{j-1} \binom{j-1}{i} a^i (x-a)^{j-i-1}(1 - F(x))dx \\ &\quad -j\int_{-\infty}^{a} \sum_{i=0}^{j-1} \binom{j-1}{i} a^i (x-a)^{j-i-1} F(x)dx \\ &= j\sum_{i=0}^{j-1} \binom{j-1}{i} a^i M_{j-i-1}(a) + a^j, \ j > 0 \end{aligned} \tag{3.85}$$

In a similar manner, moments about the centroid can be written as

$$N_j = j\sum_{i=0}^{j-1} \binom{j-1}{i} (a - \mu)^i M_{j-i-1}(a) + (a - \mu)^j, j > 0 \tag{3.86}$$

or

$$N_j(\mu) = j\sum_{i=0}^{j-1} \binom{j-1}{i} (-M_0(a))^i M_{j-i-1}(a) + (-M_0(a))^j, \ j > 0 \tag{3.87}$$

using $M_0(a)=\mu - a$.

If $a = 0$, then

$$N_j = jM_{j-1} \tag{3.88}$$

$$N_j(\mu) = j\sum_{i=0}^{j-1} \binom{j-1}{i} (-M_0)^i M_{j-i-1} + (-M_0)^j, \ j > 0 \tag{3.89}$$

It may be noted that Equations (3.87) and (3.89) possess the same coefficients independent of a.

If $a = \mu$ then Equation (3.88) gives

$$M_j(\mu) = \frac{1}{j+1} N_{j+1}(\mu) \tag{3.90}$$

Now the moment ratios can be expressed as:

$$r_3 = \frac{3M_2(a) - 6M_1(a) + 2(M_0(a))^3}{\left(2M_1(a) - (M_0(a))^2\right)^{\frac{3}{2}}} \tag{3.91}$$

$$r_4 = \frac{4M_3(a) - 12M_2(a)M_0(0) + 12M_1(a)(M_0(a))^2 - 3(M_0(a))^4}{\left(2M_1(a) - (M_0(a))^2\right)^2} \tag{3.92}$$

Generalizing it,

$$r_j = \frac{j\sum_{i=0}^{j-1}\binom{j-1}{i}(-M_0(a))^i M_{j-i-1}(a) + (-M_0(a))^j}{\left(2M_1(a) - (M_0(a))^2\right)^{\frac{j}{2}}} \tag{3.93}$$

Example 3.5 Explain the cumulative moment theory using the simplest uniform distribution: F(x) = x, x ∈ [0, 1]

Solution: The uniform distribution has no mode with equally likely outcomes. The density function is simply

$$f(x) = 1.$$

Applying Equation (3.70) to the uniform distribution we have:

$$\mathrm{M_j} = \int_0^1 x^j(1-x)\,dx = \frac{1}{j+1} - \frac{1}{j+2} = \frac{1}{(j+1)(j+2)} \tag{3.94}$$

Now, from Equation (3.94), we have:

$$\mathrm{M_0} = \frac{1}{2}, M_1 = \frac{1}{6}, M_2 = \frac{1}{12}, M_3 = \frac{1}{20}$$

Applying the relation between cumulative moments and the ordinary moments we have:

$$\mu_2 = 2M_1 - M_0^2 = \frac{2}{6} - \frac{1}{4} = \frac{1}{12}$$

$$\mu_3 = 3M_2 - 6M_1M_0 + 2M_0^3 = 3\left(\frac{1}{12}\right) - 6\left(\frac{1}{2}\right)\left(\frac{1}{6}\right) + 2\left(\frac{1}{2}\right)^2 = 0$$

$$\mu_4 = 4\mathrm{M}_3 - 12\mathrm{M}_2\mathrm{M}_0 + 12M_1M_0^2 - 3M_0^4$$

$$= 4\left(\frac{1}{20}\right) - 12\left(\frac{1}{12}\right)\left(\frac{1}{2}\right) + 12\left(\frac{1}{6}\right)\left(\frac{1}{2}\right)^2 - 3\left(\frac{1}{2}\right)^4 = 0.0125$$

Furthermore, we have:

Skewness: $r_3 = 0$
Kurtosis: $r_4 = \frac{0.0125}{\left(\frac{1}{12}\right)^2} = 1.8$

From Equation (3.67), we have: $\delta = \frac{2r_4 - 3r_3^2 - 6}{r_4 + 3} = \frac{2(1.8) - 6}{1.8 + 4} = -\frac{1}{2}$. The uniform distribution is also shown in Figure 3.1. This simple example clearly shows the relation between cumulative moment theory with ordinary moment theory.

Example 3.6 Apply cumulative moment theory to the exponential distribution

Solution: The probability density and cumulative probability distribution functions of the exponential distribution are given as: $f(x) = \lambda \exp(-\lambda x)$; $F(x) = 1 - \exp(-\lambda x)$; $\lambda > 0, x > 0$

Now applying Equation (3.70) to the exponential distribution we have:

$$M_j = \int_0^\infty x^j \exp(-\lambda x) dx = \frac{1}{\lambda^{j+1}} \Gamma(j+1) \tag{3.95}$$

Now from Equation (3.95) we have

$$M_0 = \frac{1}{\lambda}\Gamma(1) = \frac{1}{\lambda},\; M_1 = \frac{1}{\lambda^2}\Gamma(2) = \frac{1}{\lambda^2},\; M_2 = \frac{1}{\lambda^3}\Gamma(3) = \frac{2}{\lambda^3}, M_3 = \frac{1}{\lambda^4}\Gamma(4) = \frac{6}{\lambda^4}$$

Applying the relation between cumulative moments and the ordinary moments we have:

$$\mu_1 = M_0 = \frac{1}{\lambda} = \mu$$

$$\mu_2 = 2M_1 - M_0^2 = \frac{2}{\lambda^2} - \frac{1}{\lambda^2} = \frac{1}{\lambda^2} = \sigma^2$$

$$\mu_3 = 3M_2 - 6M_1M_0 + 2M_0^3 = 3\left(\frac{2}{\lambda^3}\right) - 6\left(\frac{1}{\lambda^2}\right)\left(\frac{1}{\lambda}\right) + 2\left(\frac{1}{\lambda^3}\right) = \frac{2}{\lambda^3}$$

$$\mu_4 = 4M_3 - 12M_2M_0 + 12M_1M_0^2 - 3M_0^4 = \frac{24}{\lambda^4} - \frac{24}{\lambda^4} + \frac{12}{\lambda^4} - \frac{3}{\lambda^4} = \frac{9}{\lambda^4}$$

From μ_2, μ_3, *and* μ_4 we obtain

$$skewness = \frac{\frac{2}{\lambda^3}}{\left(\frac{1}{\lambda^2}\right)^{\frac{3}{2}}} = 2;\;\; excess\ kurtosis = \frac{\frac{9}{\lambda^4}}{\left(\frac{1}{\lambda^2}\right)^2} - 3 = 6.$$

3.6 Application

In this section, the real-world data, given in the appendix, is applied to illustrate the application of the Burr system. Based on the characteristics of data, one may extend the original Burr family by introducing the scale parameter. Introducing the scale parameter into Burr III distribution [Equations 3.25–3.26], its CDF and PDF may be rewritten as:

$$F(x) = \left(1 + \left(\frac{x}{a}\right)^{-k}\right)^{-r}; a > 0, k > 0, r > 0, x \in (0, +\infty) \tag{3.96a}$$

$$f(x) = \frac{kr}{a}\left(1 + \left(\frac{x}{a}\right)^{-k}\right)^{-r-1}\left(\frac{x}{a}\right)^{-(k+1)} \tag{3.96b}$$

The log-likelihood functions for the original Burr III distribution and the Burr III distribution with scale parameter can be expressed as:

Original

$$\ln L = n \ln k + n \ln r - (k+1)\sum_{i=1}^{n} \ln x_i - (r+1)\sum_{i=1}^{n} \ln\left(1 + x_i^{-k}\right) \tag{3.97a}$$

With scale parameter

$$\ln L = n \ln k + n \ln r - n \ln a - (r+1)\sum_{i=1}^{n} \ln\left(1 + \left(\frac{x_i}{a}\right)^{-k}\right) - (k+1)\sum_{i=1}^{n} \ln\left(\frac{x_i}{a}\right) \tag{3.97b}$$

Maximizing the log-likelihood function, the parameters are estimated by solving the following set of equations for the original and Burr III distributions with scale parameter:

Original

$$\begin{cases} \frac{n}{k} - \sum_{i=1}^{n} \ln x_i - (r+1)\sum_{i=1}^{n} \left(1 + x_i^{-k}\right) x_i^{-k} \ln x_i = 0 \\ \frac{n}{r} - \sum_{i=1}^{n} \ln\left(1 + x_i^{-k}\right) = 0 \end{cases} \tag{3.97c}$$

With scale parameter

$$\begin{cases} -\dfrac{n}{a} - \dfrac{k(r+1)}{a^2}\sum_{i=1}^{n} \dfrac{x_i\left(\dfrac{x_i}{a}\right)^{-k-1}}{1+\left(\dfrac{x_i}{a}\right)^{-k}} + \dfrac{n(k+1)}{a} = 0 \\ \dfrac{n}{k} + (r+1)\sum_{i=1}^{n} \dfrac{\left(\dfrac{x_i}{a}\right)^{-k} \ln\left(\dfrac{x_i}{a}\right)}{1+\left(\dfrac{x_i}{a}\right)^{-k}} - \sum_{i=1}^{n} \ln\left(\dfrac{x_i}{a}\right) = 0 \\ \dfrac{n}{r} - \sum_{i=1}^{n} \ln\left(1+\left(\dfrac{x_i}{a}\right)^{-k}\right) = 0 \end{cases} \tag{3.97d}$$

Introducing the scale parameter into the Burr X distribution [Equations 3.56–3.57], its CDF and PDF may be rewritten as:

$$F(x) = \left(1 - \exp\left(-\left(\frac{x}{a}\right)^2\right)\right)^r; a > 0, r > 0, x \in (0, +\infty) \tag{3.98a}$$

$$f(x) = \frac{2r}{a^2} x \exp\left(-\left(\frac{x}{a}\right)^2\right)\left(1 - \exp\left(-\left(\frac{x}{a}\right)^2\right)\right)^{r-1} \tag{3.98b}$$

The log-likelihood functions for the original Burr X distribution and Burr X distribution with scale parameter can be expressed as:

Original

$$\ln L = n \ln 2 + n \ln r + \sum_{i=1}^{n} \ln x_i - \sum_{i=1}^{n} -x_i^2 + (r-1) \sum_{i=1}^{n} \ln\left(1 - \exp\left(-x_i^2\right)\right) \tag{3.99a}$$

With scale parameter

$$\ln L = n \ln 2 + n \ln r - 2n \ln a + \sum_{i=1}^{n} \ln x_i - \sum_{i=1}^{n} \left(\frac{x_i}{a}\right)^2 + (r-1)\sum_{i=1}^{n} \ln\left(1 - \exp\left(-\left(\frac{x_i}{a}\right)^2\right)\right) \tag{3.99b}$$

Maximizing the log-likelihood functions, the parameters can be estimated numerically by solving the following set of equations:

Original

$$r = -\frac{\sum_{i=1}^{n} \ln\left(1 - \exp\left(-x_i^2\right)\right)}{n} \tag{3.99c}$$

With scale parameter

$$\begin{cases} -\dfrac{2n}{a} + \dfrac{2}{a^3}\sum_{i=1}^{n} x_i^2 + \dfrac{2(r-1)}{a^3}\sum_{i=1}^{n} \dfrac{x_i^2 \exp\left(-\left(\frac{x_i}{a}\right)^2\right)}{1 - \exp\left(-\left(\frac{x_i}{a}\right)^2\right)} = 0 \\ \dfrac{n}{r} + \sum_{i=1}^{n} \ln\left(1 - \exp\left(-\left(\dfrac{x_i}{a}\right)^2\right)\right) = 0 \end{cases} \tag{3.99d}$$

Introducing the scale parameter into the Burr XII distribution [Equations 3.64–3.65], its CDF and PDF may be rewritten as:

$$F(x) = 1 - \left(1 + \left(\frac{x}{a}\right)^c\right)^{-k}; a > 0, c > 0, k > 0, x \in (0, +\infty) \tag{3.100a}$$

$$f(x) = \frac{ck\left(\frac{x}{a}\right)^{c-1}}{a\left(1 + \left(\frac{x}{a}\right)^c\right)^{k+1}} \tag{3.100b}$$

The log-likelihood functions for the original Burr XII distribution and the Burr XII distribution with scale parameter are expressed as:

Original

$$\ln L = n \ln c + n \ln k + (c-1)\sum_{i=1}^{n} \ln x_i - (k+1)\sum_{i=1}^{n} \ln\left(1 + x_i^c\right) \tag{3.101a}$$

With scale parameter

$$\begin{aligned} \ln L = {} & n \ln c + n \ln k - nc \ln a + (c-1)\sum_{i=1}^{n} \ln x_i \\ & - (k+1)\sum_{i=1}^{n} \ln\left(1 + \left(\frac{x_i}{a}\right)^c\right) \end{aligned} \tag{3.101b}$$

Maximizing the log-likelihood, the parameters can be estimated by solving the following set of equations:

Original

$$\begin{cases} \dfrac{n}{c} + \sum_{i=1}^{n} \ln x_i - (k+1)\sum_{i=1}^{n} \dfrac{x_i^c \ln x_i}{1 + x_i^c} = 0 \\ \dfrac{n}{k} - \sum_{i=1}^{n} \ln\left(1 + x_i^c\right) = 0 \end{cases} \tag{3.101c}$$

With scale parameter

$$\begin{cases} -\frac{nc}{a}+\frac{c(k+1)}{a}\sum_{i=1}^{n}\frac{\left(\frac{x_i}{a}\right)^c}{1+\left(\frac{x_i}{a}\right)^c}=0 \\ \frac{n}{c}-n\ln a+\sum_{i=1}^{n}\ln x_i-(k+1)\sum_{i=1}^{n}\frac{\left(\frac{x_i}{a}\right)^c\ln\left(\frac{x_i}{a}\right)}{1+\left(\frac{x_i}{a}\right)^c}=0 \\ \frac{n}{k}-\sum_{i=1}^{n}\ln\left(1+\left(\frac{x_i}{a}\right)^c\right)=0 \end{cases} \tag{3.101d}$$

Furthermore, for the Burr II, VII, VIII, and IX distributions, the random variable may be transformed to the range of $(-\infty, +\infty)$ using

$$x^* = \frac{x-\bar{x}}{\sigma_x} \tag{3.102}$$

and their log-likelihood functions and the parameter estimation can be expressed as:

Burr II

$$\ln L = n\ln r - \sum_{i=1}^{n} x_i - (r+1)\sum_{i=1}^{n}\ln\left(1+\exp\left(-x_i\right)\right) \tag{3.103a}$$

$$r = \frac{n}{\sum_{i=1}^{n}\ln\left(1+\exp\left(-x_i\right)\right)} \tag{3.103b}$$

Burr VII

$$\begin{aligned} \ln L = n\ln r + (r-1)\sum_{i=1}^{n}\ln\left(1+\tanh\left(x_i\right)\right) \\ +\sum_{i=1}^{n}\ln\left(1-\tanh^2(x_i)\right) - nr\ln 2 \end{aligned} \tag{3.104a}$$

$$r = \frac{1}{\ln 2 - \frac{\sum_{i=1}^{n}\ln\left(1+\tanh\left(x_i\right)\right)}{n}} \tag{3.104b}$$

Burr VIII

$$\begin{aligned} \ln L = nr\ln\left(\frac{2}{\pi}\right) + n\ln r - \sum_{i=1}^{n}\ln\left(\exp\left(-x_i\right)+\exp\left(x_i\right)\right) \\ +(r-1)\sum_{i=1}^{n}\ln\left(\text{atan}(\exp\left(x\right)\right) \end{aligned} \tag{3.105a}$$

$$r = -\frac{1}{\ln\left(\frac{2}{\pi}\right) + \sum_{i=1}^{n} \ln(\operatorname{atan}(\exp(x))} \tag{3.105b}$$

Burr IX

$$\ln L = n \ln r + \sum_{i=1}^{n} x_i - (r+1)\sum_{i=1}^{n} \ln(1 + \exp(x_i)) \tag{3.106a}$$

$$r = \frac{n}{\sum_{i=1}^{n} \ln(1 + \exp(x_i))} \tag{3.106b}$$

3.6.1 Peak Flow

According to the empirical frequency and sample statistics of peak flow at the gaging station USGS04208000 (Cuyahoga Rivera at Independence, Ohio, U.S.), the Burr II, III, X, and XII distributions (original/with scale parameter) are applied. Table 3.1 lists the parameters estimated for the peak flow with the use of different Burr distributions using the maximum likelihood estimation (MLE) method with the GA algorithm in matlab. Figure 3.13 compares the fitted frequency with the empirical frequency. It is seen that the distributions with scale parameter may be applied to study the frequency of peak flow; these distributions are the Burr III, X, and XII distributions with scale parameter. In addition, similar performance is found for the Burr III, X, and XII with scale parameters.

3.6.2 Annual Rainfall Amount

Based on the empirical frequency and sample statistics of the annual rainfall amount at the gaging station U330058 (Akron-Canton WSO AP, Ohio, U.S.), the Burr II, VII, VIII, IX, X, and XII distributions are applied to evaluate the frequency of annual rainfall amount. Table 3.2 lists the parameters estimated for the selected distributions in the Burr family. Figure 3.14 compares the

Table 3.1 *Parameters estimated for peak flows using maximum likelihood method*

	Burr II	Burr III	Burr III	Burr X	Burr X	Burr XII	Burr XII
a			61.33		225.34		246.95
c						1.17	5.61
k		0.71	3.01			0.15	0.79
r	1.26	39.54	51.24	20.11	2.7		

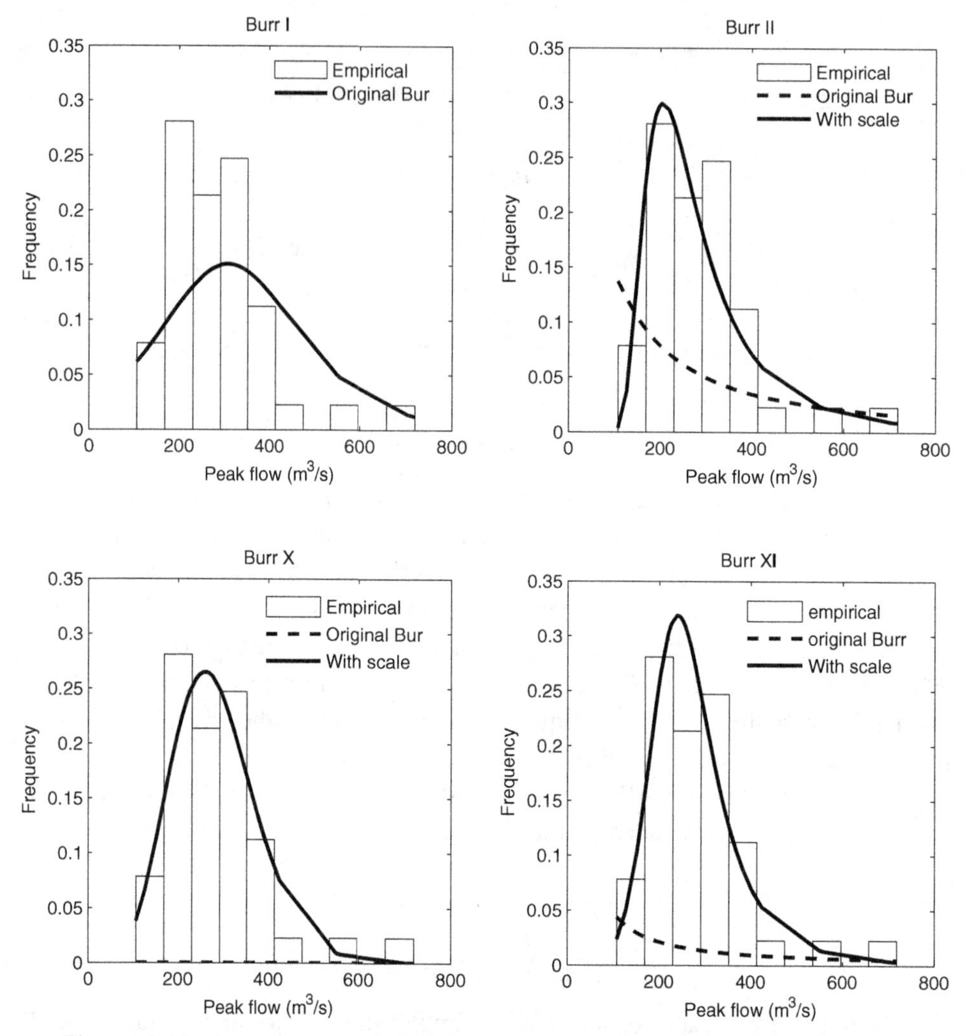

Figure 3.13 Comparisons of fitted frequency with empirical frequency of peak flow.

frequency computed from the selected parametric Burr distribution with the empirical frequency. Figure 3.14 shows that (i) with the transformation using Equation (3.84), the Burr VII and IX distributions yield better performance than does the Burr II distribution; and (ii) in case of the Burr III, X, and XII distributions, the distributions with scale parameter may be applied to study the frequency of annual rainfall amount, with the Burr X and XII distributions yielding better performances.

Table 3.2 *Parameters estimated for annual rainfall amount using maximum likelihood method*

	Burr II	Burr III	Burr III*	Burrr VII	Burr VIII	Burr IX	Burr X	Burr X*	Burr XII	Burr XII*
a			184.29					519.06		961.76
c									1.77	9.91
k		0.51	3.25			0.3			0.08	1.13
r	1.25	23.54	172.01	0.97	1.19	2.9	16.95	17.93		

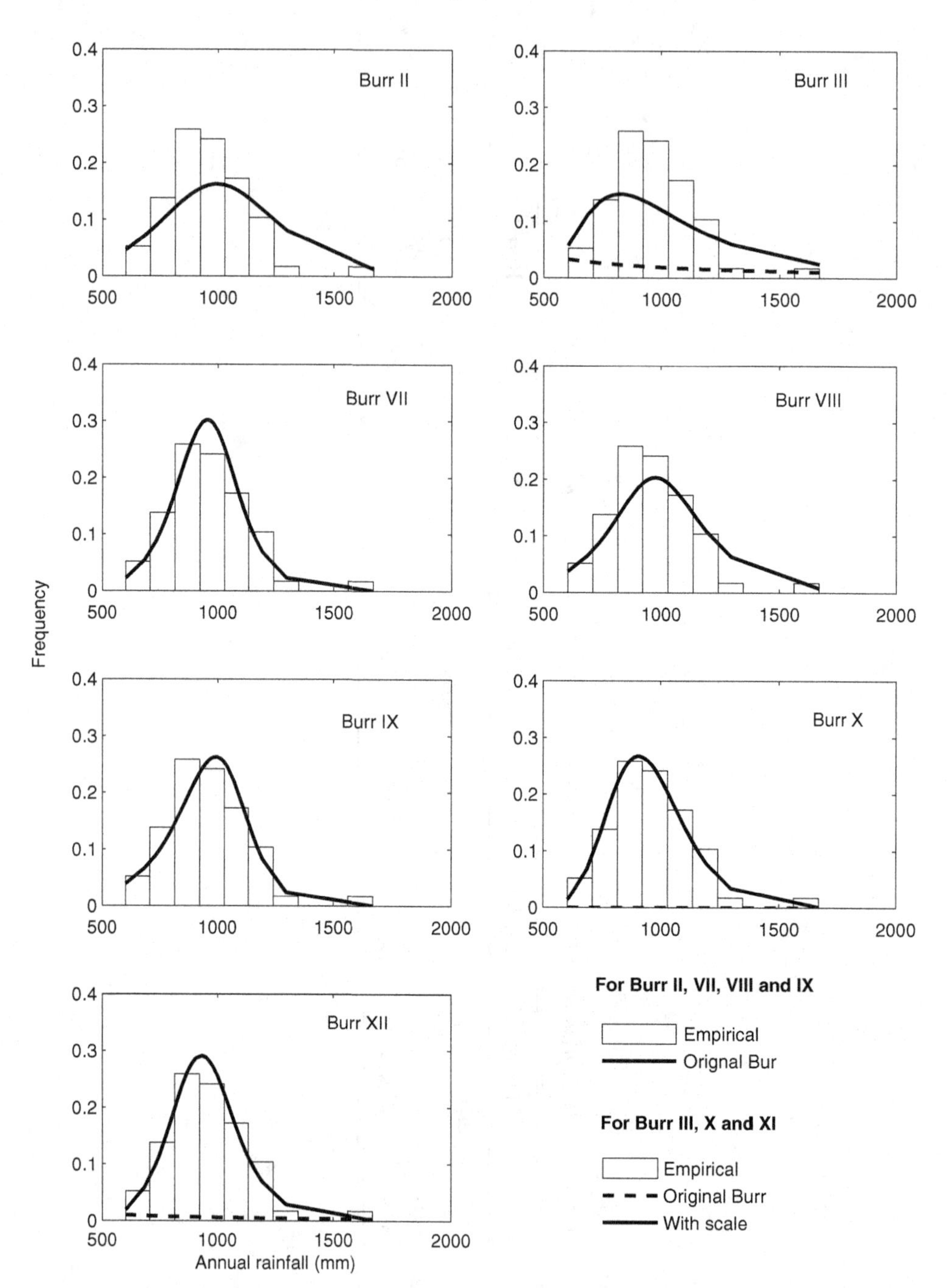

Figure 3.14 Comparisons of fitted frequency with empirical frequency of annual rainfall amount.

3.6.3 Monthly Sediment Yield

Based on the empirical frequency and the sample statistics of monthly sediment yield of May at the gaging station USGS 04208000 (Cuyahoga River at Independence, Ohio, U.S.), the Burr II, III, VI, VII, VIII, IX, X, and XII distributions are again applied for evaluation. Table 3.3 lists the parameters estimated using the MLE method. Figure 3.14 compares the fitted frequency with the empirical frequency for the selected Burr distributions. The empirical frequency of the monthly suspended sediment yield shows that it is bimodal. The fitted frequency shown in Figure 3.15 indicates that the Burr VI distribution may capture both modes of the sediment dataset. Additionally, the Burr VIII and IX distributions show similar performances. The Burr X and XII distributions with scale parameters fit the sediment data better than do the original Burr X and XII distributions. In case of the Burr III distribution, less discrepancy is found in regard to the performance of its original distribution and the distribution with scale parameter.

3.6.4 Maximum Daily Precipitation

Based on the empirical frequency and sample statistics of the maximum daily precipitation at the gaging station U330058, the Burr II, III, X, and XII distributions (original/with scale parameter) are applied. Table 3.4 lists the parameters estimated for the maximum daily precipitation with the use of different Burr distributions using the MLE method with the GA algorithm in Matlab. Figure 3.16 compares the fitted frequency distribution with the empirical frequency distribution. It is seen that the distributions with scale parameter may be applied to study the frequency distribution of maximum daily precipitation; these distributions are the Burr III, X, and XII distributions with scale parameter. Similar to the application to peak flow, similar performance is found for the Burr III, X, and XII distributions with scale parameters.

3.7 Conclusion

The discussion in the chapter shows that fundamental to deriving a Burr distribution is the definition of the $g(x)$ function. However, it is not clear how the $g(x)$ function should be specified or derived. Because the data is plotted to visualize which distribution should be fitted, it seems there should be a way to define a $g(x)$ function from the characteristics of the plot or of data. The Burr system of distributions encompasses a large variety of distribution shapes. Different Burr distributions can be useful for different fields or problems. Furthermore, the peak flow, annual rainfall amount, maximum daily precipitation amount, and monthly

Table 3.3 *Parameters estimated for monthly suspended sediment using maximum likelihood method*

	Burr II	Burr III	Burr III*	Burr VI	Burr VII	Burr VIII	Burr IX	Burr X	Burr X*	Burr XII	Burr XII*
a			15.62						118.99		244.58
c				0.99						1.39	2.06
k		0.76	1.73	0.12			0.46			0.16	5.08
r	1.25	19.83	12.44	5.83	0.95	1.17	2.46	30.04	0.98		

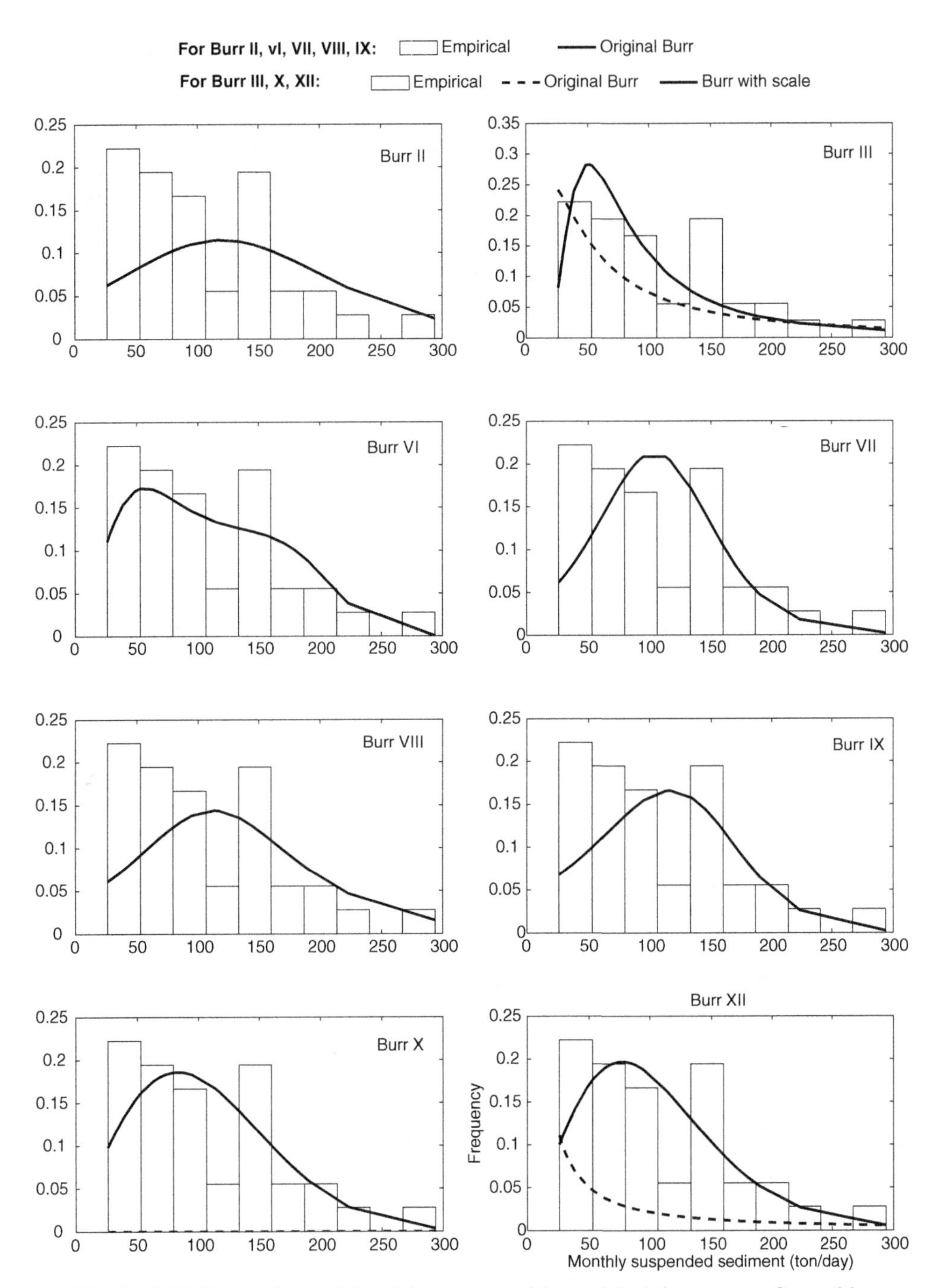

Figure 3.15 Comparison of fitted frequency with empirical frequency of monthly suspended sediment.

Table 3.4 *Parameters estimated for maximum daily precipitation with maximum likelihood method*

	Burr II	Burr III	Burr III*	Burr X	Burr X*	Burr XII	Burr XII*
a			21.08		42.15		50.66
c						1.55	5.77
k		0.99	3.72			0.16	0.85
r	1.25	37.21	18.83	36.47	3.38		

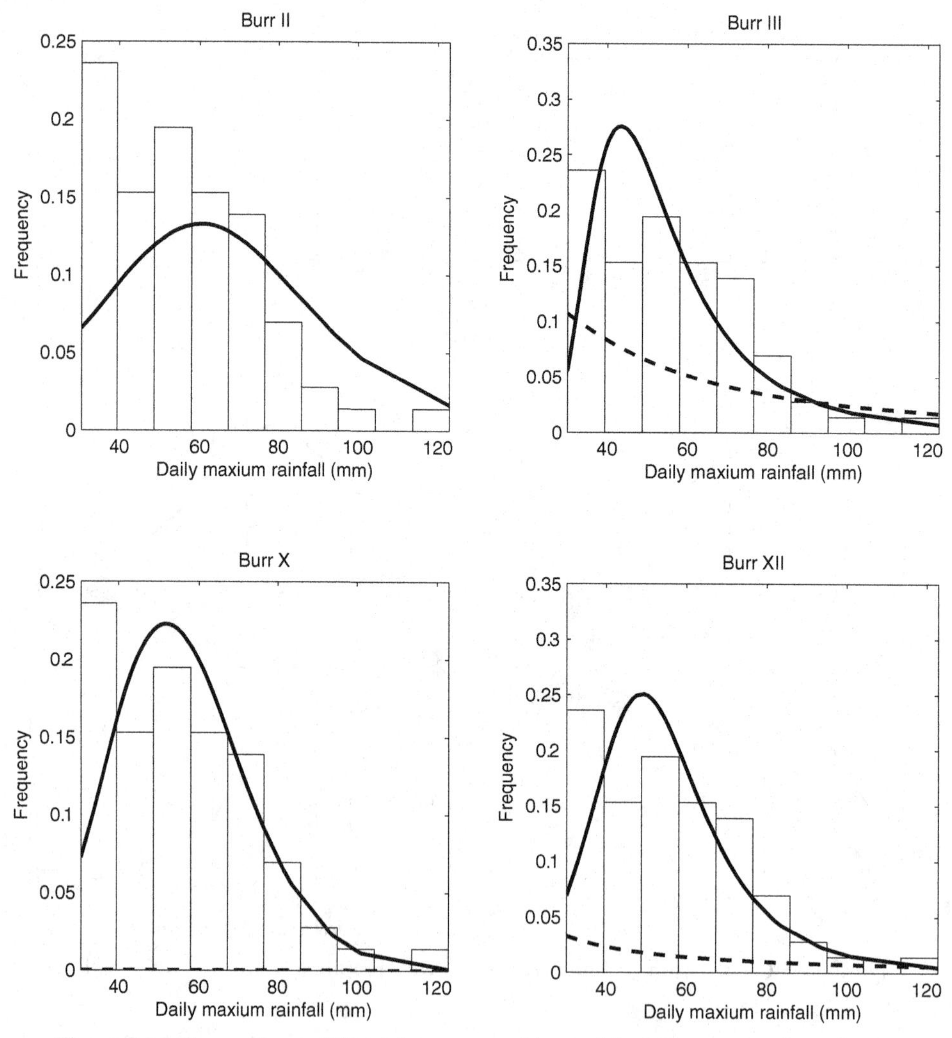

Figure 3.16 Comparison of fitted frequency with empirical frequency of maximum daily precipitation.

sediment yield are applied to evaluate the application of the Burr system. It is found: (i) in general the distributions with scale parameter outperforms the corresponding original distributions, (ii) Burr VI distribution may be applied if bimodal is detected for the dataset, and (iii) the distributions in Burr system may be properly applied to evaluate the frequency analysis in water engineering.

References

Ausloos, M., and Diricks, M. (eds.) (2006). *The Logistic Map and Route to Chaos: From the Beginnings to Modern Applications*. Berlin: Springer Science and Business Media. http://dx.doi.org/1007/3-540-32023-7.

Brouwers, F. (2015). The Burr XII distribution family and the maximum entropy principle: Power law phenomena are not necessarily "nonextensive." *Open Journal of Statistics* 5, pp. 730–741.

Burr, I. W. (1942). Cumulative frequency functions. *Annals of Mathematical Statistics* 13, pp. 215–232.

Burr, I. W. Parameters for a general system of distribution to match a grid of α_3 and α_4. *Communications in Statistics* 2, no. 1, pp. 1–21.

Burr, I. W., and Cislak, P. J. (1968). On a general system of distributions: I. Its curve-shape characteristics: II. The sample median. *Journal of the American Statistical Association* 63, no. 322, pp. 627–635.

Mielke, P. W. (1973). Another family of distributions for describing and analyzing precipitation data. *Journal of Applied Meteorology* 11, pp. 275–280.

Singh, S., and Maddala, G. (1976). A function for the size distribution of income. *Econometrica* 44, pp. 963–970.

Verhulst, P. F. (1845). Recherches mathématiques sur la loi d'accroissement de la population. *Nouveaux memoires de l'académie des Sciences, des Artes et des Beaux Arts de Belgique* 18, pp. 1–45.

4

D'Addario System of Frequency Distributions

4.1 Introduction

D'Addario (1949) integrated a probability generating function (PGF) and a transformation function (TF). He formalized the PGF by invoking a formal analogy with the most probable distribution of Brillouin's quantum statistics that generalized quantum statistics of Boltzmann, Bose and Einstein, and Fermi-Dirac. The integration led to a system of distributions that contains distributions that are commonly used in environmental and water engineering. Examples of these distributions include Pareto type I, Pareto type II, lognormal type I, lognormal type II, Amoroso, and Davis distributions. The Amoroso distribution leads to 11 special cases that include exponential, seminormal, normal, Weibull, Pearson type III, Pearson type V, Chi, Chi-square, Raleigh, Maxwell, and 3-parameter Amoroso distributions. The objective of this chapter therefore is to discuss the D'Addario system and the resulting distributions.

4.2 D'Addario System

D'Addario (1949) proposed a system of distributions that follows the concept of transformation and a PGF expressed as a differential equation. The PGF $g(z)$ of a random variable Z is expressed as

$$g(z) = \frac{A}{b + \exp\left(z^{\frac{1}{k}}\right)}; k > 0, b \in \mathbb{R} \tag{4.1}$$

where A is a normalizing constant satisfying the total probability, and b and k are parameters. The TF, connecting z to random variable x, is defined through the differential equation expressed as:

$$z^{\alpha}\frac{dz}{dx} = \frac{\beta}{x - c}; x \in [x_0, \infty), x_0 > c, \alpha, \beta \in \mathbb{R}, \beta \neq 0; \int f(z)dz = 1 \tag{4.2}$$

If $\alpha = -1$, the solution of Equation (4.2) can be expressed as:

$$\frac{1}{z}\frac{dz}{dx} = \frac{\beta}{x-c} \Rightarrow \frac{1}{z}dz = \frac{\beta}{x-c}dx$$
$$\Rightarrow z = h(x) = C_0(x-c)^{\beta}; \beta \neq 0, C_0 > 0 \tag{4.3a}$$

If $\alpha \neq -1$, the solution of Equation (4.2) can be expressed as:

$$z^{\alpha}dz = \frac{\beta}{x-c}dx \Rightarrow \frac{1}{\alpha+1}dz^{\alpha+1} = \beta d\ln(x-c)$$
$$\Rightarrow z = h(x) = ((\alpha+1)(\beta\ln(x-c)+C_0))^{\frac{1}{\alpha+1}}; C_0 \in \mathbb{R}, \beta \neq 0 \tag{4.3b}$$

In Equations (4.3a) and (4.3b), C_0 is constant of integration.

To this end, the D'Addario system is derived from the solution of Equation (4.2) in concert with Equation (4.3a) or (4.3b) as:

$$f(x) = \left|\frac{dz}{dx}\right| g(z) = \left|\frac{dh(x)}{dx}\right| \frac{A}{b + \exp\left(h(x)^{\frac{1}{k}}\right)} \tag{4.4}$$

where $\left|\frac{dz}{dx}\right| = \left|\frac{dh(x)}{dx}\right|$ is the Jacobian of the transformation given by Equation (4.3a) when $\alpha = -1$ and by Equation (4.3b) when $\alpha \neq -1$. Equation (4.4) represents the D'Addario system, which leads to several well-known distributions as special cases. The distributions resulting from the D'Addario system include Pareto type I and Pareto type II, lognormal type I and lognormal type II, Amoroso, and Davis distributions (Dagum, 1990). Each of these distributions is derived in what follows.

4.2.1 Pareto Type I Distribution

The PGF is obtained if $k = 1$, $b = 0$ in Equation (4.1) as:

$$g(z) = A[\exp(z)]^{-1} \tag{4.5a}$$

The TF may be written for Pareto Type I distribution by setting $\alpha = 0$, $c = 0$ in Equation (4.2) as:

$$\frac{dz}{dx} = \frac{\beta}{x}, \beta > 0; x \in [x_0, \infty), x_0 > 0 \tag{4.5b}$$

Integration of Equation (4.5b) yields

$$\int dz = \int_{x_0}^{x} \frac{\beta}{x} dx$$
$$\Rightarrow z = h(x) = \beta \ln x + C_0; \beta > 0, x \in [x_0, \infty), x_0 > 0 \tag{4.6}$$

Substitution of Equations (4.6) and (4.5b) into Equation (4.4) yields:

$$f(x) = \left|\frac{dz}{dx}\right| g(z) = \left|\frac{dh(x)}{dx}\right| \frac{A}{\exp(h(x))} = \frac{\beta}{x} \frac{A}{\exp(C_0)x^{\beta}} = \frac{A\beta}{\exp(C_0)x^{\beta+1}} \tag{4.7a}$$

Applying the total probability theory, the constant $\frac{A}{\exp(C_0)}$ is computed as:

$$\int_{x_0}^{\infty} f(x)dx = 1 \Rightarrow \int_{x_0}^{\infty} \left(\frac{A}{\exp(C_0)}\right) \frac{\beta}{x^{\beta+1}} dx = -\left(\frac{A}{\exp(C_0)}\right) x^{-\beta}\bigg|_{x_0}^{\infty} = \frac{A}{\exp(C_0)} x_0^{-\beta} \tag{4.7b}$$

From Equation (4.7b), we have:

$$\frac{A}{\exp(C_0)} = x_0^{\beta} \tag{4.7c}$$

Substituting Equation (4.7c) into Equation (4.7a), the density function for Pareto Type I distribution is then expressed as:

$$f(x) = \frac{\beta x_0^{\beta}}{x^{\beta+1}} \tag{4.8}$$

Frechet (1939) obtained Pareto type I distribution in a similar manner. Figure 4.1a graphs the $g(z)$ function of Pareto I distribution; Figure 4.1b graphs $z = h(x)$, the transfer function of Pareto I distribution; and Figure 4.1c graphs $f(x)$, the probability density function (PDF) of Pareto I distribution. It is shown: (i) as the exponential

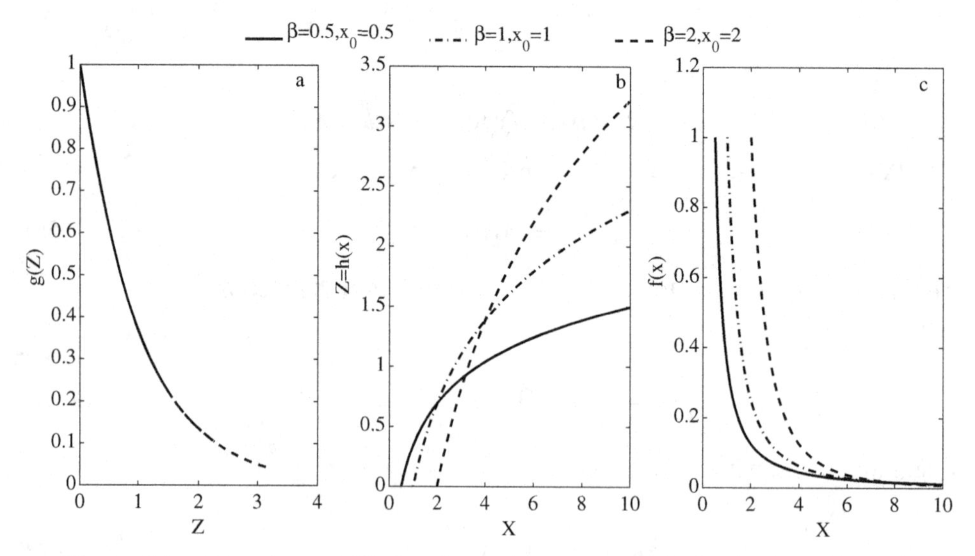

Figure 4.1 g(z) and z = h(x) functions of Pareto type I distribution. (a): g(z) function; (b): z = h(x) function; and (c): PDF function f(x).

decay function, the $g(z)$ function is a monotone decreasing function; and (ii) as the logarithm function, $z = h(x) = \beta \ln x + C_0$ is a monotone increasing function.

4.2.2 Pareto Type II Distribution

The PGF for Pareto II distribution may be written as:

$$g(z) = \frac{A}{\exp(z)} \tag{4.9a}$$

Equation (4.9a) indicates that the PGF of Pareto Type II distribution is also obtained by setting $b = 0$, $k = 1$ in Equation (4.1) as the PGF of Pareto Type I distribution.

Let $a = 0$ in Equation (4.2), the transform function of Pareto Type II distribution is obtained as:

$$\frac{dz}{dx} = \frac{\beta}{x-c}; \beta > 0, x \in [x_0, \infty), x_0 > c \neq 0 \tag{4.9b}$$

Integration of Equation (4.9b) yields

$$\begin{aligned} \int dz &= \int_{x_0}^{x} \frac{\beta}{x-c} dx \\ \Rightarrow z &= h(x) = \beta \ln(x-c) + C_0; x \in [x_0, \infty), x_0 > c \neq 0, \beta > 0 \end{aligned} \tag{4.10}$$

Substituting Equations (4.9b) and (4.10) into Equation (4.4), we have:

$$f(x) = \left(\frac{\beta}{x-c}\right) \frac{A}{\exp(C_0)(x-c)^{\beta}} \tag{4.11a}$$

In Equation (4.11a) the constant $\frac{A}{\exp(c_0)}$ may be computed from the total probability theorem as:

$$\begin{aligned} &\int_{x_0}^{\infty} f(x)dx = 1 \Rightarrow \int_{x_0}^{\infty} \frac{A}{\exp(C_0)} \frac{\beta}{(x-c)^{\beta+1}} dx = 1 \\ &\Rightarrow \frac{A}{\exp(C_0)} = \frac{1}{\int_{x_0}^{\infty} \beta (x-c)^{-\beta-1} dx} = \frac{1}{\int_{x_0}^{\infty} d(x-c)^{-\beta}} = (x_0 - c)^{\beta} \end{aligned} \tag{4.11b}$$

Using Equation (4.11b), the PDF from Equation (4.11a) is given as:

$$f(x) = \frac{\beta (x_0 - c)^{\beta}}{(x-c)^{\beta+1}} \tag{4.12}$$

which is the Pareto type II distribution, also called Lomax distribution. Compared with Pareto type I distribution, the Pareto type II distribution may also be considered

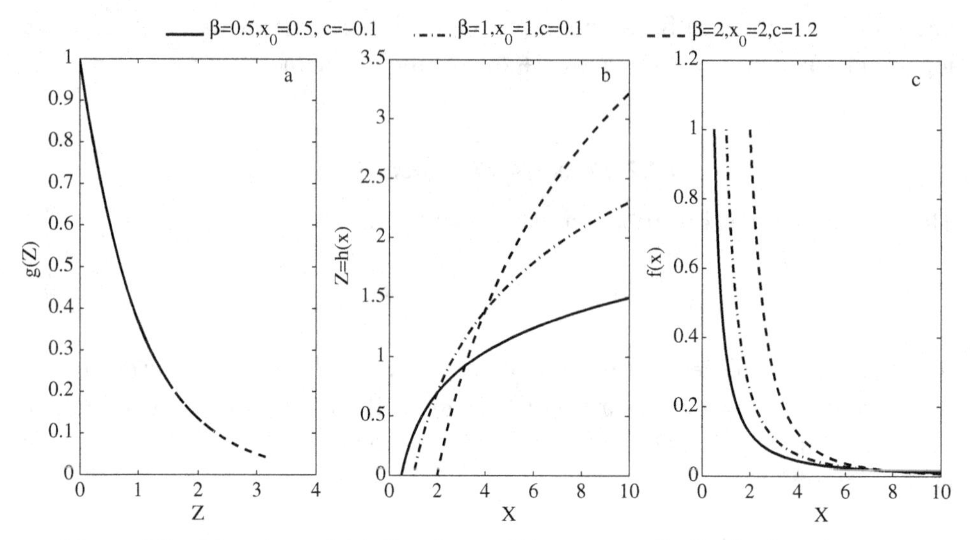

Figure 4.2 g(z) and z = h(x) functions of Pareto type II distribution. (a): g(z) function; (b): z = h(x) function; and (c): PDF f(x).

as the Pareto I distribution with location parameter. Figure 4.2 graphs $g(z), z = h(x)$, and $f(x)$ functions for Pareto type II distribution. Similar to Pareto I distribution; the generation function $g(z)$ and PDF $f(x)$ are shown as the monotone decreasing function, and the TF $z = h(x)$ is shown as monotone increasing function.

4.2.3 Lognormal (2-Parameter) Distribution

For the lognormal (2-parameter) distribution, its generating function and TF are given as:

$$g(z) = \frac{A}{\exp\left(z^2\right)} \tag{4.13}$$

$$\frac{dz}{dx} = \frac{\beta}{x}; \beta > 0, x \in (0, \infty) \tag{4.14}$$

Equations (4.13) and (4.14) indicate $b = 0, k = \frac{1}{2}$ in Equation (4.1) and $\alpha = 0, c = 0$ in Equation (4.2) for lognormal (2-parameter) distribution. Furthermore, Pareto I and lognormal 2-paramter distributions share the TF of same form. Integration of Equation (4.14) yields the same results as that in Equation (4.6) with $x \in (0, \infty)$. Equation (4.4) can be rewritten as:

$$f(x) = \frac{\beta}{x} \frac{A}{\exp\left(\left(C_0 + \beta \ln x\right)^2\right)} \tag{4.15}$$

Applying the total probability theorem, we have:

$$\int_{0^+}^{\infty} \frac{\beta}{x} \frac{A}{\exp\left((C_0 + \beta \ln x)^2\right)} dx = 1 \Rightarrow \int_{0^+}^{\infty} A \exp\left(-(C_0 + \beta \ln x)^2\right) d\beta \ln x \tag{4.16a}$$

Let $y = \beta \ln x$, the normalization factor A can be solved from Equation (4.16a) as:

$$\int_{0^+}^{\infty} A \exp\left(-(C_0 + \beta \ln x)^2\right) d \ln x = \int_{-\infty}^{\infty} A \exp\left(-(C_0 + y)^2\right) dy = 1 \tag{4.16b}$$

Applying $\int_{-\infty}^{\infty} \exp(-x^2) dx = \sqrt{\pi}$ to Equation (4.16b), the normalization factor A is solved as:

$$A = \frac{1}{\int_{-\infty}^{\infty} \exp\left(-(C_0 + y)^2\right) dy} = \frac{1}{\int_{-\infty}^{\infty} \exp\left(-(C_0 + y)^2\right) d(C_0 + y)} = \frac{1}{\sqrt{\pi}} \tag{4.16c}$$

Substituting Equation (4.16c) into Equation (4.15), the PDF is given as:

$$f(x) = \frac{1}{\sqrt{\pi}} \frac{\beta}{x} \exp\left(-(C_0 + \beta \ln x)^2\right) = \frac{1}{x\sqrt{2\pi}\left(\frac{1}{\sqrt{2}\beta}\right)} \exp\left(-\frac{\left(\ln x + \frac{C_0}{\beta}\right)^2}{2\left(\frac{1}{\sqrt{2}\beta}\right)^2}\right);$$

$$\beta > 0, x \in (0, \infty) \tag{4.17}$$

Let $\mu = -\frac{C_0}{\beta}, \sigma = \frac{1}{\sqrt{2}\beta}$, Equation (4.17) may be rewritten as:

$$f(x) = \frac{1}{x\sqrt{2\pi}\sigma} \exp\left(-\frac{(\ln x - \mu)^2}{2\sigma^2}\right) \tag{4.18}$$

Figure 4.3 graphs the $g(z), z = h(x)$, and $f(x)$ functions for the lognormal distribution with two parameters. As shown in Figure 4.3: (i) the generating function is monotone decreasing function; (ii) the density function is skewed to right; and (iii) the TF is monotone increasing function.

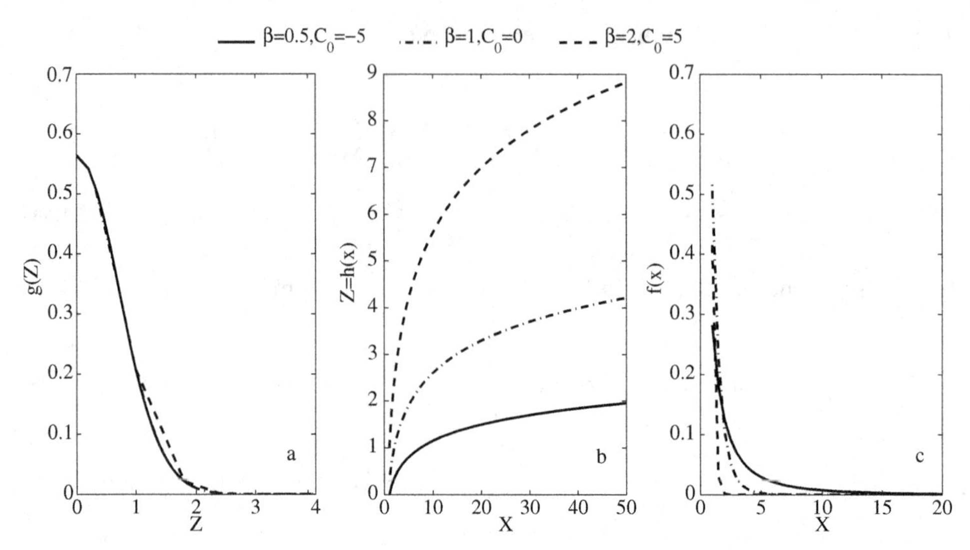

Figure 4.3 g(z) and z = h(x) functions of lognormal (2-parameter) distribution. (a): g(z) function; (b): z = h(x) function; and (c): PDF f(x).

4.2.4 Lognormal (3-Parameter) Distribution

Similar to the lognormal (2-parameter) distribution, the lognormal (3-parameter) distribution also has $b = 0, k = \frac{1}{2}$ in Equation (4.1) and $\alpha = 0,\ c \neq 0$ in Equation (4.2), i.e., the location parameter is added to the TF of the lognormal (2-parameter) distribution. To this end, the generating function and the TF are expressed for lognormal (3-parameter) as:

$$g(z) = \frac{A}{\exp\left(z^2\right)}; A = \frac{1}{\sqrt{\pi}} \tag{4.19a}$$

$$\begin{aligned} &\frac{dz}{dx} = \frac{\beta}{x - c}; \beta > 0, x \in (c, \infty) \\ &\Rightarrow z = h(x) = \beta \ln(x - c) + C_0; \beta > 0, x \in (c, \infty) \end{aligned} \tag{4.19b}$$

Substituting Equations (4.19a)–(4.19b) into Equation (4.4), the PDF can be expressed as:

$$f(x) = \left|\frac{dz}{dx}\right| \frac{A}{\mathrm{b} + \exp\left(\mathrm{z}^{\frac{1}{\mathrm{k}}}\right)} = \frac{\beta}{\sqrt{\pi}(x - c)} \exp\left(-\left(C_0 + \beta\, ln\,(x - c)\right)^2\right)$$

$$\Rightarrow f(x) = \frac{1}{(x - c)\sqrt{2\pi}\left(\frac{1}{\sqrt{2\beta}}\right)} \exp\left(-\frac{\left(\ln(x - c) + \frac{C_0}{\beta}\right)^2}{2\left(\frac{1}{\sqrt{2\beta}}\right)^2}\right); \tag{4.20a}$$

$x \in (c, \infty), \beta > 0$

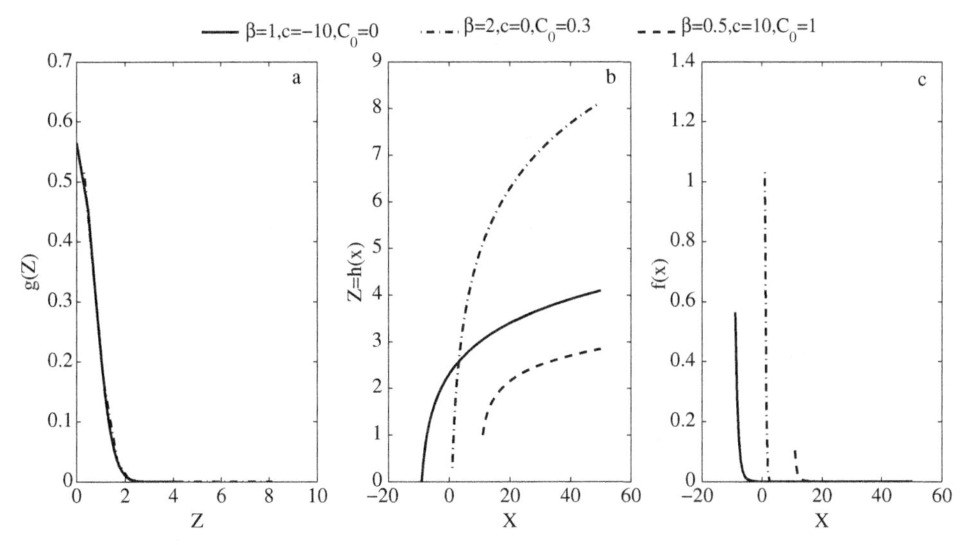

Figure 4.4 g(z) and z = h(x) functions of lognormal (3-parameter) distribution. (a): g(z) function; (b): z = h(x) function; and (c): probability density f(x) function.

Equation (4.20) represents the lognormal distribution with location parameter c, i.e., 3-parameter lognormal distribution. Again let $\mu = -\frac{C_0}{\beta}, \sigma = \frac{1}{\sqrt{2\beta}}$, Equation (4.20a) may be rewritten as:

$$f(x) = \frac{1}{(x-c)\sqrt{2\pi}\sigma} \exp\left(-\frac{(\ln(x-c)-\mu)^2}{2\sigma^2}\right), x \in (c, \infty), \sigma > 0 \tag{4.20b}$$

Figure 4.4 graphs the $g(z), z = h(x), f(x)$ functions for the lognormal distribution with three parameters. As shown in Figure 4.4: (i) the generating function is monotone decreasing; (ii) the TF is monotone increasing; and (iii) the PDF is skewed to right.

4.2.5 Davis Distribution

For Davis distribution, we have $b = -1$, $k > 0$ in the generating function [i.e., Equation 4.1] and $\alpha = -1$, $\beta = -k$, $c > 0$ in the TF [i.e., Equation 4.2]. Then the generating function and TF are given as:

$$g(z) = \frac{A}{-1 + \exp\left(z^{\frac{1}{k}}\right)} \tag{4.21a}$$

$$z^{-1}\frac{dz}{dx} = -\frac{k}{x-c} \tag{4.21b}$$

Integrating Equation (4.21b), we have:

$$\frac{1}{z}dz = -\frac{k}{x-c}dx \Rightarrow \int d\ln z = \int -kd\ln(x-c) \Rightarrow \ln z = \ln C_0(x-c)^{-k}; C_0 > 0 \tag{4.22a}$$

Equation (4.22a) can be rewritten as:

$$z = C_0(x-c)^{-k}; \left|\frac{dz}{dx}\right| = kC_0(x-c)^{-(k+1)} \tag{4.22b}$$

Substituting Equations (4.21a) and (4.22b) into Equation (4.4) we have:

$$f(x) = \frac{AkC_0(x-c)^{-k-1}}{-1+\exp\left(C_0^{\frac{1}{k}}(x-c)^{-1}\right)} \tag{4.23a}$$

Applying total probability into Equation (4.23a), we have:

$$\int_{x_0}^{\infty} f(x)dx = 1 \Rightarrow \int_{x_0}^{\infty} \frac{AkC_0(x-c)^{-k-1}}{-1+\exp\left(C_0^{\frac{1}{k}}(x-c)^{-1}\right)}dx = 1$$

$$\Rightarrow A = \frac{1}{\int_{x_0}^{\infty} \frac{kC_0(x-c)^{-k-1}}{\exp\left(C_0^{\frac{1}{k}}(x-c)^{-1}\right)-1}dx}. \tag{4.23b}$$

The integral in Equation (4.23b) can be evaluated using Riemann zeta function when $x_0 \to c^+$ as:

$$\int_{x_0}^{\infty} \frac{kC_0(x-c)^{-k-1}}{\exp\left(C_0^{\frac{1}{k}}(x-c)^{-1}\right)-1}dx = \int_{c^+}^{\infty} \frac{kC_0(x-c)^{-k-1}}{\exp\left(C_0^{\frac{1}{k}}(x-c)^{-1}\right)-1}dx$$

$$= -\int_{c^+}^{\infty} \frac{dC_0(x-c)^{-k}}{\exp\left(C_0^{\frac{1}{k}}(x-c)^{-1}\right)-1} \tag{4.23c}$$

Let $t = C_0^{\frac{1}{k}}(x-c)^{-1}$ in Equation (4.23c), Equation (4.23c) can be rewritten as:

$$-\int_{c^+}^{\infty} \frac{dC_0(x-c)^{-k}}{\exp\left(C_0^{\frac{1}{k}}(x-c)^{-1}\right)-1} = \int_0^{\infty} \frac{kt^{k-1}dt}{\exp(t)-1} = k\Gamma(k)\zeta(k) \tag{4.23d}$$

Substituting Equation (4.23d) into Equation (4.23b), we have:

$$A = \frac{1}{\int_{x_0}^{\infty} \frac{kC_0(x-c)^{-k-1}}{\exp\left(C_0^{\frac{1}{k}}(x-c)^{-1}\right)-1}dx} = \frac{1}{k\Gamma(k)\zeta(k)} \tag{4.23e}$$

Thereby, the PDF [i.e., Equation 4.23a] can be expressed as:

$$f(x) = \frac{C_0}{\Gamma(k)\zeta(k)} \frac{1}{(x-c)^{k+1}} \left(-1 + \exp\left(C_0^{\frac{1}{k}}(x-c)^{-1}\right)\right)^{-1}; \quad x \in [x_0, \infty), x_0 > c, k > 0 \tag{4.24a}$$

Let $C_0^{\frac{1}{k}} = \lambda > 0$. Then, Equation (4.24a) may be rewritten as:

$$f(x) = \frac{\lambda^k}{\Gamma(k)\zeta(k)(x-c)^{k+1}} \left(-1 + \exp\left(\lambda(x-c)^{-1}\right)\right)^{-1}; \quad x \in [x_0, \infty), x_0 > c, k > 0 \tag{4.24b}$$

$$\Gamma(k) = \int_0^{\infty} x^{k-1} \exp(-k)dx \tag{4.24c}$$

$$\zeta(k) = \sum_{n=1}^{\infty} n^{-k} \tag{4.24d}$$

In the preceding equations, $\Gamma(k)$ and $\zeta(k)$ are gamma and Riemann Zeta functions expressed in Equations (4.24c)–(4.24d), respectively. Figure 4.5 plots $g(z), z = h(x)$, $f(x)$ functions of Davis distribution. It is shown that: (i) both the generating function $g(z)$ and the TF $z = h(x)$ are decreasing function; (ii) the PDF is skewed to the right.

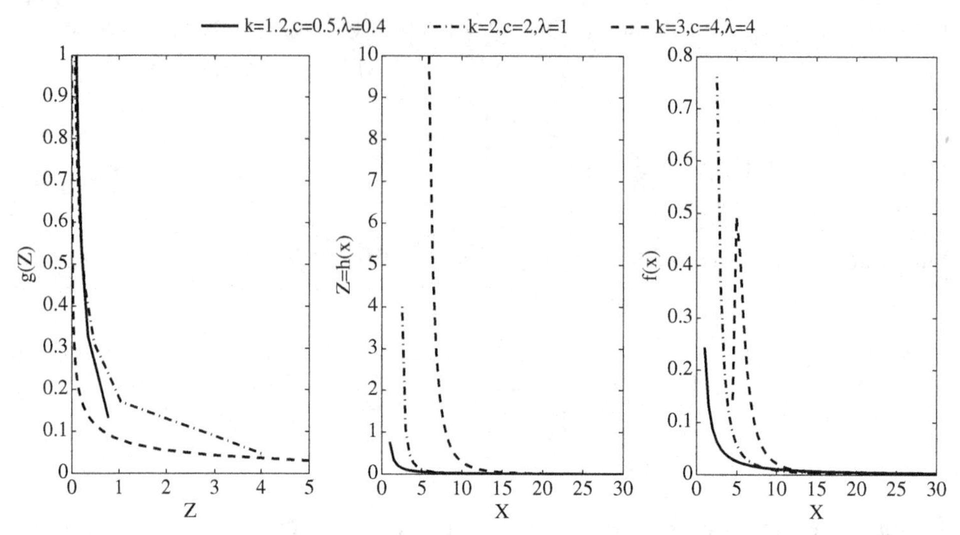

Figure 4.5 g(z), z = h(x) and f(x) functions of Davis distribution. (a): g(z) function; (b): z = h(x) function; and (c): probability density f(x) function.

4.2.6 Amoroso Distribution

For Amoroso distribution, we have $b = 0$, $k > 0$ in generating function [Equation 4.1] and $\alpha = -1, \beta \neq 0, c \neq 0$ in the TF [Equation 4.2]. Then, the generating function and TF are given as:

$$g(z) = A/\exp\left(z^{\frac{1}{k}}\right) \tag{4.25}$$

$$\begin{aligned} &\frac{1}{z}\frac{dz}{dx} = \frac{\beta}{x-c} \\ &\Rightarrow z = h(x) = C_0(x-c)^{\beta}; \left|\frac{dz}{dx}\right| = C_0|\beta|(x-c)^{\beta-1}; x \in [x_0, \infty), x_0 > c \end{aligned} \tag{4.26}$$

Substituting Equations (4.25)–(4.26) into Equation (4.4), we have:

$$f(x) = \left|\frac{dz}{dx}\right| g(z) = AC_0|\beta|(x-c)^{\beta-1}\exp\left(-\left(C_0(x-c)^{\beta}\right)^{\frac{1}{k}}\right) \tag{4.27a}$$

Applying the total probability to Equation (4.27a), the normalization factor can then be solved as following:

$$\begin{aligned} &\int_{x_0}^{\infty} f(x)dx = 1 \Rightarrow \int_{x_0}^{\infty} A|\beta|C_0(x-c)^{\beta-1}\exp\left(-\left(C_0(x-c)^{\beta}\right)^{\frac{1}{k}}\right)dx = 1 \\ &\Rightarrow A = \frac{1}{\int_{x_0}^{\infty} C_0|\beta|(x-c)^{\beta-1}\exp\left(-\left(C_0(x-c)^{\beta}\right)^{\frac{1}{k}}\right)dx} \end{aligned} \tag{4.27b}$$

Let $y = \left(C_0(x-c)^{\beta}\right)^{\frac{1}{k}}$, we have: $C_0(x-c)^{\beta} = y^k \Rightarrow d[C_0(x-c)^{\beta}] = ky^{k-1}dy$. Equation (4.27b) can now be solved as:

(i) if $\beta > 0$ and $x_0 \to c^+$:

$$\begin{aligned}
&\int_{x_0}^{\infty} C_0|\beta|(x-c)^{\beta-1}\exp\left(-\left(C_0(x-c)^{\beta}\right)^{\frac{1}{k}}\right)dx \\
&= \int_{c^+}^{\infty} C_0\beta(x-c)^{\beta-1}\exp\left(-\left(C_0(x-c)^{\beta}\right)^{\frac{1}{k}}\right)dx \\
&= \int_0^{\infty} ky^{k-1}\exp(-y)dy = k\Gamma(k) = \Gamma(k+1)
\end{aligned} \tag{4.27c}$$

(ii) if $\beta < 0$ and $x_0 \to c^+$:

$$\begin{aligned}
&\int_{x_0}^{\infty} C_0|\beta|(x-c)^{\beta-1}\exp\left(-\left(C_0(x-c)^{\beta}\right)^{\frac{1}{k}}\right)dx \\
&= -\int_{c^+}^{\infty} C_0\beta(x-c)^{\beta-1}\exp\left(-\left(C_0(x-c)^{\beta}\right)^{\frac{1}{k}}\right)dx \\
&= -\int_{\infty}^{0} ky^{k-1}\exp(-y)dy = \int_0^{\infty} ky^{k-1}\exp(-y)dy = k\Gamma(k) = \Gamma(k+1)
\end{aligned} \tag{4.27d}$$

From Equations (4.27c) and (4.27d), we have: $A = \frac{1}{k\Gamma(k)} = \frac{1}{\Gamma(k+1)}$. The PDF [i.e., Equation 4.27a] can now be rewritten as:

$$\begin{aligned}
f(x) &= \frac{C_0|\beta|(x-c)^{\beta-1}}{k\Gamma(k)}\exp\left(-\left(C_0(x-c)^{\beta}\right)^{\frac{1}{k}}\right) \\
&= \frac{C_0}{\Gamma(k)}\frac{|\beta|}{k}(x-c)^{\beta-1}\exp\left(-C_0^{\frac{1}{k}}(x-c)^{\frac{\beta}{k}}\right)
\end{aligned} \tag{4.28a}$$

Let $C_0^{\frac{1}{k}} = \lambda, s = \frac{\beta}{k}$, Equation (4.28a) can be rewritten as:

$$f(x) = \frac{\lambda^k|s|}{\Gamma(k)}(x-c)^{sk-1}\exp(-\lambda(x-c)^s); \lambda > 0, k > 0, s \neq 0, c \neq 0, \quad x \in [x_0, \infty), x_0 > c \tag{4.28b}$$

Both Equation (4.28a) and Equation (4.28b) represent the 4-parameter Amoroso distribution.

The Amoroso distribution includes several distributions as special cases, which are presented in the following text. Figure 4.6 graphs $g(z), z = h(x)$, and $f(x)$ functions for Amoroso distribution.

From the 4-parameter Amoroso distribution, one may obtain the special cases as follows. Figure 4.7 graphs the density functions for each special case.

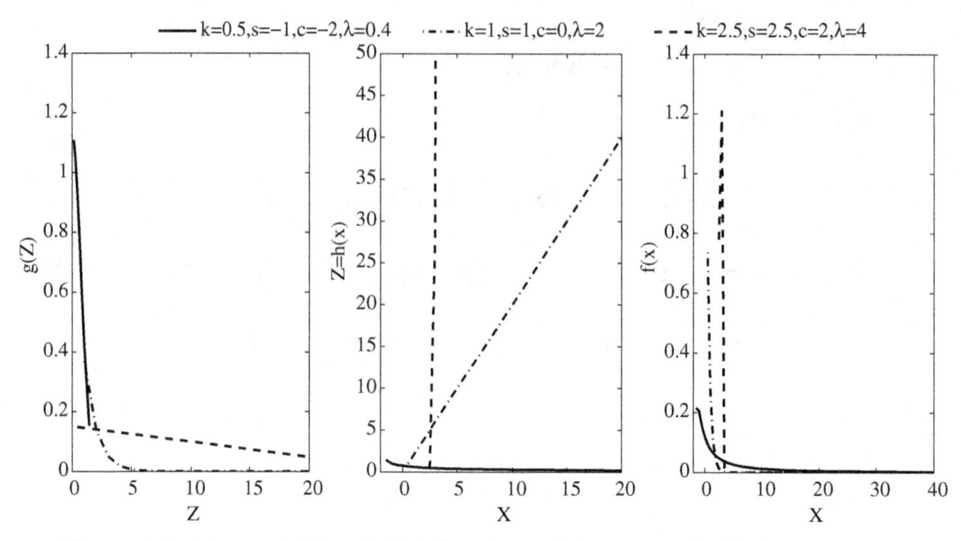

Figure 4.6 g(z), z = h(x) and f(x) functions of Amoroso distribution. (a): g(z) function; (b): z = h(x) function; and (c): probability density f(x) function.

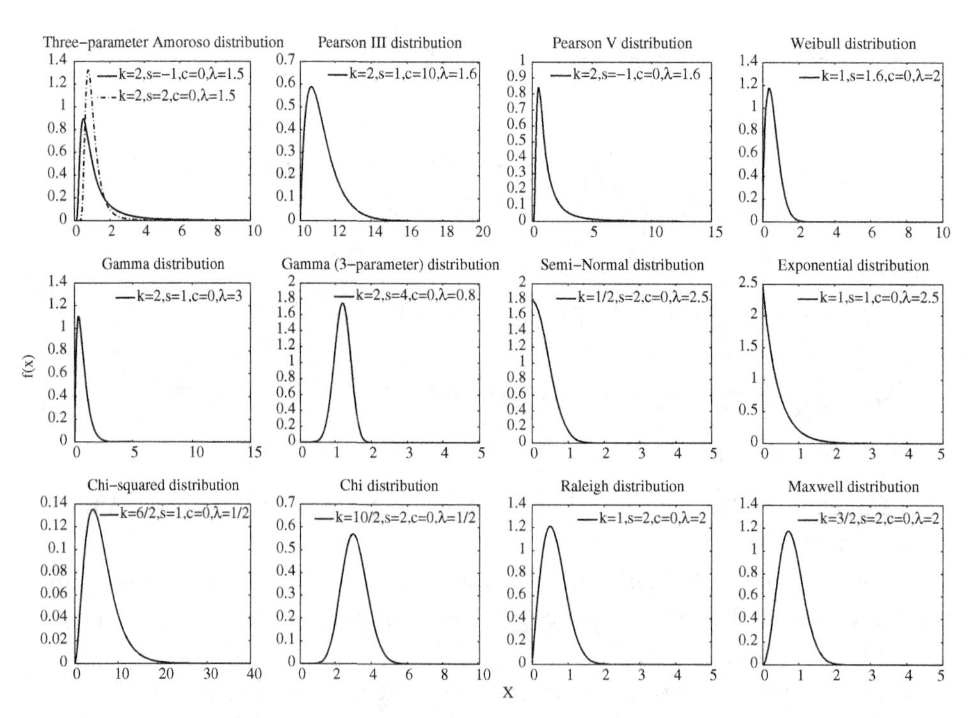

Figure 4.7 PDFs for the special cases of Amoroso distribution.

Three-parameter Amoroso Distribution: The 4-parameter Amoroso distribution reduces to the 3-parameter Amoroso distribution if $c = 0$. Thus, Equation (4.28b) may be rewritten as:

$$f(x) = \frac{\lambda^k |s|}{\Gamma(k)} x^{sk-1} \exp\left(-\lambda x^s\right); \lambda > 0, k > 0, s \neq 0, \quad x \in (0, \infty) \tag{4.29}$$

Alternatively, the 3-parameter Amoroso distribution may be directly derived from D'Addario system with the generating function and the TF expressed as:

$$g(z) = \frac{A}{\exp\left(z^{\frac{1}{k}}\right)} \tag{4.30a}$$

$$\frac{1}{z}\frac{dz}{dx} = \frac{\beta}{x}; \beta > 0, x \in (0, \infty) \tag{4.30b}$$

Comparing Equations (4.30a)–(4.30b) to Equations (4.1)–(4.2), it is shown that $g(z)$ and $z = h(x)$ for 3-parameter Amoroso distribution can be obtained by setting $\alpha = -1$ and $c = 0$ in Equations (4.1)–(4.2).

From Equation (4.30b), we obtain:

$$z = h(x) = C_0 x^\beta \Rightarrow \left|\frac{dz}{dx}\right| = \beta C_0 x^{\beta-1}; \beta > 0, C_0 > 0, x \in (0, \infty) \tag{4.31}$$

Then, the PDF can be expressed as:

$$f(x) = \left|\frac{dz}{dx}\right| g(z) \Rightarrow f(x) = \frac{A\beta C_0 x^{\beta-1}}{\exp\left(\left(C_0 x^\beta\right)^{\frac{1}{k}}\right)}; \beta, C_0, k > 0, x \in (0, \infty) \tag{4.32}$$

Applying total probability to Equation (4.32), we can solve for the normalization factor A as:

$$\begin{aligned} \int_0^\infty f(x)dx = 1 &\Rightarrow \int_0^\infty A\beta C_0 x^{\beta-1} \exp\left(-\left(C_0 x^\beta\right)^{\frac{1}{k}}\right) dx = 1 \\ &\Rightarrow \int_0^\infty A \exp\left(-\left(C_0 x^\beta\right)^{\frac{1}{k}}\right) d\left(C_0 x^\beta\right) = 1 \end{aligned} \tag{4.33a}$$

$$A = \frac{1}{\int_0^\infty \exp\left(-\left(C_0 x^\beta\right)^{\frac{1}{k}}\right) d\left(C_0 x^\beta\right)} \tag{4.33b}$$

Let $y = \left(C_0 x^\beta\right)^{\frac{1}{k}} \Rightarrow d\left(C_0 x^\beta\right) = k y^{k-1} dy$, and the integral in Equation (4.33b) can be solved as:

$$\int_0^\infty \exp\left(-\left(C_0 x^\beta\right)^{\frac{1}{k}}\right) d\left(C_0 x^\beta\right) = \int_0^\infty k y^{k-1} \exp(-y) dy = k\Gamma(k) = \Gamma(k+1) \tag{4.33c}$$

Thereby, we have $A = \frac{1}{k\Gamma(k)} = \frac{1}{\Gamma(k+1)}$.

Equation (4.32) can now be rewritten as:

$$f(x) = \left|\frac{dz}{dx}\right| g(z) \Rightarrow f(x) = \frac{\beta C_0 x^{\beta-1}}{\mathrm{k}\Gamma(\mathrm{k}) \exp\left(\left(C_0 x^{\beta}\right)^{\frac{1}{k}}\right)}; \beta, C_0, k > 0, x \in (0, \infty) \quad (4.34)$$

In Equation (4.34), let $\lambda = C_0^{\frac{1}{k}}, s = \frac{\beta}{k}$; we can rewrite Equation (4.34) in the exactly same form as the 3-parameter Amoroso distribution expressed by Equation (4.29).

Pearson type III distribution: The 4-parameter Amoroso distribution becomes a 3-parameter Pearson III distribution if $c \neq 0$, $s = 1$. Thus, Equation (4.28b) may be rewritten as:

$$f(x) = \frac{\lambda^k}{\Gamma(k)}(x-c)^{k-1} \exp\left(-\lambda(x-c)\right); x \in [x_0, \infty); x_0 > c \quad (4.35)$$

The Pearson type III distribution can be directly derived from the D'Addario system with the generating function and TFs expressed as:

$$g(z) = \frac{A}{\exp\left(z^{\frac{1}{k}}\right)}; k > 0 \quad (4.36a)$$

$$\frac{1}{z}\frac{dz}{dx} = \frac{k}{x-c}; \beta = k; c \neq 0 \quad (4.36b)$$

From Equation (4.36b), we have:

$$z = h(x) = C_0(x-c)^k \text{ and } \left|\frac{dz}{dx}\right| = C_0 k (x-c)^{k-1}; x \in [x_0, \infty), x_0 > c \quad (4.36c)$$

The PDF can now be expressed as:

$$f(x) = \left|\frac{dz}{dx}\right| g(z) = AC_0 k (x-c)^{k-1} \exp\left(-\left(C_0(x-c)^k\right)^{\frac{1}{k}}\right) \quad (4.36d)$$

Applying the total probability, the normalization factor A is solved as:

$$\begin{aligned} &\int_{x_0}^{\infty} AC_0 k (x-c)^{k-1} \exp\left(-\left(C_0(x-c)^k\right)^{\frac{1}{k}}\right) dx = 1 \\ &\Rightarrow \int_{x_0}^{\infty} A \exp\left(-\left(C_0(x-c)^k\right)^{\frac{1}{k}}\right) dC_0(x-c)^k = 1 \\ &\Rightarrow A = \frac{1}{\int_{x_0}^{\infty} \exp\left(-\left(C_0(x-c)^k\right)^{\frac{1}{k}}\right) dC_0(x-c)^k} \end{aligned} \quad (4.37)$$

Let $y = \left(C_0(x-c)^k\right)^{\frac{1}{k}}$ and $x_0 \to c^+$, we solve $A = \frac{1}{k\Gamma(k)} = \frac{1}{\Gamma(k+1)}$. Equation (4.36d) can now be rewritten as:

$$f(x) = \frac{1}{\Gamma(k)} C_0(x-c)^{k-1} \exp\left(-C_0^{\frac{1}{k}}(x-c)\right) \tag{4.38}$$

In Equation (4.38) let $\lambda = C_0^{\frac{1}{k}}$, the Equation (4.38) can be expressed in the same form as Pearson type III distribution given as Equation (4.35).

It is known Pearson type III distribution is commonly applied in flood frequency analysis.

Pearson type V distribution: The 4-parameter Amoroso distribution becomes a 2-parameter Pearson V distribution if $c = 0$, $s = -1$. Thus, Equation (4.28b) may be rewritten as:

$$f(x) = \frac{\lambda^k}{\Gamma(k)} x^{-k-1} \exp\left(-\lambda x^{-1}\right) \tag{4.39}$$

This is also known as Vinci (1921) distribution.

To derive directly from D'Addario system, the Pearson type V distribution may be obtained in what follows. The generating function and the TF are expressed as:

$$g(z) = \frac{A}{\exp\left(z^{\frac{1}{k}}\right)}, k > 0 \tag{4.40a}$$

$$\frac{1}{z}\frac{dz}{dx} = \frac{\beta}{x}; \beta = -k; \mathrm{x} \in (0, \infty) \tag{4.40b}$$

Solving differential Equation (4.40b), we have:

$$z = C_0 x^\beta \Rightarrow z = C_0 x^{-k} \Rightarrow \left|\frac{dz}{dx}\right| = C_0 k x^{-k-1} \tag{4.40c}$$

Substituting Equations (4.40c) into Equation (4.4) we have:

$$f(x) = \left|\frac{dz}{dx}\right| g(z) = AC_0 k x^{-k-1} \exp\left(-C_0^{\frac{1}{k}} x^{-1}\right) \tag{4.41a}$$

Applying the total probability to Equation (4.41a), the normalization factor A can be solved as follows:

$$\int_0^\infty f(x)dx = 1 \Rightarrow \int_0^\infty AC_0 k x^{-k-1} \exp\left(-C_0^{\frac{1}{k}} x^{-1}\right) dx = 1$$

$$\Rightarrow A = \frac{1}{\int_0^\infty C_0 k x^{-k-1} \exp\left(-C_0^{\frac{1}{k}} x^{-1}\right) dx} \tag{4.41b}$$

$$= \frac{1}{-\int_0^\infty \exp\left(-C_0^{\frac{1}{k}} x^{-1}\right) d\left(C_0 x^{-k}\right)}$$

let $y = C_0^{\frac{1}{k}} x^{-1}$, we have: $d(C_0 x^{-k}) = dy^k = ky^{k-1}dy$; and the integral in Equation (4.41b) can be evaluated as:

$$-\int_0^{\infty} \exp\left(-C_0^{\frac{1}{k}} x^{-1}\right) d\left(C_0 x^{-k}\right) = \int_0^{\infty} \mathrm{k}y^{\mathrm{k}-1} \exp(-y)dy = k\Gamma(k) = \Gamma(k+1) \tag{4.41c}$$

$$A = \frac{1}{k\Gamma(k)} = \frac{1}{\Gamma(k+1)} \tag{4.41d}$$

Now, Equation (4.41a) may be rewritten as:

$$f(x) = \frac{C_0 x^{-k-1}}{\Gamma(k)} \exp\left(-C_0^{\frac{1}{k}} x^{-1}\right) \tag{4.42}$$

Let $\lambda = C_0^{\frac{1}{k}}$ in Equation (4.42). We obtain the PDF in the same form as Equation (4.39), i.e., the Pearson type V distribution.

Weibull distribution: The 4-parameter Amoroso distribution becomes a 2-parameter Weibull distribution if $c = 0$, $k = 1$. Thus, Equation (4.28b) may be rewritten as:

$$f(x) = \lambda s x^{s-1} \exp(-\lambda x^s);\ \lambda,\ s > 0,\ x \in [0, \infty) \tag{4.43a}$$

Similarly, the 3-parameter Weibull distribution (i.e., Weibull distribution with location) is rewritten using Equation (4.28b) as:

$$f(x) = \lambda s(x-c)^{s-1} \exp(-\lambda(x-c)^s);\ \lambda > 0,\ k > 0,\ s \neq 0,\ c \neq 0,\ x \in (c, \infty) \tag{4.43b}$$

The 2- and 3-parameter Weibull distribution may be derived directly from the D'Addario system. The generating function and TF can be given as:

$$g(z) = \frac{A}{\exp(z)} \tag{4.44a}$$

$$\frac{1}{\mathrm{z}}\frac{dz}{dx} = \frac{\beta}{x-c}; \beta > 0, x \in (\mathrm{c}, \infty) \tag{4.44b}$$

Solution of differential Equation (4.44b) yields

$$\mathrm{z} = \mathrm{h(x)} = \mathrm{C_0}(x-c)^{\beta}; \left|\frac{dz}{dx}\right| = C_0\beta(x-c)^{\beta-1}; \mathrm{C_0} > 0 \tag{4.44c}$$

The PDF can then be expressed as:

$$\mathrm{f(x)} = \left|\frac{\mathrm{dz}}{\mathrm{dx}}\right| g(z) = AC_0\beta(x-c)^{\beta-1} \exp\left(-C_0(x-c)^{\beta}\right) \tag{4.45}$$

Applying the total probability to Equation (4.45), the normalization factor A can be solved as:

$$\int_c^{\infty} f(x)dx = 1 \Rightarrow \int_c^{\infty} AC_0\beta(x-c)^{\beta-1}\exp\left(-C_0(x-c)^{\beta}\right)dx = 1$$

$$\Rightarrow A = \frac{1}{\int_c^{\infty} C_0\beta(x-c)^{\beta-1}\exp\left(-C_0(x-c)^{\beta}\right)dx} \tag{4.46a}$$

$$\Rightarrow A = \frac{1}{\int_c^{\infty} \exp\left(-C_0(x-c)^{\beta}\right)d\left(c_0(x-c)^{\beta}\right)} \Rightarrow A = 1$$

and Equation (4.45) can be rewritten as:

$$\mathrm{f(x)} = C_0\beta(x-c)^{\beta-1}\exp\left(-C_0(x-c)^{\beta}\right) \tag{4.46b}$$

Equation (4.46b) has exactly the same form as of the Weibull distribution [i.e., Equation 4.43a] by setting $\lambda = C_0$, $s = \beta$, $c = 0$. Additionally, Equation (4.46b) has exactly the same form of 3-parameter Weibull distribution [i.e., Equation 4.43b] by setting $\lambda = C_0$, $s = \beta$, $c \neq 0$.

Two-parameter gamma distribution: The 4-parameter Amoroso distribution becomes a 2-parameter gamma distribution if $c = 0$, $s = 1$. Thus, Equation (4.28b) may be rewritten as:

$$f(x) = \frac{\lambda^k}{\Gamma(k)}x^{k-1}\exp(-\lambda x); \lambda > 0, k > 0, x \in (0, \infty) \tag{4.47}$$

The 2-parameter gamma distribution can also be derived from the D'Addario system. The generating function and TFs can be given as:

$$g(z) = \frac{A}{\exp\left(z^{\frac{1}{k}}\right)}; \mathrm{k} > 0 \tag{4.48a}$$

$$\frac{1}{z}\frac{dz}{dx} = \frac{k}{x}; x > 0 \tag{4.48b}$$

The solution of Equation (4.48b) yields

$$z = h(x) = C_0x^k; \left|\frac{dz}{dx}\right| = C_0kx^{k-1} \tag{4.48c}$$

Then the PDF may be written as:

$$f(x) = \left|\frac{dz}{dx}\right|g(z) = AC_0kx^{k-1}\exp\left(-\left(C_0x^k\right)^{\frac{1}{k}}\right) \tag{4.49}$$

Applying the total probability to Equation (4.49), the normalization factor A can be solved as:

$$\int_0^\infty f(x)dx = 1 \Rightarrow \int_0^\infty AC_0 kx^{k-1} \exp\left(-\left(C_0 x^k\right)^{\frac{1}{k}}\right)dx = 1$$

$$\Rightarrow A = \frac{1}{\int_0^\infty C_0 kx^{k-1} \exp\left(-\left(C_0 x^k\right)^{\frac{1}{k}}\right)dx} = \frac{1}{k\Gamma(k)} = \frac{1}{\Gamma(k+1)} \tag{4.50a}$$

and Equation (4.49) can be rewritten as:

$$f(x) = \frac{C_0}{\Gamma(k)} x^{k-1} \exp\left(-C_0^{\frac{1}{k}} x\right) \tag{4.50b}$$

Equation (4.50b) is exactly of the same form as Equation (4.47) by setting $\lambda = C_0^{\frac{1}{k}}$.

Three-parameter gamma distribution: The 4-parameter Amoroso distribution becomes a 3-parameter gamma distribution if $c = 0$. Thus, Equation (4.28b) may be rewritten as:

$$f(x) = \frac{\lambda^k s}{\Gamma(k)} (x)^{sk-1} \exp\left(-\lambda x^s\right); \lambda > 0, k > 0, s > 0,\ \ x \in (0, \infty) \tag{4.51}$$

Three-parameter gamma distribution may be derived from the D'Addario system. The generating function and TF can be given as:

$$g(z) = \frac{A}{\exp\left(z^{\frac{1}{k}}\right)} \tag{4.52a}$$

$$\frac{1}{z}\frac{dz}{dx} = \frac{\beta}{x}; \beta > 0, x \in (0, \infty) \tag{4.52b}$$

The solution of Equation (4.52b) yields

$$z = h(x) = C_0 x^\beta; \left|\frac{dz}{dx}\right| = C_0 \beta x^{\beta-1} \tag{4.52c}$$

and the PDF can be given as:

$$f(x) = \left|\frac{dz}{dx}\right| g(z) = AC_0 \beta x^{\beta-1} \exp\left(-\left(C_0 x^\beta\right)^{\frac{1}{k}}\right) \tag{4.53}$$

Applying the total probability to Equation (4.53), the normalization factor A can be solved as:

$$\int_0^\infty f(x)dx = 1 \Rightarrow \int_0^\infty AC_0\beta x^{\beta-1}\exp\left(-\left(C_0x^\beta\right)^{\frac{1}{k}}\right)dx = 1$$
$$\Rightarrow A = \frac{1}{\int_0^\infty C_0\beta x^{\beta-1}\exp\left(-\left(C_0x^\beta\right)^{\frac{1}{k}}\right)dx} = \frac{1}{k\Gamma(k)} = \frac{1}{\Gamma(k+1)} \tag{4.54a}$$

and Equation (4.53) is rewritten as:

$$f(x) = \frac{C_0\beta x^{\beta-1}}{k\Gamma(k)}\exp\left(-C_0^{\frac{1}{k}}x^{\frac{\beta}{k}}\right) \tag{4.54b}$$

Equation (4.54b) is exactly of the same form as Equation (4.51) by setting $\lambda = C_0^{\frac{1}{k}}$, $s = \frac{\beta}{k}$.

Seminormal distribution: The 4-parameter Amoroso distribution becomes a 1-parameter seminormal distribution if $c = 0, s = 2, k = \frac{1}{2}$. Thus, Equation (4.28b) may be rewritten as:

$$f(x) = \frac{2\sqrt{\lambda}}{\sqrt{\pi}}\exp\left(-\lambda x^2\right) = \frac{\sqrt{2}}{\sqrt{\frac{1}{2\lambda}}\sqrt{\pi}}\exp\left(-\frac{x^2}{2\left(\frac{1}{2\lambda}\right)}\right); \lambda > 0, x \in [0,\infty) \tag{4.55}$$

In Equation (4.55) $\sqrt{\frac{1}{2\lambda}}$ represents the standard deviation.

The seminormal distribution may be derived from the D'Addario system. The generating function and TFs are:

$$g(z) = \frac{A}{\exp\left(z^2\right)} \tag{4.56a}$$

$$\frac{1}{z}\frac{dz}{dx} = \frac{1}{x} \tag{4.56b}$$

Solution of Equation (4.56b) yields

$$z = h(x) = C_0x \Rightarrow \left|\frac{dz}{dx}\right| = C_0 \tag{4.56c}$$

The PDF is given as:

$$f(x) = \left|\frac{dz}{dx}\right|g(z) = AC_0\exp\left(-C_0^2x^2\right) \tag{4.57}$$

Applying the total probability to Equation (4.57), the normalization factor A can be solved as:

$$\int_0^\infty AC_0\exp\left(-C_0^2x^2\right)dx = 1 \Rightarrow A = \frac{1}{\int_0^\infty C_0\exp\left(-C_0^2x^2\right)dx} \Rightarrow A = \frac{2}{\sqrt{\pi}} \tag{4.58}$$

Then Equation (4.57) can be rewritten as:

$$f(x) = \frac{2}{\sqrt{\pi}} C_0 \exp\left(-C_0^2 x^2\right) \tag{4.59}$$

Equation is exactly the same form as Equation (4.55) by setting $\lambda = C_0^2$.

Exponential distribution: The exponential distribution is obtained if $s = 1$, $k = 1$, $c = 0$. Thus, Equation (4.28b) may be rewritten as:

$$f(x) = \lambda \exp(-\lambda x); \lambda > 0, x \in [0, \infty) \tag{4.60}$$

The exponential distribution may be derived from the D'Addario system. The generating function and TFs are:

$$g(z) = A / \exp(z) \tag{4.61a}$$

$$\frac{1}{z}\frac{dz}{dx} = \frac{1}{x}; x \in (0, \infty) \tag{4.61b}$$

The solution of Equation (4.61b) yields:

$$z = h(x) = C_0 x; \left|\frac{dz}{dx}\right| = C_0 \tag{4.62}$$

and the PDF can be rewritten as:

$$f(x) = \left|\frac{dz}{dx}\right| g(z) = AC_0 \exp(-C_0 x) \tag{4.63}$$

Applying the total probability into Equation (4.63), the normalization factor A can be obtained as:

$$\int_0^\infty f(x)dx = 1 \Rightarrow \int_0^\infty AC_0 \exp(-C_0 x)dx = 1 \Rightarrow A = \frac{1}{\int_0^\infty C_0 \exp(-C_0 x)dx} = 1 \tag{4.64}$$

and Equation (4.63) can be rewritten as:

$$f(x) = C_0 \exp(-C_0 x) \tag{4.65}$$

Equation (4.65) is exactly of the same form as of Equation (4.60) by setting $\lambda = C_0$.

Chi-squared distribution: The chi-square distribution is obtained if $\lambda = \frac{1}{2}$, $k = \frac{n}{2}, s = 1, c = 0$. Thus, Equation (4.28b) may be rewritten as:

$$f(x) = \frac{1}{2^{\frac{n}{2}}\Gamma\left(\frac{n}{2}\right)} x^{\frac{n}{2}-1} \exp\left(-\frac{x}{2}\right); n \in \mathbb{N}^+, x \in [0, \infty) \tag{4.66}$$

Chi distribution: The chi distribution is obtained if $\lambda = \frac{1}{2}, c = 0, k = \frac{K}{2} > 0$, $s = 2$. Thus Equation (4.28b) may be rewritten as:

$$f(x) = \frac{1}{2^{\frac{K}{2}-1}\Gamma\left(\frac{K}{2}\right)} x^{K-1} \exp\left(-\frac{1}{2}x^2\right); K > 0, x \in [0, \infty) \tag{4.67}$$

The Chi distribution can also be derived directly from the D'Addario system. The generating function and TFs are given as:

$$g(z) = \frac{A}{\exp\left(z^{\frac{1}{k}}\right)}; \mathrm{k} > 0 \tag{4.68a}$$

$$\frac{1}{z}\frac{dz}{dx} = \frac{2k}{x}; x \in (0, \infty) \tag{4.68b}$$

Solving the differential Equation (4.68b), we have:

$$z = C_0 x^{2k} \Rightarrow \left|\frac{dz}{dx}\right| = 2C_0 k x^{2k-1} \tag{4.69}$$

and the PDF can be written as:

$$f(x) = \left|\frac{dz}{dx}\right| g(z) = 2C_0 k x^{2k-1} A \exp\left(-\left(C_0 x^{2k}\right)^{\frac{1}{k}}\right) \tag{4.70}$$

Applying the total probability to Equation (4.70), we can solve for the normalization factor A as:

$$\begin{aligned} \int_0^\infty f(x)dx = 1 &\Rightarrow \int_0^\infty 2C_0 k x^{2k-1} A \exp\left(-\left(C_0 x^{2k}\right)^{\frac{1}{k}}\right) dx = 1 \\ &\Rightarrow A = \frac{1}{\int_0^\infty 2C_0 k x^{2k-1} \exp\left(-\left(C_0 x^{2k}\right)^{\frac{1}{k}}\right) dx} \\ &= \frac{1}{\int_0^\infty \exp\left(-\left(C_0 x^{2k}\right)^{\frac{1}{k}}\right) dC_0 x^{2k}} \end{aligned} \tag{4.71}$$

In Equation (4.71), let $y = \left(C_0 x^{2k}\right)^{\frac{1}{k}} \Rightarrow d\left(C_0 x^{2k}\right) = d\left(y^k\right) = k y^{k-1} dy$ and Equation (4.71) may be solved as:

$$A = \frac{1}{\int_0^\infty \exp(-y) k y^{k-1} dy} = \frac{1}{k\Gamma(k)} \tag{4.72}$$

Now Equation (4.70) can be rewritten as:

$$f(x) = \frac{2C_0 x^{2k-1}}{\Gamma(k)} \exp\left(-C_0^{\frac{1}{k}} x^2\right); \mathrm{x} \in (0, \infty) \tag{4.73}$$

In Equation (4.73) let $C_0^{\frac{1}{k}} = \frac{1}{2} \Rightarrow C_0 = \left(\frac{1}{2}\right)^k$ and $k = \frac{K}{2}, K \in (0, \infty)$; Equation (4.73) may be expressed in the same form as Equation (4.67).

Raleigh distribution: The Raleigh distribution is obtained from Amoroso distribution if $c = 0$, $k = 1$, $s = 2$. Thus Equation (4.28b) may be rewritten as:

$$f(x) = 2\lambda x \exp(-\lambda x^2); x \in [0, \infty) \tag{4.74}$$

let $\sigma^2 = \frac{1}{2\lambda}$, Equation (4.74) can be rewritten as:

$$f(x) = \frac{x}{\sigma^2} \exp\left(-\frac{x^2}{2\sigma^2}\right), x \in [0, \infty) \tag{4.75}$$

The Raleigh distribution can also be derived directly from the D'Addario system. The generating function and TF can be given as:

$$g(z) = A/\exp(z) \tag{4.76a}$$

$$\frac{1}{z}\frac{dz}{dx} = \frac{2}{x} \Rightarrow z = C_0 x^2 \Rightarrow \left|\frac{dz}{dx}\right| = 2C_0 x \tag{4.76b}$$

The PDF can then be given as:

$$f(x) = \left|\frac{dz}{dx}\right| g(z) = 2C_0 x A \exp(-C_0 x^2) \tag{4.77}$$

Applying the total probability to Equation (4.77), the normalization factor A can be solved as:

$$\int_0^\infty f(x)dx = 1 \Rightarrow \int_0^\infty 2C_0 x A \exp(-C_0 x^2)dx = 1$$
$$\Rightarrow A = \frac{1}{\int_0^\infty \exp(-C_0 x^2) dC_0 x^2} = 1 \tag{4.78}$$

And Equation (4.77) can be rewritten as:

$$f(x) = 2C_0 x \exp(-C_0 x^2); x \in [0, \infty), C_0 > 0 \tag{4.79}$$

In Equation (4.79) let $\lambda = C_0$. Equation (4.79) can be expressed in the same form as Equation (4.74). Additionally, let $\sigma^2 = \frac{1}{2C_0}$. Equation (4.79) can be expressed in the same form as Equation (4.75).

Maxwell distribution: The Maxwell distribution is obtained from the Amoroso distribution if $c = 0, k = \frac{3}{2}, s = 2$. Thus, Equation (4.28b) may be rewritten as:

$$f(x) = \frac{2\lambda^{3/2}}{\frac{1}{2}\sqrt{\pi}} x^2 \exp(-\lambda x^2); x \in [0, \infty) \tag{4.76}$$

Let $\lambda = \frac{1}{2\sigma^2}$. Equation (4.76) can be rewritten as:

$$f(x) = \frac{x^2\sqrt{2}}{\sigma^3\sqrt{\pi}} \exp\left(-\frac{x^2}{2\sigma^2}\right); x \in [0, \infty) \tag{4.77}$$

The Maxwell distribution may also be derived directly from the D'Addario system. The generating function and TFs can be given as:

$$g(z) = A \exp\left(-z^{\frac{2}{3}}\right) \tag{4.78a}$$

$$\frac{1}{z}\frac{dz}{dx} = \frac{3}{x} \Rightarrow z = C_0 x^3 \Rightarrow \left|\frac{dz}{dx}\right| = 3C_0 x^2; x \in [0, \infty) \tag{4.78b}$$

The PDF can then be written as:

$$f(x) = \left|\frac{dz}{dx}\right| g(z) = 3C_0 x^2 A \exp\left(-\left(C_0 x^3\right)^{\frac{2}{3}}\right) \tag{4.79}$$

Applying the total probability to Equation (4.79), we have:

$$\begin{aligned} \int_0^\infty f(x)dx = 1 &\Rightarrow \int_0^\infty 3C_0 x^2 A \exp\left(-\left(C_0 x^3\right)^{\frac{2}{3}}\right) dx = 1 \\ &\Rightarrow A \int_0^\infty \exp\left(-\left(C_0 x^3\right)^{\frac{2}{3}} d\left(C_0 x^3\right) = 1\right. \end{aligned} \tag{4.80}$$

From Equation (4.80), the normalization factor A can be solved as:

$$A = \frac{1}{\int_0^\infty \exp\left(-C_0 x^3\right)^{\frac{2}{3}}\big) d\left(C_0 x^3\right)} = \frac{1}{\int_0^\infty \left(\frac{3}{2}\right) y^{\frac{1}{2}} \exp(-y) dy} = \frac{1}{\frac{3}{2}\Gamma\left(\frac{3}{2}\right)} = \frac{4}{3\sqrt{\pi}} \tag{4.81}$$

Now, Equation (4.79) can be rewritten as:

$$f(x) = \frac{4C_0 x^2}{\sqrt{\pi}} \exp\left(-C_0^{\frac{2}{3}} x^2\right) \tag{4.82}$$

In Equation (4.82) let $\lambda = C_0^{\frac{2}{3}}$. Equation (4.82) can be expressed in the same form as Equation (4.76). Additionally, if setting $\sigma^2 = \frac{1}{2} C_0^{-\frac{2}{3}}$, Equation (4.82) can be expressed in the same form as Equation (4.77).

4.3 Application

In this section, the peak flow, monthly discharge, monthly sediment, Total Persulfate Nitrogen (TPN), and maximum daily precipitation amount are applied to evaluate the applicability of the D'Addario system in water engineering. The following distributions are applied for evaluation: log-normal (2-/3-parameter distribution), 3- and 4-parameter Amoroso distribution, Maxwell distribution, Rayleigh distribution, and seminormal distribution. The maximum likelihood method for the random variable of sample size n is applied for the parameter estimation that are briefly discussed as follows.

Log-normal (2-parameter)

$$\ln L = -n \ln\left(\sqrt{2\pi}\sigma\right) - \sum_{i=1}^{n} \ln x_i - \sum_{i=1}^{n} \frac{(\ln x_i - \mu)^2}{2\sigma^2} \tag{4.83a}$$

Maximizing Equation (4.83a), the parameters can be expressed as:

$$\begin{cases} \sum_{i=1}^{n} \dfrac{\ln x_i - \mu}{\sigma^2} = 0 \Rightarrow \hat{\mu} = \dfrac{1}{n}\sum_{i=1}^{n} \ln x_i = \overline{\ln x} \\ \dfrac{n}{\sigma} - \dfrac{1}{\sigma^3}\sum_{i=1}^{n} (\ln x - \mu)^2 = 0 \Rightarrow \hat{\sigma}^2 = \dfrac{1}{n}\sum_{i=1}^{n} (\ln x - \mu)^2 = Var(\ln x) \end{cases} \tag{4.83b}$$

Lognormal (3-parameter)

$$\ln L = -n \ln\left(\sqrt{2\pi}\sigma\right) - \sum_{i=1}^{n} \ln (x_i - c) - \sum_{i=1}^{n} \frac{(\ln (x_i - c) - \mu)^2}{2\sigma^2} \tag{4.84a}$$

Maximizing Equation (4.84a), we have:

$$\begin{cases} \sum_{i=1}^{n} \dfrac{\ln (x_i - c) - \mu}{\sigma^2} = 0 \Rightarrow \hat{\mu} = \dfrac{1}{n}\sum_{i=1}^{n} \ln (x_i - c) = \overline{\ln (x - c)} \\ \dfrac{n}{\sigma} - \dfrac{1}{\sigma^3}\sum_{i=1}^{n} (\ln (x_i - c) - \mu)^2 = 0 \Rightarrow \hat{\sigma}^2 = Var(\ln (x - c)) \\ \sum_{i=1}^{n} \dfrac{1}{x_i - c} + \dfrac{1}{\sigma^2}\sum_{i=1}^{n} \dfrac{\ln (x_i - c) - \mu}{x_i - c} = 0 \end{cases} \tag{4.84b}$$

Then the parameters may be estimated numerically by solving the system of equations given as Equation (4.84b).

Amoroso (3-parameter)

$$\ln L = kn \ln \lambda + n \ln |s| - n \ln \Gamma(k) + (sk - 1)\sum_{i=1}^{n} \ln x_i - \lambda \sum_{i=1}^{n} x_i^s \tag{4.85a}$$

Maximizing Equation (4.85a), we have:

$$\begin{cases} n\ln\lambda - n\psi(k) + s\sum_{i=1}^{n} \ln x_i = 0 \\ \dfrac{kn}{\lambda} - \sum_{i=1}^{n} x_i^s = 0 \\ n\dfrac{sign(s)}{|s|} + k\sum_{i=1}^{n} \ln x_i - \lambda s\sum_{i=1}^{n} x_i^{s-1} = 0 \end{cases} \tag{4.85b}$$

Then the parameters may be estimated numerically by solving the system of equations given as Equation (4.85b).

Amoroso (4-parameter)

$$\begin{aligned} \ln L = {} & kn\ln\lambda + n\ln |s| - n\ln\Gamma(k) + (sk-1)\sum_{i=1}^{n} \ln(x_i - c) \\ & - \lambda\sum_{i=1}^{n} (x_i - c)^s \end{aligned} \tag{4.86a}$$

$$\begin{cases} n\ln\lambda - n\psi(k) + s\sum_{i=1}^{n} \ln(x_i - c) = 0 \\ \dfrac{kn}{\lambda} - \sum_{i=1}^{n} (x_i - c)^s = 0 \\ n\dfrac{sign(s)}{|s|} + k\sum_{i=1}^{n} \ln(x_i - c) - \lambda s\sum_{i=1}^{n} (x_i - c)^{s-1} = 0 \\ -(sk-1)\sum_{i=1}^{n} \dfrac{1}{x_i - c} - \lambda\sum_{i=1}^{n} (x_i - c)^s \ln(x_i - c) = 0 \end{cases} \tag{4.86b}$$

Then the parameters may be estimated by solving the system of equations given as Equation (4.86b).

Maxwell

$$\ln L = n\ln\left(\frac{2}{\pi}\right)^{0.5} - 3n\ln\sigma + 2\sum_{i=1}^{n} \ln x_i - \sum_{i=1}^{n} \frac{x_i^2}{2\sigma^2} \tag{4.87a}$$

Maximizing Equation (4.87a), we have:

$$-\frac{3n}{\sigma} + \frac{1}{\sigma^3}\sum_{i=1}^{n} x_i^2 = 0 \Rightarrow \hat{\sigma} = \sqrt{\frac{\sum_{i=1}^{n} x_i^2}{3n}} \tag{4.87b}$$

Rayleigh

$$\ln L = \sum_{i=1}^{n} \ln x_i - n\ln\sigma^2 - \frac{1}{2\sigma^2}\sum_{i=1}^{n} x_i^2 \tag{4.88a}$$

Maximizing Equation (4.88a), we have:

$$-\frac{n}{\sigma^2}+\frac{1}{2\sigma^4}\sum_{i=1}^{n}x_i^2=0\Rightarrow\hat{\sigma}^2=\frac{n}{2}\sum_{i=1}^{n}x_i^2 \tag{4.88b}$$

Seminormal

$$\ln L=\frac{n}{2}\ln\lambda+n\ln\left(\frac{2}{\sqrt{\pi}}\right)-\lambda\sum_{i=1}^{n}x_i^2 \tag{4.89a}$$

Maximizing Equation (4.89a), we have:

$$\frac{n}{2\lambda}-\sum_{i=1}^{n}x_i^2=0\Rightarrow\hat{\lambda}=\frac{n}{2\sum_{i=1}^{n}x_i^2} \tag{4.89b}$$

4.3.1 Peak Flow

Here we will illustrate whether the distributions belonging to the D'Addario system may be applied for flood frequency analysis. The lognormal and 3- and 4-parameter Amoroso distributions are applied as candidate distributions. Table 4.1 lists the parameters estimated for these three distributions using the maximum likelihood estimation method. Figure 4.8 compares the fitted frequency distributions with the empirical frequency distribution. Figure 4.8 indicates that all three distribution candidates may be applied for flood frequency analysis.

4.3.2 Monthly Discharge

In this section, the monthly discharge of June at the gaging station USGS09239500 (Yampa River at Steamboat Springs, Colorado, U.S.) is selected for analysis. The 2- and 3-parameter lognormal, 4-parameter Amoroso, Maxwell, Rayleigh, and seminormal distributions are applied to illustrate their application. Table 4.2 lists the parameters estimated using the maximum likelihood estimation method and

Table 4.1 *Parameters estimated for the peak flow*

Parameters	LN (2)	Amoroso (4)	Amoroso (3)
μ	5.58		
σ	0.35		
c		7.15	
λ		0.1	0.014
k		9.31	5.03
s		0.82	1.05

Table 4.2 *Parameters estimated for the monthly discharge (month of June)*

Parameters	LN (2)	LN (3)	Amoroso (4)	Maxwell	Rayleigh	Seminormal
μ	3.76	3.798				
σ	0.63	0.59				
c		−1.47	0.38			
λ			0.09		39.24	8.49E-04
k			3.53			
s			0.93	32.04		

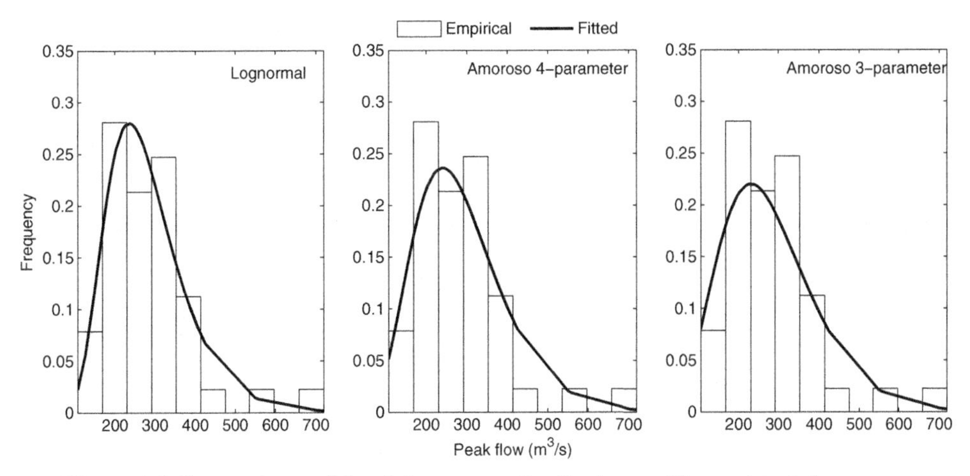

Figure 4.8 Comparison of fitted frequency distributions with empirical frequency distributions of peak flow.

Figure 4.9 compares the fitted frequency distributions with the empirical frequency distribution. Comparison in Figure 4.9 indicates that the Maxwell distribution may be applied to evaluate the frequency distribution of monthly discharge.

4.3.3 Deseasonalized TPN

Due to the obvious periodicity appearing in the observed TPN shown in Figure 4.10, the observed TPN are deseasonalized using the full deseasonalization as:

$$DTPN_{i,j} = \frac{TPN_{i,j} - \hat{\mu}_i}{\hat{\sigma}_i}, i = 1, 2, \ldots, s \tag{4.90}$$

In Equation (4.90), $\hat{\mu}_i, \hat{\sigma}_i$ are the sample mean and sample standard deviation for i-th season, respectively, and s is total number of seasons ($s = 12$).

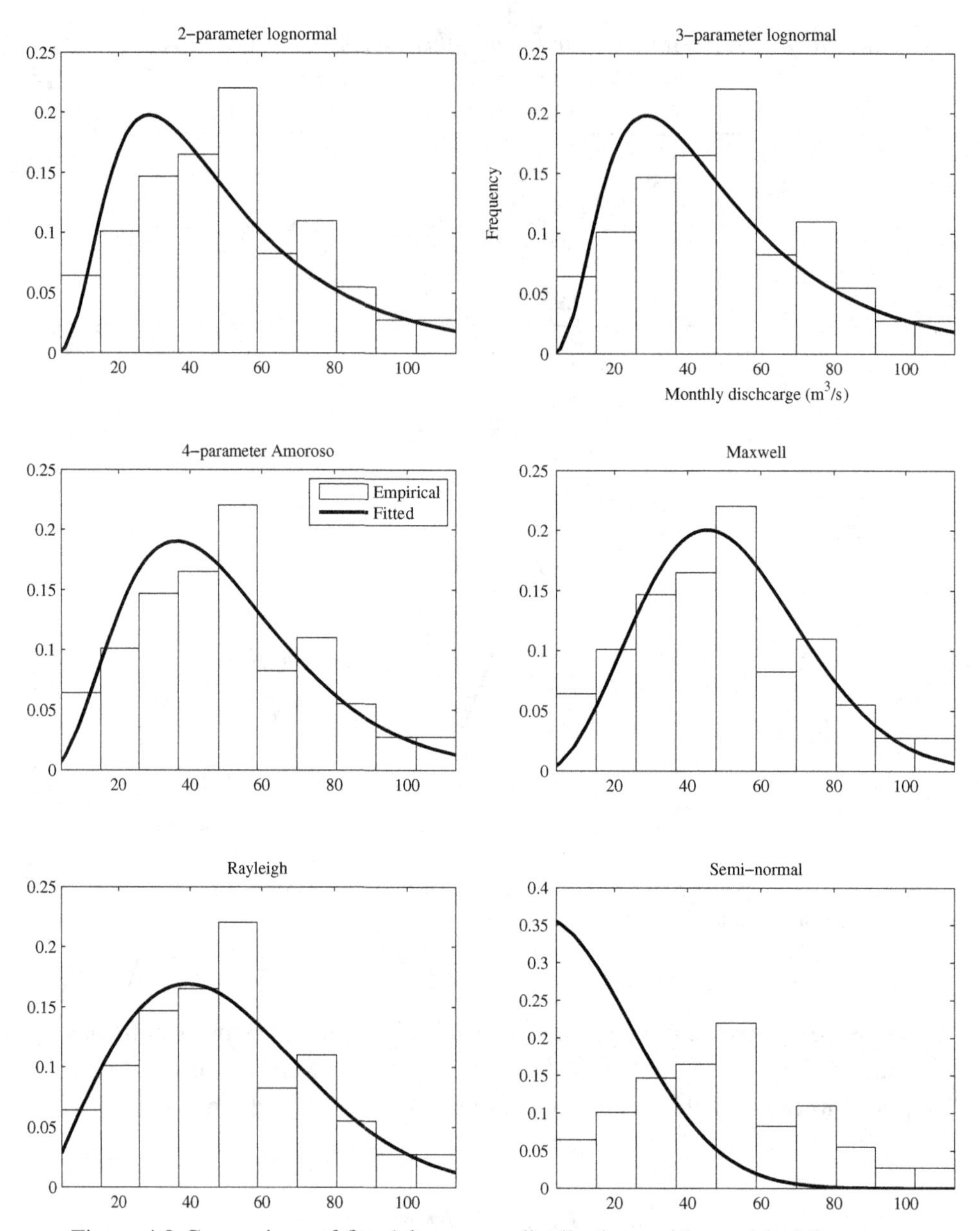

Figure 4.9 Comparison of fitted frequency distributions with empirical frequency distribution for monthly discharge.

To eliminate the negative values for TPN after full deseasonalization, the linear transformation is applied to the deseasonlized TPN as:

$$DTPN^* = \frac{DTPN - (1+d)\min(DTPN)}{(1+d)[\max(DTPN) - \min(DTPN)]}; d = 0.1 \tag{4.91}$$

Table 4.3 *Parameters estimated for the deseasonalized TPN with linear transformation*

Parameters	LN (2)	LN (3)	Amoroso (3)	Maxwell	Rayleigh	Seminormal
μ	−1.19	−0.71				
σ	0.54	0.31				
c		−0.17				
λ			8.92		0.27	19.39
k			1.60			
s			1.73	0.22		

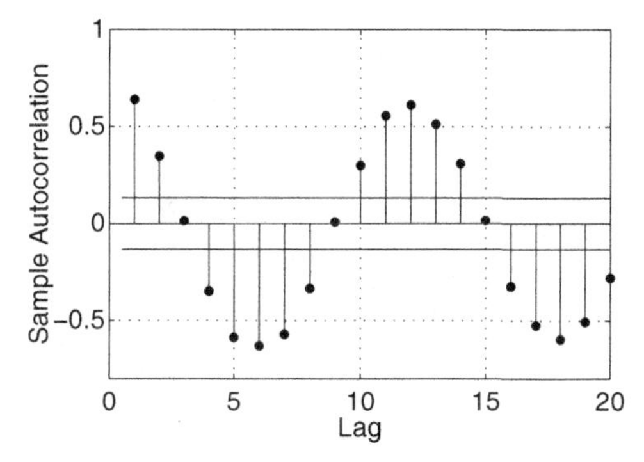

Figure 4.10 Sample autocorrelation of observed TPN.

Now, the deseasonalized TPN after linear transformation is selected for analysis. The 2- and 3-parameter lognormal, 3-parameter Amoroso, Maxwell, Rayleigh, and seminormal distributions are applied to illustrate their application. Table 4.3 lists the parameters estimated using the maximum likelihood estimation method and Figure 4.11 compares the fitted frequency distributions with the empirical frequency distribution. Comparison in Figure 4.11 indicates that all the distribution candidates except the seminormal distribution may be applied to model the deseasonalized TPN after linear transformation.

4.3.4 Daily Maximum Precipitation

The distribution candidates to study daily maximum precipitation are: the 2- and 3-parameter lognormal, 3- and 4-parameter Amoroso, Maxwell, Rayleigh, and seminormal distributions. Applying the previously mentioned distribution candidates to the observed daily maximum precipitation, Table 4.4 lists the parameters estimated

Table 4.4 *Parameters estimated for maximum daily precipitation*

Parameters	LN (2)	LN (3)	Amoroso (3)	Amoroso (4)	Maxwell	Rayleigh	Seminormal
μ	3.98	3.99					
σ	0.32	0.31					
c		−0.63		0.03			
λ			0.08	0.04		41.86	0.001
k			6.08	5.69			
s			1.08	1.20	34.17		

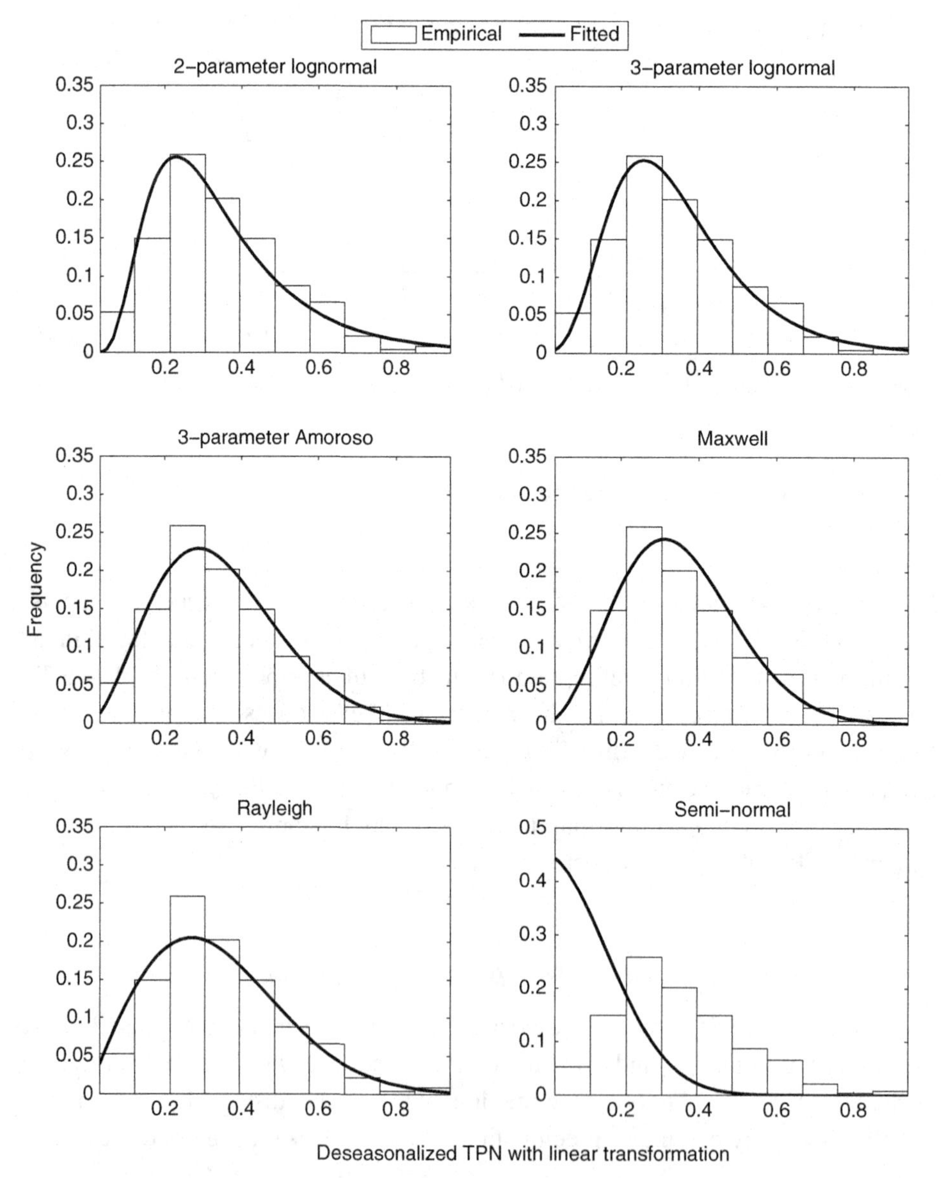

Figure 4.11 Comparison of fitted frequency distribution with empirical frequency distribution for the deseasonalized TPN with linear transformation.

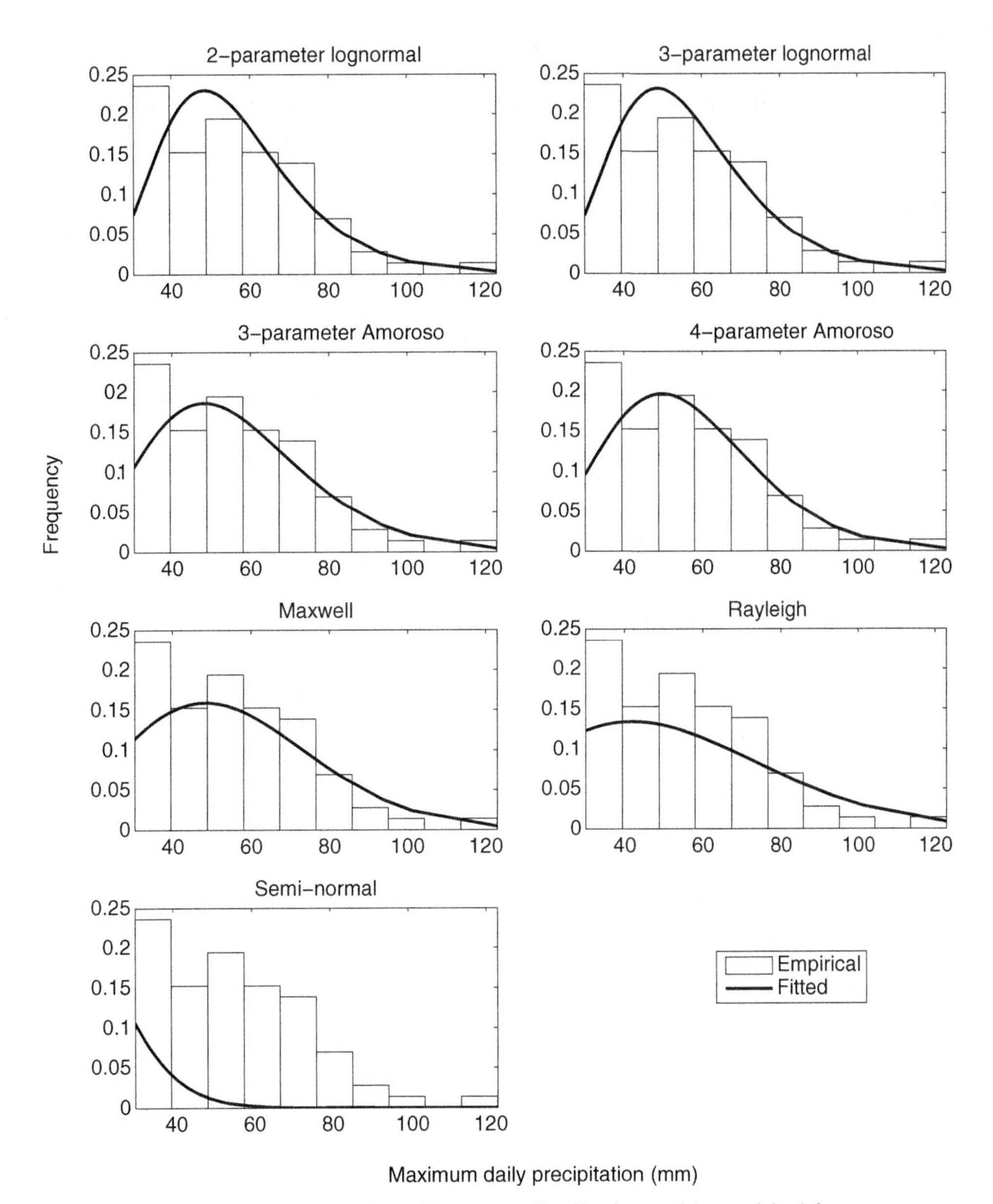

Figure 4.12 Comparison of fitted frequency distributions with empirical frequency distribution for maximum daily precipitation.

with the maximum likelihood estimation method. Figure 4.12 compares the fitted frequency distributions with the empirical frequency distribution. Figure 4.12 indicates lognormal (2- and 3-paramter) and Amoroso distribution (3- and 4-parameter) yield similar results and may be applied to study the maximum daily precipitation values.

4.4 Conclusion

Although the D'Addario system yields several useful distributions, it is not popular in environmental and water engineering. It is also possible to derive other distributions following the same line of thinking as the D'Addario system. The application examples show that the distributions belonging to the D'Addario system may be applied for frequency analysis in environmental and water engineering.

References

D'Addario, R. (1949). Richerche sulla curva dei redditi. *Giornale Degli Economisti e Annali di Economia* 8, pp. 91–114.

Dagum, C. (1990). Generation and properties of income distribution functions. In *Income and Wealth Distribution, Inequality and Poverty*, edited by C. Dagum and M. Zenga, pp. 1–17. Berlin: Springer.

Frechet, M. (1939). Sur les formules de repartition des renenus. *Revue de l'Institut International de Statistique* 7, pp. 32–38.

Vinci, F. (1921). Nuovi contribute allo American della distribuzione dei redditi. *Giornale degli Economisti e Riviste di Statistics* 61, pp. 365–369.

5

Dagum System of Frequency Distributions

5.1 Introduction

Dagum (1977) discussed a set of logical-empirical postulates that should be used for deriving a particular distribution that include parsimony, interpretation of parameters, efficiency of parameter estimation, model flexibility, and goodness of fit. Ease of computation and algebraic manipulation can also be added to this set. Using these postulates, Dagum (1980a, 1980b, 1983, 1990, 2006) derived a set of 11 distributions, called the Dagum family or system, for modeling income distribution. However, some of these distributions are also commonly used in hydrology, hydrometeorology, hydraulics, and environmental engineering. The objective of this chapter therefore is to discuss the Dagum system and its properties, and derive its frequency distributions.

5.2 Dagum System of Distributions

The Dagum system of distributions is derived based on the concept of elasticity. In physics, the elasticity indicates the ability of an object to resist the deformation and to return to its original size and shape. In economics, the elasticity is defined as the response of one variable to the change of another variable, for example, how a quantity may change if one drops the selling price. In mathematics, the elasticity of a positive continuous (differentiable) function with positive input variable [let us say x] and positive output function [let us say $y(x)$] at a given point [i.e., x_0] may be written following Sydsaeter and Hammond (1995) as:

$$\eta(y(x_0)) = \frac{x_0}{y(x_0)} y'(x_0) = \lim_{x \to x_0} \frac{x_0}{y(x_0)} \frac{y(x) - y(x_0)}{x - x_0} \approx \frac{\%\Delta y(x_0)}{\%\Delta x_0} \tag{5.1a}$$

In Equation (5.1a), η denotes elasticity for positive continuous (differentiable) function at point x_0. This equation clearly shows that the elasticity is a measure of how the output variable changes with the change of input variable. Equation (5.1a) may also be written as a form of the differentiation in the logarithm domain as:

$$\eta(y(x)) = \frac{d \ln y(x)}{d \ln x} \tag{5.1b}$$

With the observation of the change of continuous cumulative distribution function (CDF) with respect to the change of positive random variable x, Dagum (1980a, 1983) found that the function $F(x)$ started from a finite and positive value as $F(x)$ tended to zero and decreased toward zero as $F(x)$ tended to unity, and hence random variable X tended to infinity and was a decreasing and a concave function of $F(x)$. Based on these considerations as well as the concept of elasticity, Dagum (1980a, 1983) defined the elasticity for the CDF of continuous random variable, $\eta(x, F(x))$, as

$$\eta(x, F(x)) = \frac{d \ln (F(x) - a)}{d \ln x}; a < 1, x \in [x_0, \infty), x_0 > 0 \tag{5.2}$$

where a is a constant.

Hypothesizing Equation (5.2) as a multiplication of functions x and $F(x)$ leads to:

$$\eta(x, F(x)) = \frac{d \ln (F(x) - a)}{d \ln x} = g(x)G(F) \leq k; k, g(x), F(x) > 0, \frac{d(g(x)F(x))}{dx} < 0 \tag{5.3}$$

Equation (5.3) is the differential equation for generating frequency distributions. $\frac{d(g(x)F(x))}{dx} < 0$ leads to the elasticity function of $F(x)$ being a decreasing function bounded by $x \geq x_0$. Different distributions can be obtained, based on the specification of functions $g(x)$ and $G(F)$.

Dagum (1990) obtained 11 distributions as members of his system that included Pareto types I, type II, and type III (Pareto, 1897, 1985; Benini, 1906; Weibull, 1951; Fisk, 1961; Arnold, 1980, 1983); Singh-Maddala (1976); log-Gompertz (Dagan, 1980a); and Dagum type I, II, and III (Dagum, 1980a, 1983, 1990) distributions. Within these 11 distributions, Fisk distribution (Champernowne, 1952; Fisk, 1961) is also called log-logistic distribution with Burr III and Burr XII (Burr, 1942) as its generalizations. Hence, the Dagum system is also regarded as the generalized logistic-Burr system of distributions and can be further extended by introducing location and scale parameters and using transformations (Dagum,

1983). Dagum (1977) analyzed a set of 14 properties that a probability distribution should satisfy to be a relevant income distribution. Many of these properties are relevant for hydrologic frequency distributions.

5.3 Derivation of Frequency Distributions

The individual distributions of the Dagum system are discussed in what follows.

5.3.1 Pareto Type I Distribution

For the Pareto type I distribution, the random variable X is bounded as: $0 < x_0 \leq x < \infty$. The elasticity function is expressed as:

$$\eta(x, F(x)) = \frac{d \ln F(x)}{d \ln x} = g(x)G(F) = c\frac{1-F}{F} \tag{5.4}$$

where: $g(x) = c, G(F) = \frac{1-F}{F}$, with $a = 0$ in Equation (5.2).

Equation (5.4) may be rewritten as:

$$\frac{1}{1-F(x)}dF = \frac{c}{x}dx \tag{5.5}$$

Integration of Equation (5.5) yields

$$\ln(1 - F(x)) = -c \ln x + const = \ln (Cx^{-c}) \tag{5.6}$$

where $const = \ln C$ is constant of integration. Setting the initial condition as: $F(x_0) = 0$, $x = x_0$, the constant C may then be expressed as:

$$\ln\left(Cx_0^{-c}\right) = 0 \Rightarrow C = x_0^c; \quad x_0 > 0, c > 0 \tag{5.7}$$

Substituting Equation (5.7) into Equation (5.6), Equation (5.6) can be rewritten as

$$F(x) = 1 - \left(\frac{x}{x_0}\right)^{-c} = 1 - \left(\frac{x_0}{x}\right)^c; \quad x \geq x_0, x_0 > 0, c > 0 \tag{5.8}$$

Equation (5.8) is known as the Pareto I distribution. Its probability density function (PDF) can be expressed as

$$f(x) = \frac{c}{x_0}\left(\frac{x}{x_0}\right)^{-c-1} = \frac{cx_0^c}{x^{c+1}}; \quad x \geq x_0, x_0 > 0, c > 0 \tag{5.9}$$

where x_0 and c are the scale and shape parameters of Pareto type I distribution, respectively. Both $g(F)$ and PDF of Pareto type I distribution are always L-shaped

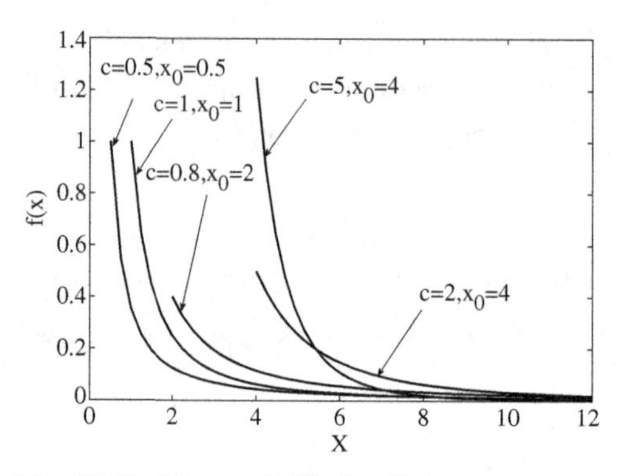

Figure 5.1 Plot of the PDF of Pareto I distribution.

(i.e., strictly decreasing nonmodal). Figure 5.1 plots the PDF of Pareto type I distribution with different parameter values.

5.3.2 Pareto Type II Distribution

For the Pareto II distribution, the random variable X is bounded as: $0 < x_0 \le x < \infty$. The elasticity function of Pareto II distribution can be expressed as:

$$\eta(x, F(x)) = \frac{d \ln F(x)}{d \ln x} = g(x)G(F) = \frac{cx}{x-d}\frac{1-F}{F} \tag{5.10}$$

where: $g(x) = \frac{cx}{x-d}, c > 0$, and $G(F) = \frac{1-F}{F}$ with $a = 0$ in Equation (5.2).

Furthermore, Equation (5.10) can be rewritten as:

$$\frac{1}{1-F}dF = \frac{c}{x-d}dx \tag{5.11}$$

Integration of Equation (5.11) may be written as:

$$\begin{aligned} \int \frac{1}{1-F}dF = \int \frac{c}{x-d}dx \Rightarrow -\ln(1-F) = c\ln(x-d) + const \\ \Rightarrow F = 1 - C(x-d)^{-c} \end{aligned} \tag{5.12}$$

In Equation (5.12), C is the integration constant. Setting the initial condition: $F(x_0) = 0$; $x = x_0 > d$, the constant C can be solved for as:

$$1 - C(x_0 - d)^{-c} = 0 \Rightarrow C = (x_0 - d)^c \tag{5.13}$$

Substituting Equation (5.13) into Equation (5.12), the CDF of Pareto II distribution is then written as:

$$F = 1 - \left(\frac{x-d}{x_0-d}\right)^{-c}; x \ge x_0 > d, c > 0, b = x_0 - d \tag{5.14}$$

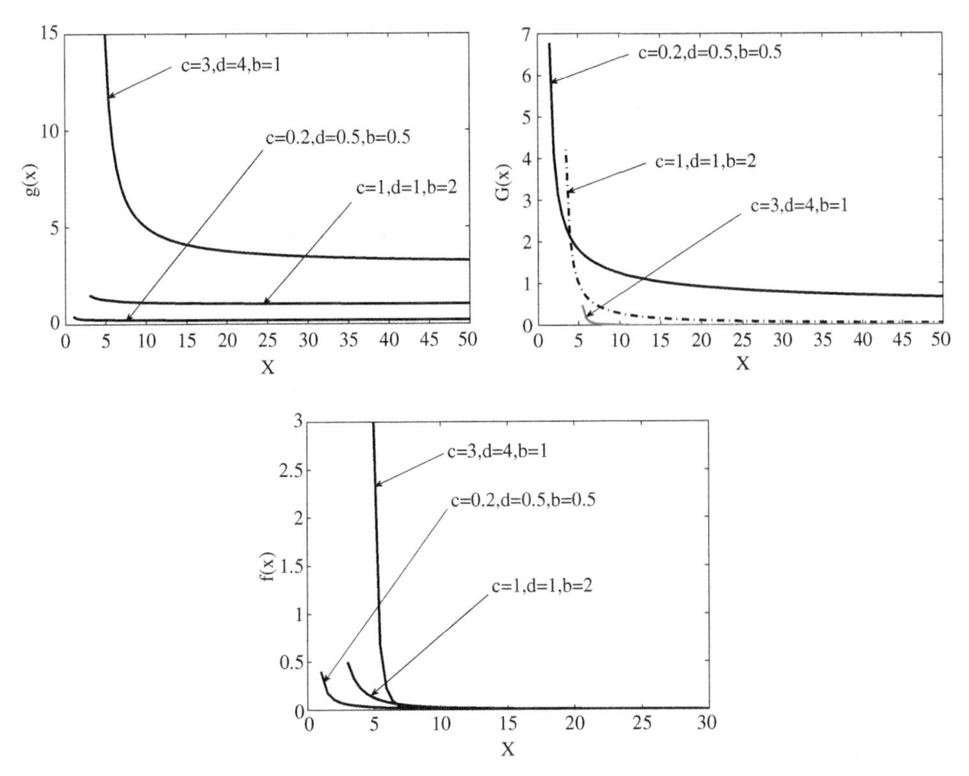

Figure 5.2 (a) Plot of the g(x) function of Pareto II distribution; (b) plot of the G(F) function of Pareto II distribution; and (c) plot of the PDF of Pareto II distribution.

Its PDF can be written as

$$f(x) = \frac{c}{x_0 - d}\left(\frac{x-d}{x_0-d}\right)^{-c-1} = \frac{c}{b}\left(\frac{x-d}{b}\right)^{-c-1} \tag{5.15}$$

The functions $g(x), G(F)$, and PDF of Pareto II distribution are plotted in Figures 5.2a–5.2c. $g(x), G(F)$, and PDF are decreasing functions.

5.3.3 Pareto Type III Distribution

For the Pareto type III distribution, the random variable X is bounded as: $x \in [x_0, \infty)$, $x_0 > 0$. Its elasticity function can be given as:

$$\eta(x, F(x)) = \frac{d \ln F(x)}{d \ln x} = g(x)G(F) = \left(bx + \frac{cx}{x-d}\right)\frac{1-F}{F} \tag{5.16}$$

where: $g(x) = bx + \frac{cx}{x-d}, b > 0, c > 0$, and $G(F) = \frac{1-F}{F}$ with $a = 0$ in Equation (5.2).

Integration of Equation (5.16) yields

$$\frac{1}{1-F}dF = \left(b + \frac{c}{x-d}\right)dx \Rightarrow d\ln(1-F) = -\left(b + \frac{c}{x-d}\right)dx$$
$$\Rightarrow F = 1 - C\exp(-bx)(x-d)^{-c} \tag{5.17a}$$

The integration constant C can be calculated by substituting $F(x_0) = 0$, $x = x_0$ into Equation (5.17a) as:

$$1 - C\exp(-bx_0)(x_0 - d)^{-c} = 0 \Rightarrow C = \exp(bx_0)(x_0 - d)^c \tag{5.17b}$$

Hence, the CDF of the Pareto type III distribution is written as:

$$F(x) = 1 - \exp(bx_0)(x_0 - d)^c \exp(-bx)(x-d)^{-c}$$
$$\Rightarrow F(x) = 1 - \exp(-b(x - x_0))\left(\frac{x-d}{x_0-d}\right)^{-c}; x > x_0 > d \tag{5.18}$$

The PDF of the Pareto type III distribution is written as:

$$f(x) = b\exp(-b(x-x_0))\left(\frac{x-d}{x_0-d}\right)^{-c} + \frac{c}{x_0-d}\exp(-b(x-x_0))\left(\frac{x-d}{x_0-d}\right)^{-c-1} \tag{5.19}$$

Figures 5.3a and 5.3b plot the $g(x)$ function and the PDF of Pareto type III distribution. Similar to Pareto type I and type II distributions, the PDF of Pareto type III distribution is L-shaped and skewed. However, the $g(x)$ function of Pareto type III distribution is an increasing function in general. The $G(F)$ function is a decreasing function as that for Pareto type I and type II distributions. Figure 5.3a

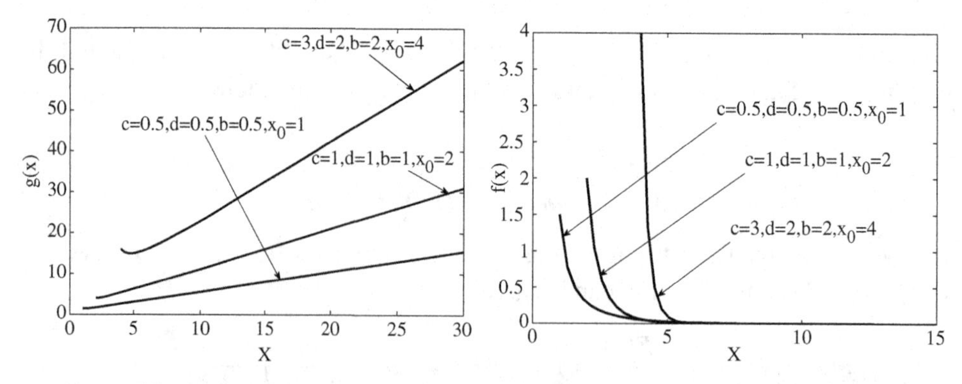

Figure 5.3 (a) Plot of the g(x) function of Pareto III distribution; and (b) plot of the PDF of Pareto III distribution.

graphs the $g(x)$ function of Pareto type III distribution and Figure 5.3b. graphs the PDF of Pareto type III distribution.

5.3.4 Benini Distribution

In the Benini distribution, the random variable X is bounded as: $x \in [x_0, \infty)$, $x_0 > 0$. Its elasticity function can be expressed as:

$$\eta(x, F(x)) = \frac{d \ln F}{d \ln x} = g(x)G(F) = 2c \ln x \frac{1-F}{F} \tag{5.20}$$

where: $g(x) = 2c \ln x, c > 0$ and $G(F) = \frac{1-F}{F}$ with $a = 0$ in Equation (5.1).

Integration of Equation (5.20) yields:

$$\begin{aligned} &\frac{1}{1-F} dF = 2c \ln x d \ln x \Rightarrow -d \ln (1-F) = 2c \ln x d \ln x \\ &\Rightarrow F(x) = 1 - \exp\left(-\int_{x_0}^{x} 2c \ln t d \ln t\right) \end{aligned} \tag{5.21}$$

In Equation (5.21), using the bound for the random variable X and set $y = x/x_0$, Equation (5.21) may be rewritten as:

$$\begin{aligned} F(x) &= F\left(\frac{x}{x_0}\right) = F(y) = 1 - \exp\left(-\int_1^y 2c \ln t d \ln t\right) \\ &= 1 - \exp\left(-c(\ln t)^2\Big|_1^y\right) = 1 - \exp\left(-c(\ln y)^2\right) \\ &= 1 - \exp\left(-c\left(\ln\left(\frac{x}{x_0}\right)\right)^2\right) \end{aligned} \tag{5.22a}$$

Equation (5.22a) may be rewritten as:

$$F(x) = 1 - \exp\left(\left(-c \ln\left(\frac{x}{x_0}\right)\right) \ln\left(\frac{x}{x_0}\right)\right) = 1 - \left(\frac{x}{x_0}\right)^{-c \ln\left(\frac{x}{x_0}\right)}; \tag{5.22b}$$

$x \geq x_0 > 0, c > 0$

Equations (5.22a)–(5.22b) is the Benini distribution. Its PDF can be written as

$$f(x) = \frac{2c}{x} \ln\left(\frac{x}{x_0}\right) \exp\left(-c\left(\ln\left(\frac{x}{x_0}\right)\right)^2\right); x \geq x_0 > 0, c > 0 \tag{5.23}$$

The $g(x)$ function is a monotone increasing function. Figure 5.4 plots the PDF of the Benini distribution with different parameter values. It is shown that the PDF of Benini distribution is bell shaped and skewed.

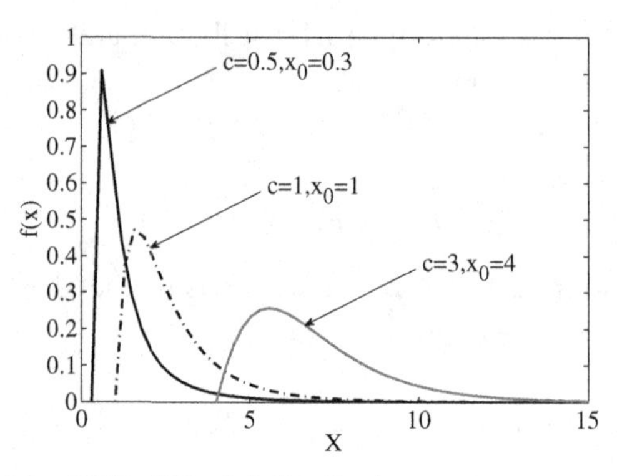

Figure 5.4 Plot of the PDF of Benini distribution.

5.3.5 Weibull Distribution

In the Weibull distribution, the random variable X is bounded. The elasticity function of Weibull distribution can be expressed as:

$$\eta(x, F(x)) = \frac{d\ln F}{d\ln x} = g(x)F(x) = bx(x-d)^{c-1}\frac{1-F}{F}; c, b > 0, x \in [d, \infty) \tag{5.24}$$

where: $g(x) = bx(x-d)^{c-1}, G(F) = \frac{1-F}{F}$, with $a = 0$ in Equation (5.2).

Integration Equation (5.24) leads to:

$$\begin{aligned} \frac{1}{1-F}dF = b(x-d)^{c-1}dx \Rightarrow 1 - F = C\exp\left(-\frac{b}{c}(x-d)^c\right) \\ \Rightarrow F = 1 - C\exp\left(-\frac{b}{c}(x-d)^c\right) \end{aligned} \tag{5.25a}$$

The integration constant C can be evaluated by setting $F(x) = 0$ for $x = d$ as:

$$1 - C\exp\left(-\frac{b}{c}(d-d)^c\right) = 1 \Rightarrow C = 1 \tag{5.25b}$$

Thus, the CDF of the Weibull distribution can be written as:

$$F(x) = 1 - \exp\left(-\frac{b}{c}(x-d)^c\right) \tag{5.26}$$

Let $k = \left(\frac{c}{b}\right)^{\frac{1}{c}}$ in Equation (5.26). Equation (5.26) may be rewritten as:

$$F(x) = 1 - \exp\left(-\left(\frac{x-d}{k}\right)^c\right) \tag{5.27}$$

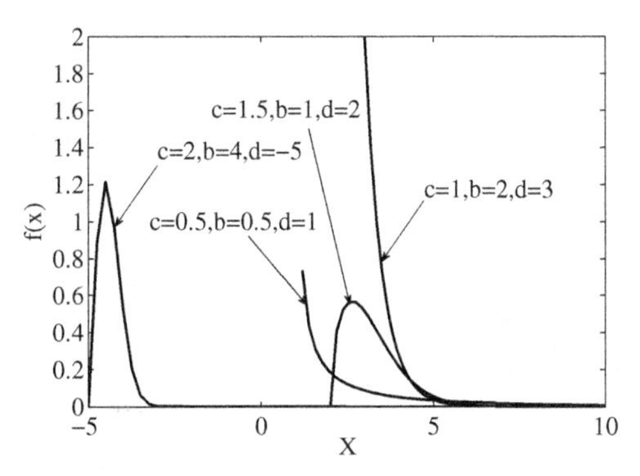

Figure 5.5 Plot of the PDF of Weibull distribution.

Equation (5.27) is a Weibull distribution with location parameter in which d, k, and c are location, scale, and shape parameters, respectively.

The PDF of the Weibull distribution may be written as:

$$f(x) = b(x-d)^{c-1}\exp\left(-\frac{b}{c}(x-d)^c\right) = \frac{c}{k}\left(\frac{x-d}{k}\right)^{c-1}\exp\left(-\left(\frac{x-d}{k}\right)^c\right) \tag{5.28}$$

The PDF of Weibull distribution is L-shaped if $c \leq 1$. It is bell shaped and skewed if $c > 1$. Figure 5.5 plots the PDF of Weibull distribution.

5.3.6 Log-Gompertz Distribution

For the log-Gompertz distribution, the random variable X is bounded as: $x \in [0, \infty)$. The elasticity function of log-Gompertz distribution can be expressed as:

$$\eta(x, F(x)) = \frac{d \ln F}{d \ln x} = (-\ln c)(-\ln F); c \in (0,1) \tag{5.29}$$

where: $g(x) = -\ln c, G(F) = -\ln F$, with $a = 0$ in Equation (5.2).

Integration of Equation (5.29) leads to:

$$\begin{aligned} &\frac{1}{-\ln F} d\ln F = (-\ln c) d\ln x \Rightarrow -\left(\frac{1}{-\ln F}\right) d(-\ln F) = -\ln c\, d\ln x \\ &\Rightarrow d\ln(-\ln F) = d\ln x^{\ln c} \Rightarrow F = C\exp(-\exp(\ln c(\ln x))) \\ &= C\exp\left(-x^{\ln c}\right) \end{aligned} \tag{5.30a}$$

Setting $F(x) = 0$ for $x = 0$, we can easily obtain the value of integration constant C, i.e., $C = 1$. Equation (5.30a) may be rewritten as:

$$F = \exp(-\exp(\ln c(\ln x))) = \exp(-x^{\ln c});\ c \in (0,1),\ x \in [0,\infty) \tag{5.30b}$$

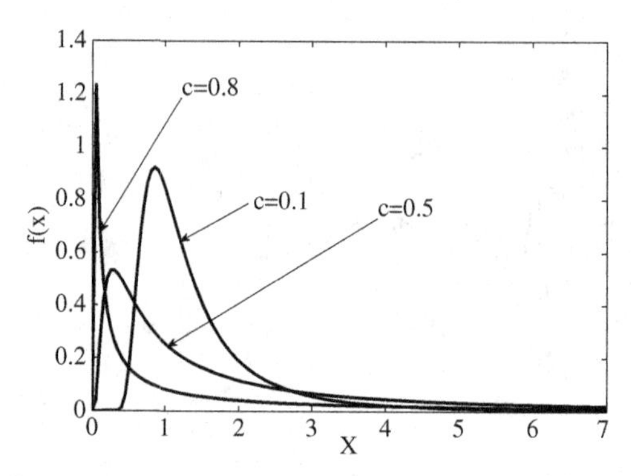

Figure 5.6 Plot of the PDF of log-Gompertz distribution.

Let $b = -\ln c$, $y = \ln x$. Equation (5.30b) may be rewritten as:

$$F(y) = \exp(-\exp(-by));\ b > 0,\ y \in (-\infty, \infty) \tag{5.31}$$

From Equation (5.31), it is seen that log-Gompertz distribution is also an extreme value I distribution. The PDF of log-Gompertz distribution is then given, using Equation (5.30b), as:

$$f(x) = \frac{x^{\ln c}(-\ln c)}{x} \exp\left(-x^{\ln c}\right); c \in (0,1), x \in [0, \infty) \tag{5.32}$$

The PDF of log-Gompertz distribution is bell shaped and skewed. It is generalized so that the L-shape distribution may be obtained if $c \rightarrow 1^{-}$. Figure 5.6 plots the PDF of log-Gompertz distribution.

5.3.7 Fisk Distribution

The Fisk distribution is also called the log-logistic distribution, where the random variable X is bounded as $x \in [0, \infty)$. The elasticity function of Fisk distribution can be expressed as:

$$\eta(x, F(x)) = \frac{d \ln F}{d \ln x} = c(1 - F); c > 0 \tag{5.33}$$

where: $g(x) = c, G(F) = 1 - F$, with $a = 0$ in Equation (5.2).

Integration of Equation (5.33) yields

$$\begin{aligned} &\frac{1}{F(1-F)} dF = c\, d \ln x \Rightarrow \left(\frac{1}{F} + \frac{1}{1-F}\right) dF = c\, d \ln x \\ &\Rightarrow \frac{F}{1-F} = C \exp\left(\ln x^{c}\right) \Rightarrow F = \frac{Cx^{c}}{1 + Cx^{c}}; c > 0 \end{aligned} \tag{5.34a}$$

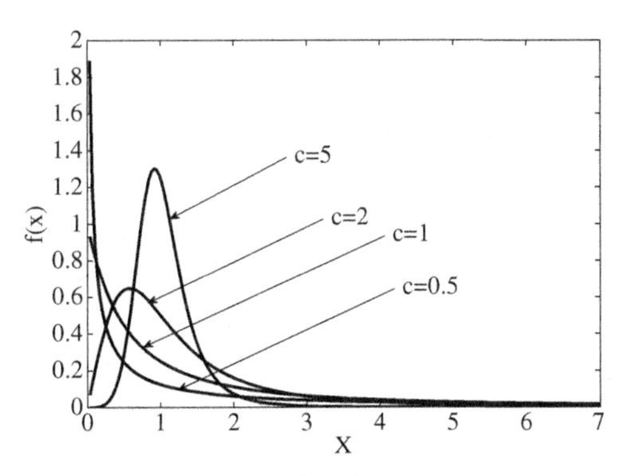

Figure 5.7 Plot of the PDF of Fisk distribution.

Setting $F(x) = 0$ for $x \to 0^+$, the integration constant C may be evaluated as a free parameter such that $C > 0$. Simply setting C = 1, Equation (5.34a) can be rewritten as:

$$F = \frac{x^c}{1 + x^c} = \frac{1}{1 + x^{-c}}; c > 0, x \in (0, \infty) \tag{5.34b}$$

Then, the PDF of the Fisk distribution can be given as:

$$f(x) = cx^{-(c+1)}(1 + x^{-c})^{-2} \tag{5.35}$$

Figure 5.7 plots the PDF of Fisk distribution.

5.3.8 Singh-Maddala Distribution

For the Singh-Maddala distribution, the random variable X is bounded as $x \in [0, \infty)$. Its elasticity function can be expressed as:

$$\eta(x, F(x)) = \frac{d \ln F}{d \ln x} = \frac{c\left(1 - (1 - F)^b\right)}{F(1 - F)^{-1}}; c > 0, b > 0 \tag{5.36}$$

where: $g(x) = c, G(F) = \frac{1-(1-F)^b}{F(1-F)^{-1}}$, with $a = 0$ in Equation (5.2).

Integration of Equation (5.36) yields:

$$\left(\frac{1}{1-F} + \frac{(1-F)^{b-1}}{1-(1-F)^b}\right) dF = c\, d \ln x \Rightarrow \frac{1-(1-F)^b}{(1-F)^b} = Cx^{bc}$$
$$\Rightarrow F = 1 - \left(1 + Cx^{bc}\right)^{-\left(\frac{1}{b}\right)};$$
$$b, c > 0, x \in [0, \infty) \tag{5.37a}$$

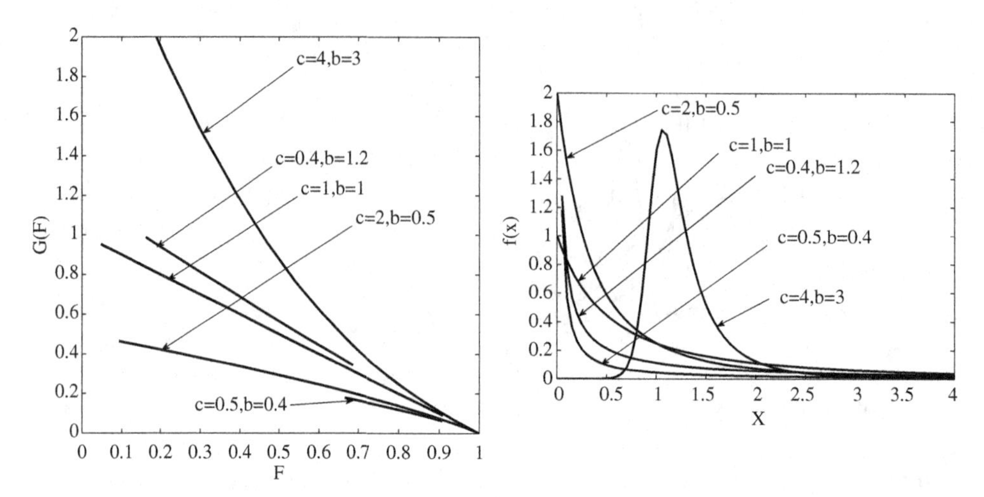

Figure 5.8 (a) Plot of the G(F) function of Singh-Maddala distribution; and (b) plot of the PDF of Singh-Maddala distribution.

Setting $F(x) = 0$ for $x = 0$, the integration constant C in Equation (5.37a) can be evaluated as a free parameter such that $C > 0$. Simply setting $C = 1$, Equation (5.37a) can be rewritten as:

$$F(x) = 1 - \left(1 + x^{bc}\right)^{-\frac{1}{b}} \tag{5.37b}$$

The PDF of Singh-Maddala distribution may then be written as:

$$f(x) = cx^{bc-1}\left(1 + x^{bc}\right)^{-\frac{1}{b}-1}; x \in [0, \infty), b, c > 0 \tag{5.38}$$

The $G(F)$ function is a decreasing function. The PDF of Singh-Maddala distribution is L-shaped if $b \leq 1$ or $c \leq 1$ and is bell shaped if $b > 1$ and $c > 1$. Figure 5.8a graphs the G(F) function and Figure 5.8b graphs the PDF of Singh-Maddala distribution.

5.3.9 Dagum I Distribution

For Dagum I distribution, the random variable X is bounded as: $x \in [0, \infty)$. Its elasticity function can be expressed as:

$$\eta(x, F(x)) = \frac{d \ln F}{d \ln x} = g(x)G(F) = c\left(1 - F^{\frac{1}{b}}\right); b, c > 0 \tag{5.39}$$

where: $g(x) = c, G(F) = 1 - F^{\frac{1}{b}}$, with $a = 0$ in Equation (5.2).

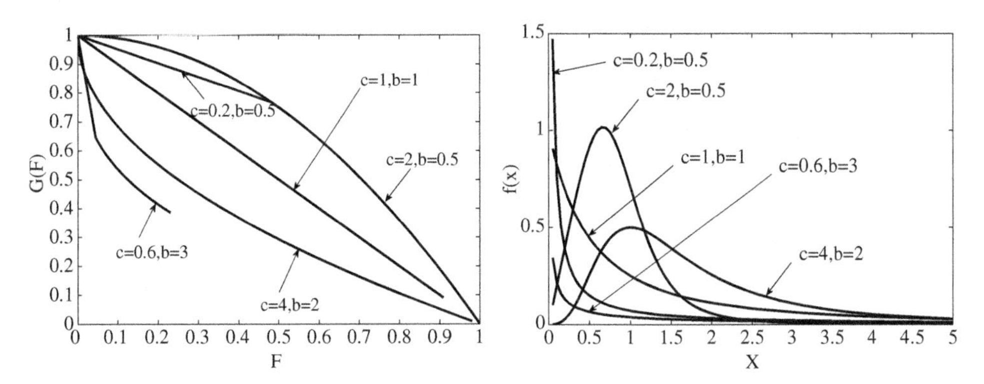

Figure 5.9 (a) Plot of the G(F) function of Dagum I distribution; and (b) graph of the PDF of Dagum I distribution.

Integration of Equation (5.39) yields

$$\frac{1}{F\left(1-F^{\frac{1}{b}}\right)}dF = c\,d\ln x \Rightarrow \left(\frac{1}{F}+\frac{F^{\frac{1}{b}-1}}{1-F^{\frac{1}{b}}}\right)dF = c\,d\ln x = \frac{F^{\frac{1}{b}}}{1-F^{\frac{1}{b}}} \tag{5.40a}$$

$$= Cx^{\frac{c}{b}} \Rightarrow F = \left(1+Cx^{-\frac{c}{b}}\right)^{-b}$$

Setting $F = 0$ for $x = 0$, we again find that integration constant C is a free parameter. Simply setting $C = 1$, Equation (5.40a) can be rewritten as:

$$F(x) = \left(1+x^{-\frac{c}{b}}\right)^{-b} \tag{5.40b}$$

The PDF of Dagum I distribution is given as:

$$f(x) = cx^{-\frac{c}{b}-1}\left(1+x^{-\frac{c}{b}}\right)^{-b-1}; x \in [0,\infty), b, c > 0 \tag{5.41}$$

The $G(F)$ of Dagum I distribution is a decreasing function. The PDF of Dagum I distribution is L-shaped if shape parameter is $c \leq 1$, and is bell shaped if the shape parameter is $c > 1$. Figure 5.9a plots the G(F) function and Figure 5.9b plots the PDF of Dagum I distribution.

5.3.10 Dagum II Distribution

For the Dagum II distribution, the random variable X is bounded as $x \in [0,\infty)$. Its elasticity function can be expressed as:

$$\eta(x, F(x)) = \frac{d\ln(F-a)}{d\ln x} = c\left(1-\left(\frac{F-a}{1-a}\right)^{b}\right); c > 1, a \in (0,1), b > 0 \tag{5.42}$$

where: $g(x) = c, G(F) = 1-\left(\frac{F-a}{1-a}\right)^{\frac{1}{b}}$.

Integration of Equation (5.42) yields

$$\frac{1}{(F-a)\left(1-\left(\frac{F-a}{1-a}\right)^{b}\right)}dF = c\,d\ln x \Rightarrow \left(\frac{1}{F-a}+\frac{1}{1-a}\frac{\left(\frac{F-a}{1-a}\right)^{b-1}}{1-\left(\frac{F-a}{1-a}\right)^{b}}\right)dF = c\,d\ln x$$

$$\Rightarrow (F-a)\left(1-\left(\tfrac{F-a}{1-a}\right)^{b}\right)^{-\frac{1}{b}} = Cx^{c}$$

$$\Rightarrow F = a+\left((1-a)^{-b}+Cx^{-bc}\right)^{-\frac{1}{b}}$$

$$\Rightarrow F = a+(1-a)\left(1+\tfrac{Cx^{-bc}}{(1-a)^{-b}}\right)^{-\frac{1}{b}}$$

(5.43a)

Setting $F(x) = a$ at $x = 0$, the integration constant C may be evaluated as:

$$a+(1-a)\left(1+\frac{Cx^{-bc}}{(1-a)^{-b}}\right)^{-\frac{1}{b}} = 0 \tag{5.43b}$$

From Equation (5.43b) we see that when $x = 0$, $1+\frac{Cx^{-bc}}{(1-a)^{-b}} \to \infty \Rightarrow \left(1+\frac{Cx^{-\frac{c}{b}}}{(1-a)^{-b}}\right)^{-\frac{1}{b}} \to 0 \Rightarrow F(0) = a$. Now, we have again shown that the integration constant C may be considered as a free parameter. Simply setting $C = 1$, Equation (5.43a) can be rewritten as:

$$F = a+(1-a)\left(1+\frac{x^{-bc}}{(1-a)^{-b}}\right)^{-\frac{1}{b}};\quad a \in (0,1), c > 1, b > 0, x \in (0,\infty)$$

(5.43c)

In Equation (5.43c) let $\beta = (1-a)^{-\frac{b}{c}}; \alpha = c; p = \frac{1}{b}$; Equation (5.43c) is rewritten as:

$$F = a+(1-a)\left(1+\left(\frac{x}{\beta}\right)^{-\alpha}\right)^{-p} \tag{5.43d}$$

The PDF of Dagum II distribution can be given as:

$$f(x) = c(1-a)^{b+1}x^{-(bc+1)}\left(1+\frac{x^{-bc}}{(1-a)^{-b}}\right)^{-\frac{1}{b}-1} \tag{5.44}$$

The G(F) function for Dagum II distribution is a monotone decreasing function. The PDF of Dagum distribution is bell-shaped skewed. Figure 5.10a plots the G(F) function and Figure 5.10b plots the PDF of Dagum II distribution.

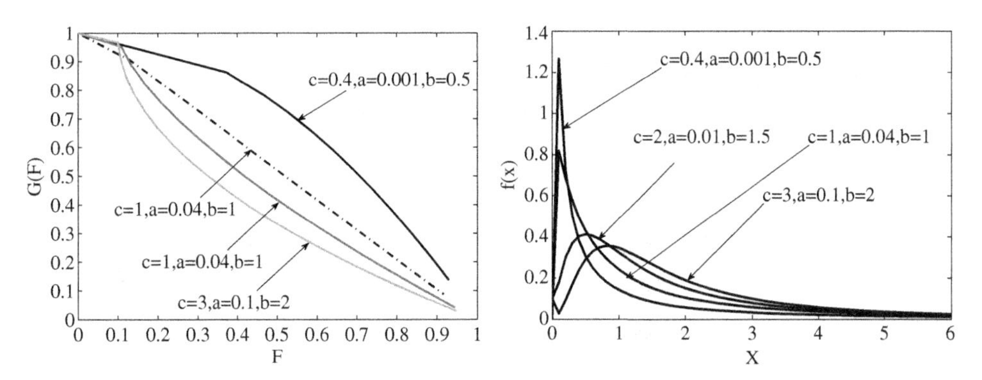

Figure 5.10 (a) Plot of the G(F) function of Dagum II distribution; and (b) graph of the PDF of Dagum II distribution.

5.3.11 Dagum III Distribution

The elasticity function for Dagum III distribution can be given as:

$$\eta(x, F(x)) = g(x)G(F) = c\left(1 - \left(\frac{F-a}{1-a}\right)^{\frac{1}{b}}\right), a < 0 \tag{5.45}$$

where $g(x) = c, G(F) = 1 - \left(\frac{F-a}{1-a}\right)^{\frac{1}{b}}$ with $a < 0$. Comparing to the elasticity function for Dagum II distribution, it is seen that the only difference is the parameter constraint in the *G(F)* functions, i.e., a < 0 for Dagum III distribution while $a \in (0, 1)$ for Dagum II distribution. The difference in *G(F)* function leads to the variable *X* bounded as $x \in [x_0, \infty)$, $x_0 > 0$. With the same form of the elasticity function as Dagum II distribution, the CDF and PDF of Dagum III distribution may be expressed exactly as those of Dagum II distribution as Equations (5.43c–5.43d) for CDF and Equation (5.44) for PDF, respectively. The lower bound x_0 can be evaluated using $F(x) = 0$, *if* $x = x_0$ from Equation (5.43c) or Equation (5.43d) as:

From Equation (5.43c): $a + (1-a)\left(1 + \frac{x_0^{-bc}}{(1-a)^{-b}}\right)^{-\frac{1}{b}} = 0$

$$\Rightarrow x_0 = (-a)^{-\frac{1}{c}}\left((1-a)^b - 1\right)^{\left(-\frac{1}{bc}\right)} \tag{5.46a}$$

or

From Equation (5.43d): $a + (1-a)\left(1 + \left(\frac{x_0}{\beta}\right)^{-\alpha}\right)^{-p} = 0$

$$\Rightarrow x_0 = \beta\left(\left(-\frac{a}{1-a}\right)^{-\frac{1}{p}} - 1\right)^{-\frac{1}{\alpha}} = \beta\left(\left(1 - \frac{1}{a}\right)^{\frac{1}{p}} - 1\right)^{-\frac{1}{\alpha}} \tag{5.46d}$$

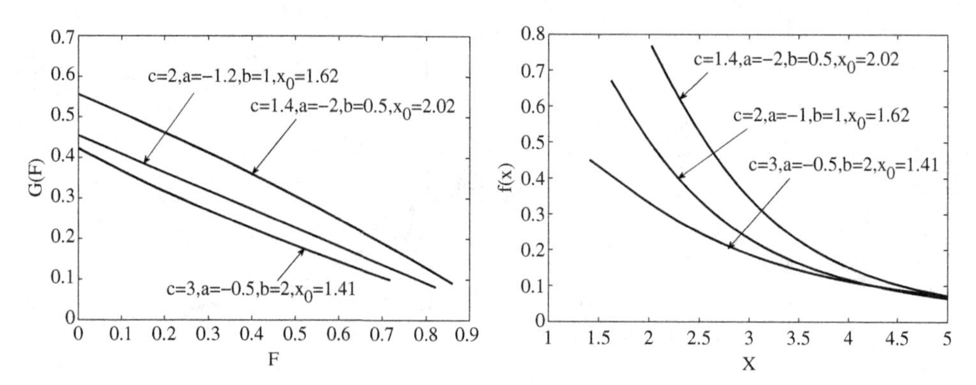

Figure 5.11 (a) Plot of the G(F) function of Dagum III distribution; and (b) plot of the PDF of Dagum III distribution.

Similar to Dagum II distribution, the *G*(*F*) function is a monotone decreasing function for Dagum III distribution and the PDF is positively skewed. Figure 5.11a plots the *G*(*F*) function and Figure 5.11b graphs the PDF of Dagum III distribution.

5.4 Application

In this section, monthly suspended sediment yield, peak flow, maximum daily precipitation, and total flow deficit for drought are selected to evaluate the applicability of frequency distributions of the Dagum system. The distribution candidates are Fisk, Singh-Maddala, Weibull, Dagum I, and Dagum II distributions. The parameters are estimated using maximum likelihood method, which is briefly discussed as following:

Fisk Distribution

$$\ln L = n \ln c - (c+1)\sum_{i=1}^{n} \ln x_i - 2\sum_{i=1}^{n} \ln\left(1 + x_i^{-c}\right) \tag{5.47a}$$

Maximizing Equation (5.47a), the parameter c may be estimated by solving:

$$\frac{n}{c} - \sum_{i=1}^{n} \ln x_i + 2\sum_{i=1}^{n} \frac{\ln x_i}{1 + x_i^c} = 0 \tag{5.47b}$$

Singh-Maddala Distribution

$$\ln L = n \ln c + (bc-1)\sum_{i=1}^{n} \ln x_i - \left(\frac{1}{b} + 1\right)\sum_{i=1}^{n} \ln\left(1 + x_i^{bc}\right) \tag{5.48a}$$

Maximizing Equation (5.48a), we have:

$$\begin{cases} c\sum_{i=1}^{n} \ln x_i - c\left(1+\frac{1}{b}\right)\sum_{i=1}^{n} (\ln x_i)\frac{x_i^{bc}}{1+x_i^{bc}} + \frac{1}{b^2}\sum_{i=1}^{n} \ln\left(1+x_i^{bc}\right) = 0 \\ \frac{n}{c} + b\sum_{i=1}^{n} \ln x_i - (b+1)\sum_{i=1}^{n} (\ln x_i)\frac{x_i^{bc}}{1+x_i^{bc}} = 0 \end{cases} \tag{5.48b}$$

Now the parameters may be estimated by solving the system of equations given as Equation (5.48b).

Weibull Distribution (3-Parameter)

$$\ln L = n\ln c - n\ln k + (c-1)\sum_{i=1}^{n} \ln\left(\frac{x_i-d}{k}\right) - \sum_{i=1}^{n}\left(\frac{x_i-d}{k}\right)^{c} \tag{5.49a}$$

Maximizing Equation (5.49a) we have:

$$\begin{cases} \frac{n}{c} + \sum_{i=1}^{k} \ln\left(\frac{x_i-d}{k}\right) - \sum_{i=1}^{n}\left(\frac{x_i-d}{k}\right)^{c} \ln\left(\frac{x_i-d}{k}\right) = 0 \\ -(c-1)\sum_{i=1}^{n}\frac{1}{x_i-d} + \sum_{i=1}^{n}\frac{c(x_i-d)^{c-1}}{k^c} = 0 \\ -\frac{nc}{k} + \frac{c}{k^c}\sum_{i=1}^{n}(x_i-d)^{c-1} = 0 \end{cases} \tag{5.49b}$$

Now, the parameters may be estimated by solving the system of equations given as Equation (5.49b).

Dagum I Distribution (Original: 2-Parameter)

$$\ln L = n\ln c - \left(1+\frac{c}{b}\right)\sum_{i=1}^{n} \ln x_i - (b+1)\sum_{i=1}^{n} \ln\left(1+x_i^{-\frac{c}{b}}\right) \tag{5.50a}$$

Maximizing Equation (5.50a), we have:

$$\begin{cases} \frac{c}{b^2}\sum_{i=1}^{n} \ln x_i - \sum_{i=1}^{n} \ln\left(1+x_i^{-\frac{c}{b}}\right) - (b+1)\sum_{i=1}^{n}\frac{c\ln x_i}{b^2\left(1+x_i^{\frac{c}{b}}\right)} = 0 \\ \frac{n}{c} - \frac{1}{b}\sum_{i=1}^{n} \ln x_i + (b+1)\sum_{i=1}^{n}\frac{\ln x_i}{b\left(1+x_i^{\frac{c}{b}}\right)} = 0 \end{cases} \tag{5.50b}$$

Now, the parameters may be estimated by solving the system of equations given as Equation (5.50b).

Dagum I Distribution with Location Parameter (3-Parameter)

Introducing the location parameter, the density function of Dagum I distribution with location parameter may be expressed as:

$$f(x;b,c,\mu) = c(x-\mu)^{-\left(\frac{c}{b}+1\right)}\left(1+(x-\mu)^{-\frac{c}{b}}\right)^{-b-1};\mu \leq \min(x) \tag{5.51a}$$

The log-likelihood function is then given as:

$$\ln L = n\ln c - \left(\frac{c}{b}+1\right)\sum_{i=1}^{n}\ln(x_i-\mu) - (b+1)\sum_{i=1}^{n}\ln\left(1+(x_i-\mu)^{-\frac{c}{b}}\right) \tag{5.51b}$$

Maximizing Equation (5.51b) we have:

$$\begin{cases} \frac{c}{b^2}\sum_{i=1}^{n}\ln(x_i-\mu) - \sum_{i=1}^{n}\ln\left(1+(x_i-\mu)^{-\frac{c}{b}}\right) \\ \quad -(b+1)\sum_{i=1}^{n}\frac{c\ln(x_i-\mu)}{b^2\left(1+(x_i-\mu)^{\frac{c}{b}}\right)} = 0 \\ \frac{n}{c} - \frac{1}{b}\sum_{i=1}^{n}\ln(x_i-\mu) + (b+1)\sum_{i=1}^{n}\frac{\ln(x_i-\mu)}{b\left(1+(x_i-\mu)^{\frac{c}{b}}\right)} = 0 \\ \left(\frac{c}{b}+1\right)\sum_{i=1}^{n}\frac{1}{x_i-\mu} - \frac{c(b+1)}{b}\sum_{i=1}^{n}\frac{1}{(x_i-\mu)^{\frac{c}{b}+1}+(x_i-\mu)} = 0 \end{cases} \tag{5.51c}$$

Now the parameters may be estimated by solving the system of equations given as Equation (5.51c).

Dagum II Distribution (Original: 3-Parameter)

$$\begin{aligned} \ln L = {} & n\ln c + n\left(\frac{1}{b}+1\right)\ln(1-a) \\ & -\left(\frac{c}{b}+1\right)\sum_{i=1}^{n}\ln x_i - (b+1)\sum_{i=1}^{n}\ln\left(1+(1-a)^{\frac{1}{b}}x_i^{-\frac{c}{b}}\right) \end{aligned} \tag{5.52a}$$

Maximizing Equation (5.51a), we have:

$$\begin{cases} -n\left(1+\frac{1}{b}\right)\left(\frac{1}{1-a}\right)+\frac{(b+1)(1-a)^{\frac{1}{b}-1}}{b}\sum_{i=1}^{n}\frac{1}{(1-a)^{\frac{1}{b}}+x_i^{\frac{c}{b}}}=0 \\ -nb^{-2}\ln(1-a)+cb^{-2}\sum_{i=1}^{n}\ln x_i-\sum_{i=1}^{n}\ln\left(1+(1-a)^{\frac{1}{b}}x_i^{-\frac{c}{b}}\right) \\ \quad +\frac{(b+1)(1-a)^{\frac{1}{b}}}{b^2}\sum_{i=1}^{n}\frac{\ln(1-a)-c\ln x_i}{(1-a)^{\frac{1}{b}}+x_i^{\frac{c}{b}}}=0 \\ \frac{n}{c}-\frac{1}{b}\sum_{i=1}^{n}\ln x_i+\frac{(b+1)(1-a)^{\frac{1}{b}}}{b}\sum_{i=1}^{n}\frac{\ln x_i}{(1-a)^{\frac{1}{b}}+x_i^{\frac{c}{b}}}=0 \end{cases} \tag{5.52b}$$

Now the parameters may be estimated by solving the system of equations given as Equation (5.52b).

Dagum II Distribution with Location Parameter (4-Parameter)

Introducing the location parameter, the PDF of Dagum II distribution may be rewritten as:

$$f(x)=c(1-a)^{1+\frac{1}{b}}(x-\mu)^{-\frac{c}{b}-1}\left(1+(1-a)^{\frac{1}{b}}(x-\mu)^{-\frac{c}{b}}\right)^{-b-1} \tag{5.53a}$$

The log-likelihood function can then be given as:

$$\begin{aligned} \ln L = n\ln c+n\left(1+\frac{1}{b}\right)\ln(1-a)-\left(\frac{c}{b}+1\right)\sum_{i=1}^{n}\ln(x_i-\mu) \\ -(b+1)\sum_{i=1}^{n}\ln\left(1+(1-a)^{\frac{1}{b}}(x_i-\mu)^{-\frac{c}{b}}\right) \end{aligned} \tag{5.53b}$$

Maximizing Equation (5.53b), we have:

$$\begin{cases} -n\left(1+\frac{1}{b}\right)\left(\frac{1}{1-a}\right)+\frac{(b+1)(1-a)^{\frac{1}{b}-1}}{b}\sum_{i=1}^{n}\frac{1}{(1-a)^{\frac{1}{b}}+(x_i-\mu)^{\frac{c}{b}}}=0 \\ -nb^{-2}\ln(1-a)+cb^{-2}\sum_{i=1}^{n}\ln(x_i-\mu)-\sum_{i=1}^{n}\ln\left(1+(1-a)^{\frac{1}{b}}(x_i-\mu)^{-\frac{c}{b}}\right) \\ \quad +\frac{(b+1)(1-a)^{\frac{1}{b}}}{b^2}\sum_{i=1}^{n}\frac{\ln(1-a)-c\ln(x_i-\mu)}{(1-a)^{\frac{1}{b}}+(x_i-\mu)^{\frac{c}{b}}}=0 \\ \frac{n}{c}-\frac{1}{b}\sum_{i=1}^{n}\ln(x_i-\mu)+\frac{(b+1)(1-a)^{\frac{1}{b}}}{b}\sum_{i=1}^{n}\frac{\ln(x_i-\mu)}{(1-a)^{\frac{1}{b}}+(x_i-\mu)^{\frac{c}{b}}}=0 \\ \left(\frac{c}{b}+1\right)\sum_{i=1}^{n}\frac{1}{x_i-\mu}-\frac{c(b+1)(1-a)^{\frac{1}{b}}}{b}\sum_{i=1}^{n}\frac{1}{(1-a)^{\frac{1}{b}}(x_i-\mu)+(x_i-\mu)^{1+\frac{c}{b}}}=0 \end{cases} \tag{5.53c}$$

Now, the parameters may be estimated by solving the system of equations given as Equation (5.53c).

Pareto II Distribution

The log-likelihood function of Pareto II distribution is given as:

$$\ln L = n \ln c - n \ln b - (c+1) \sum_{i=1}^{n} \ln \left(\frac{x_i - d}{b} \right) \tag{5.54a}$$

Maximizing Equation (5.54a), we have:

$$\begin{cases} \dfrac{n}{c} - \sum_{i=1}^{n} \ln \left(\dfrac{x_i - d}{b} \right) = 0 \\ -\dfrac{n}{b} + \dfrac{(c+1)}{b^2} \sum_{i=1}^{n} \dfrac{b}{x_i - d} = 0 \\ (c+1) \sum_{i=1}^{n} \dfrac{b}{x_i - d} = 0 \end{cases} \tag{5.54b}$$

now the parameters may be estimated by solving the system of equations given as Equation (5.54b).

5.4.1 Monthly Sediment Yield

Based on the empirical frequency of the monthly sediment yield, the Fisk, Singh-Maddala, Dagum I (with/without location parameter), and Dagum II (with/without location parameter) are applied. The maximum likelihood estimation method is applied to estimate the parameters with the use of the genetic algorithm. Table 5.1 lists the parameters estimated for monthly suspended sediment. Figures 5.12–5.14 compare the fitted frequency distribution with the empirical frequency. The results

Table 5.1 *Parameters estimated for monthly suspended sediment*

Parameters	Fisk	Singh-Maddala	Dagum I	Dagum I[1]	Dagum II	Dagum II[1]
a					0.002	0.002
b		8.74	204.87	82.55	47.42	19.59
c	0.34	0.23	260.12	93.86	44.72	16.22
μ				13.00		21.11

[1] Dagum distributions with location parameter.

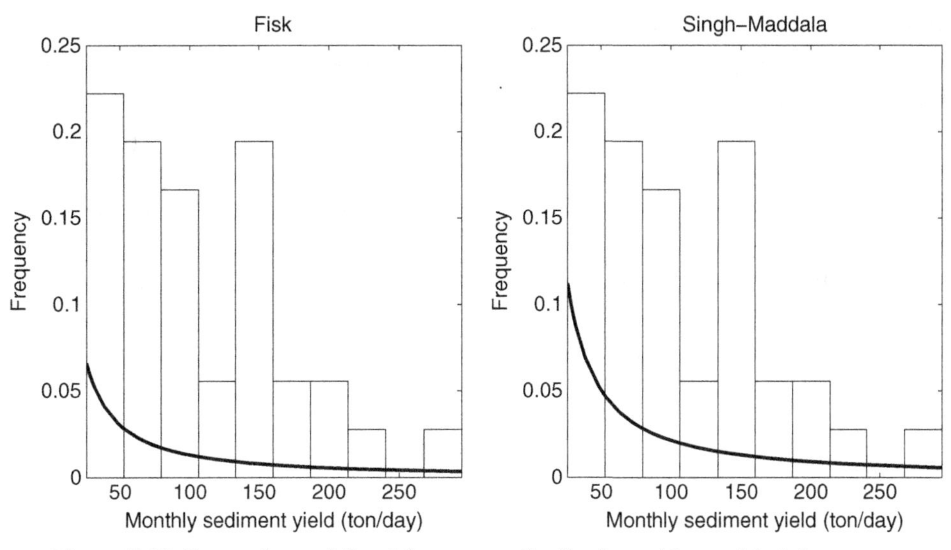

Figure 5.12 Comparison of fitted frequency distribution with empirical frequency distribution for monthly suspended sediment: Fisk and Singh-Maddala distribution.

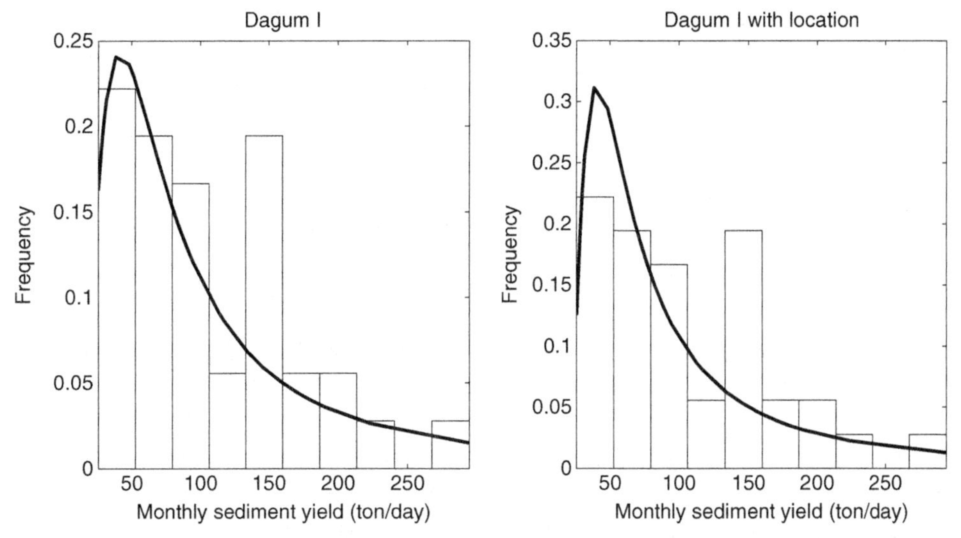

Figure 5.13 Comparison of fitted frequency distribution with empirical frequency distribution for monthly suspended sediment: Dagum I distribution with/without location parameter.

of monthly suspended sediment indicate that the Dagum I distribution performs better than do the Fisk, Singh-Maddala, and Dagum II distributions. Additionally, the Dagum I and II distributions without location parameter perform better than the corresponding distributions with the location parameter. However, the

Table 5.2 *Parameters estimated for peak flow*

Parameters	Dagum I	Dagum II	Dagum I[1]	Dagum II[1]
a		0.001		0.006
b	192.52	17.28	272.33	76.88
c	189.06	10.01	315.61	70.75
μ			89.79	98.12

[1] Dagum distributions with location parameter.

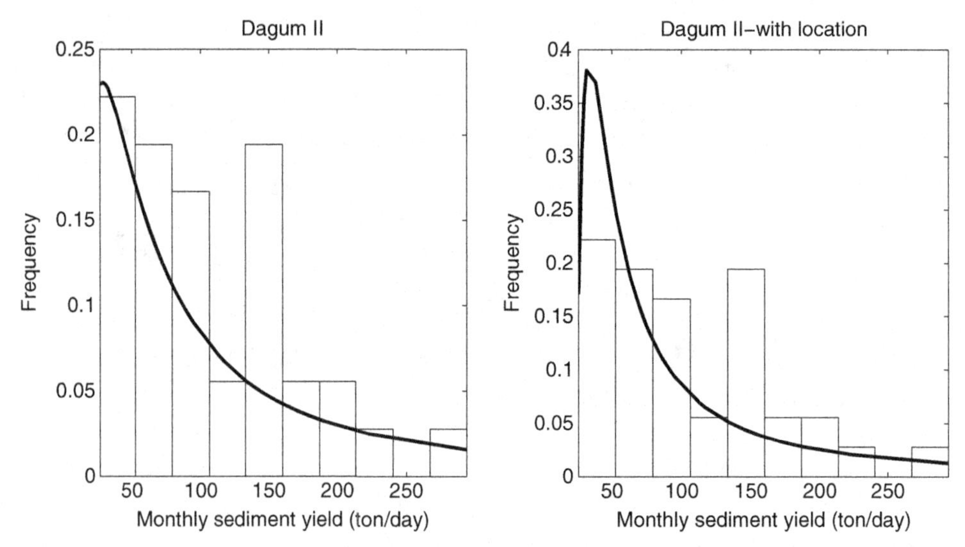

Figure 5.14 Comparison of fitted frequency distribution with empirical frequency distribution for monthly suspended sediment: Dagum II distribution with/without location parameter.

performances are inferior to the Stoppa III and generalized 4-parameter Pareto distributions (given in Chapter 6).

5.4.2 Peak Flow

Dagum I and II distributions with/without location parameter are selected to evaluate the peak distribution flow. Table 5.2 lists the parameters estimated for peak flow distribution using the maximum likelihood estimation method with genetic algorithm. Figures 5.15–5.16 compare the fitted frequency distribution with empirical frequency distribution for peak flow. The results of peak flow distribution indicate that the Dagum I and II distributions with location parameter outperform their original distribution. Again, similar to monthly suspended sediment, the performances are inferior to the Pearson system (i.e., log-Pearson III in Chapter 2),

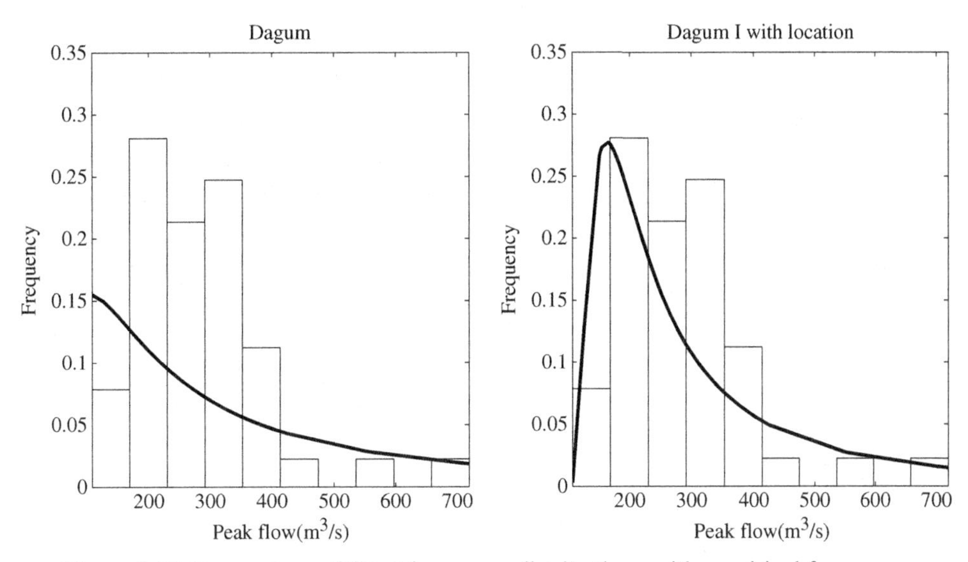

Figure 5.15 Comparison of fitted frequency distributions with empirical frequency distribution for peak flow: Dagum I and Dagum I distributions with location parameter.

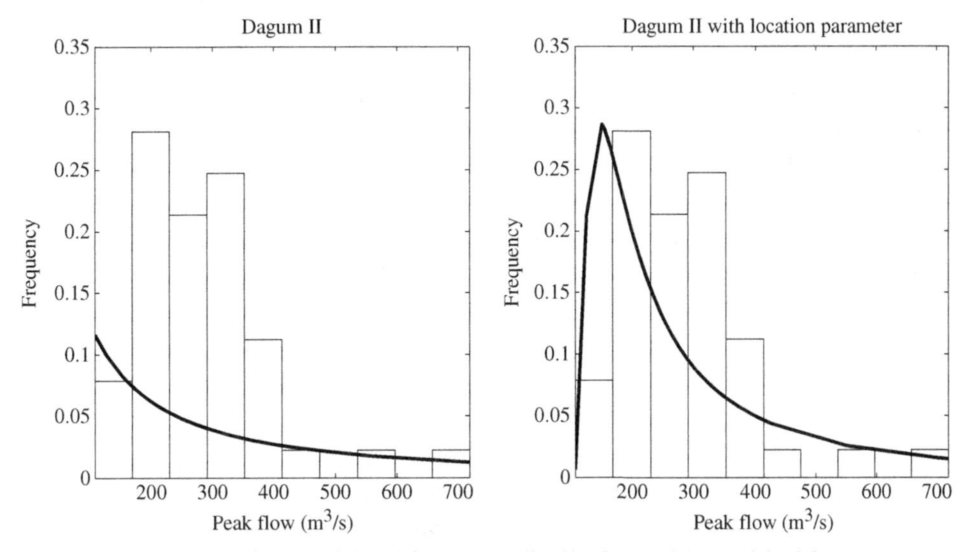

Figure 5.16 Comparison of fitted frequency distributions with empirical frequency distribution for peak flow: Dagum II and Dagum II distributions with location parameter.

Burr system (i.e., Burr III and XII with scale parameter in Chapter 3), the D'Addario system (i.e., log-normal, 3- and 4-parameter Amoroso distributions in Chapter 4), Esteban system (i.e., 3-parameter gamma distribution and generalized Beta distribution of second kind in Chapter 7).

5.4.3 Maximum Daily Precipitation

Singh-Maddala, Weibull (3-parameter), Dagum I, and Dagum II distributions with and without location parameter are applied to evaluate maximum daily precipitation. Table 5.3 lists the parameters estimated using the maximum likelihood estimation method with genetic algorithm. Figures 5.17–5.19 compare the fitted frequency to the empirical frequency. The results of comparison show that the Dagum I and II distributions with location parameters fit the maximum daily precipitation better than other distribution candidates belonging to the Dagum family. Additionally, compared with the gamma and log-Pearson distributions of Pearson family applied in Chapter 2, and 2- and 3-parameter lognormal distributions of Burr family applied in Chapter 3, the distribution candidates in Chapters 2 and 3 perform better than do the distribution candidates of Dagum family.

Table 5.3 *Parameters estimated for maximum daily precipitation*

Parameters	Singh-Maddala	Weibull (3-parameter)	Dagum I	Dagum I[1]	Dagum II	Dagum II[1]
a					0.0001	0.0064
b	12.42	0.002	428.74	48.27	33.85	20.34
c	0.26	1.68	678.72	64.71	32.68	22.79
d		6.05				
μ				25.61		27.70

[1] Dagum distributions with location parameter.

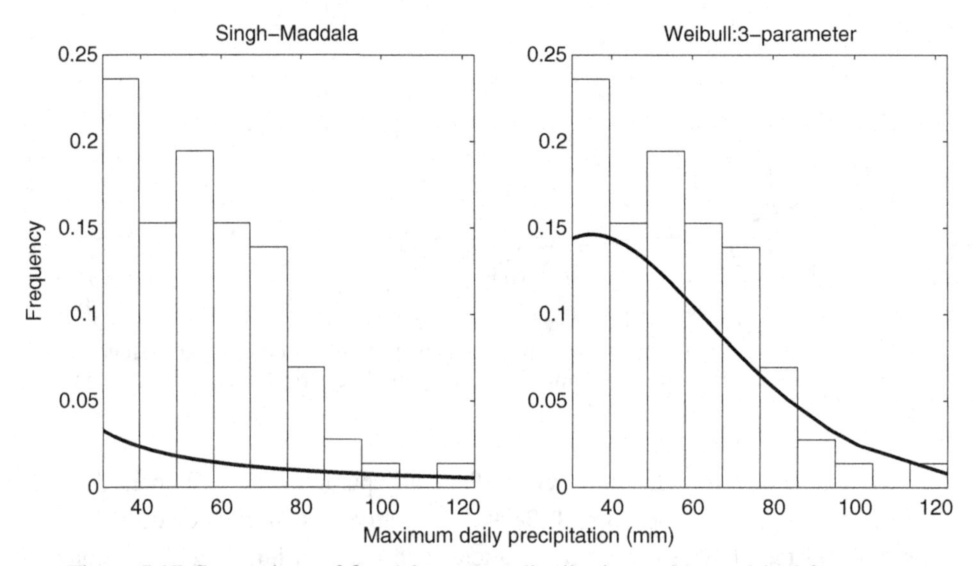

Figure 5.17 Comparison of fitted frequency distributions with empirical frequency distribution for maximum daily precipitation: Singh-Maddala and Weibull (3-parameter) distributions.

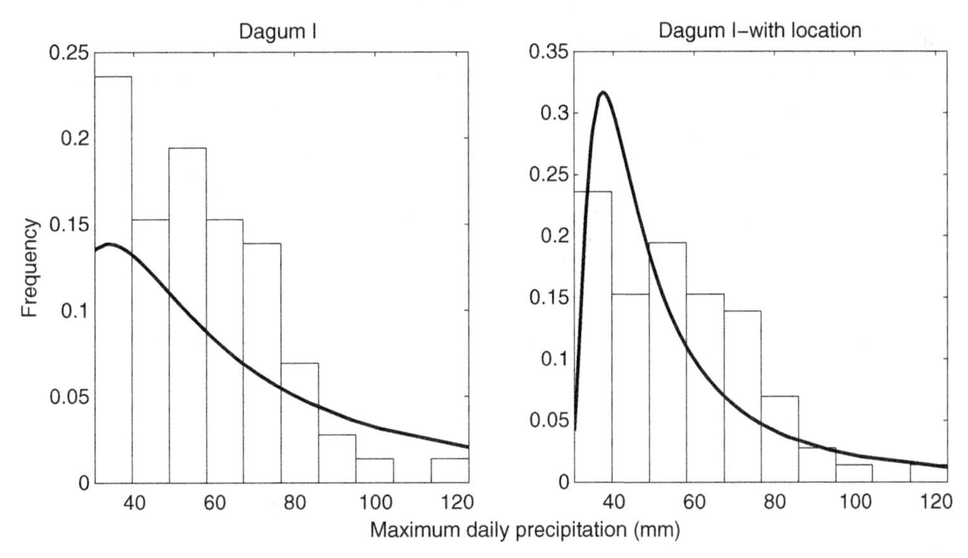

Figure 5.18 Comparison of fitted frequency distributions with empirical frequency distribution for maximum daily precipitation: Dagum I and Dagum I distributions with location parameter.

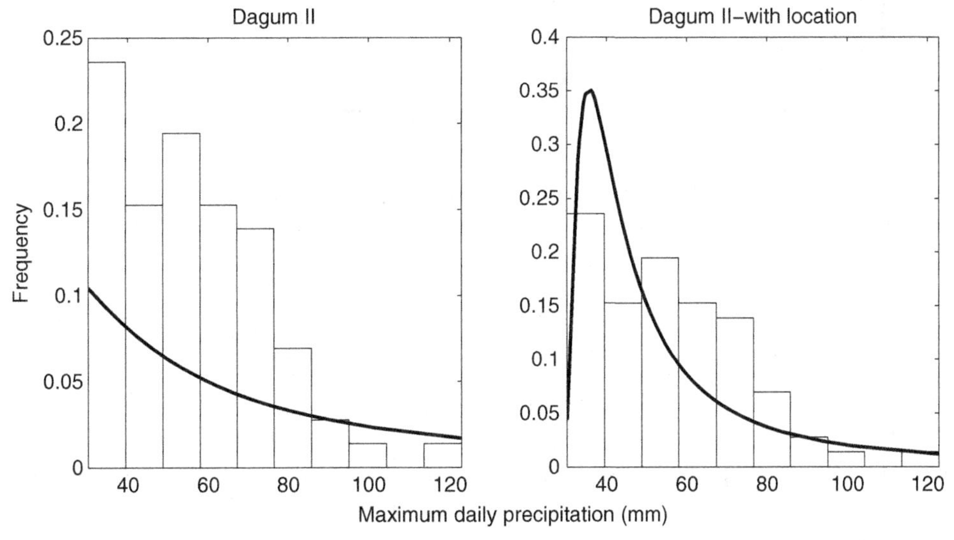

Figure 5.19 Comparison of fitted frequency distributions with empirical frequency distribution for maximum daily precipitation: Dagum II and Dagum II distributions with location parameter.

5.4.4 Drought (Total Flow Deficit)

The drought at Tilden, Texas, is applied for analysis. Amongst the flow deficit (total flow deficit less than the average monthly flow), drought duration, drought interarrival time, and average drought intensity, the total flow deficit is selected for

Table 5.4 *Parameters estimated for total flow deficit*

Parameters	Fisk	Singh-Maddala	Pareto II (3-parameter)	Dagum I	Dagum I[1]	Dagum II	Dagum II[1]
a						0.0001	0.007
b		4.87	132.58	156.48	153.84	6.42	74.99
c	0.15	0.10	0.19	83.46	81.63	1.53	34.39
d			67.31				
μ					25.89		25.89

[1] Dagum distributions with location parameter.

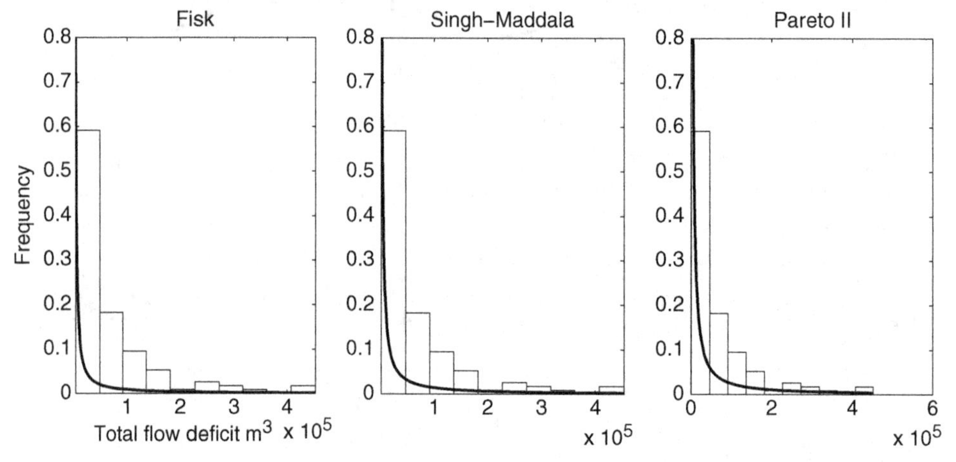

Figure 5.20 Comparison of fitted frequency distributions with empirical frequency distribution for total flow deficit: Fisk, Singh-Maddala, and Pareto II distributions.

analysis. According to its empirical frequency, the following distribution candidates are selected: Pareto II, Fisk distribution, Singh-Maddala distribution, and Dagum I and II distributions with and without location parameter. Table 5.4 lists the parameters estimated for the distribution candidates. Figures 5.20–5.22 compare the fitted frequency distributions to the empirical frequency distribution. Comparison shows that for the total flow deficit, all the distribution candidates show similar performances and may be applied for the investigation.

5.5 Conclusion

Based on the elasticity of CDF, the Dagum system of distributions is revisited in this chapter. Although the practitioners and engineers may not be familiar with the system, the special cases of the system have been widely applied in the field of water resources and environmental engineering, such as Pareto I, Weibull, Fisk

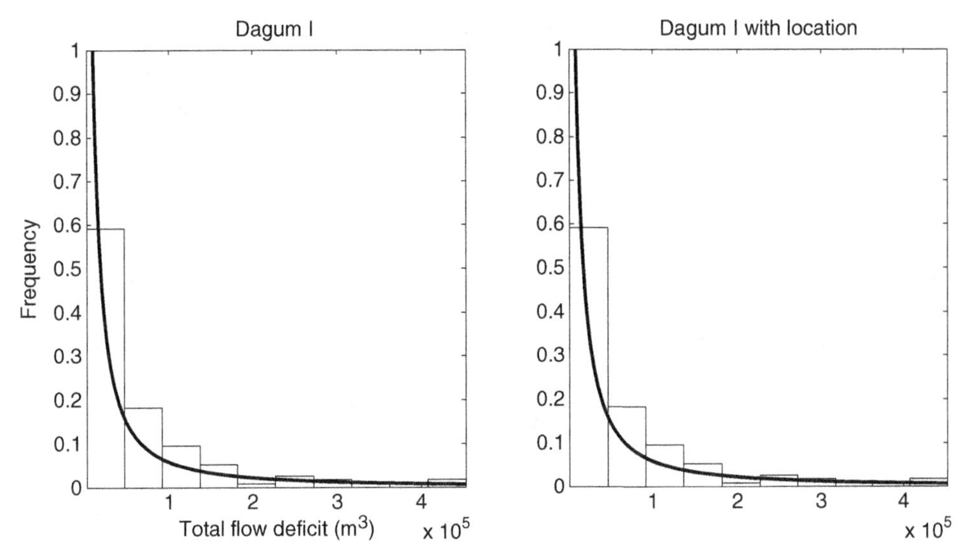

Figure 5.21 Comparison of fitted frequency distributions with empirical frequency distribution for total flow deficit: Dagum I distribution with/without location parameter.

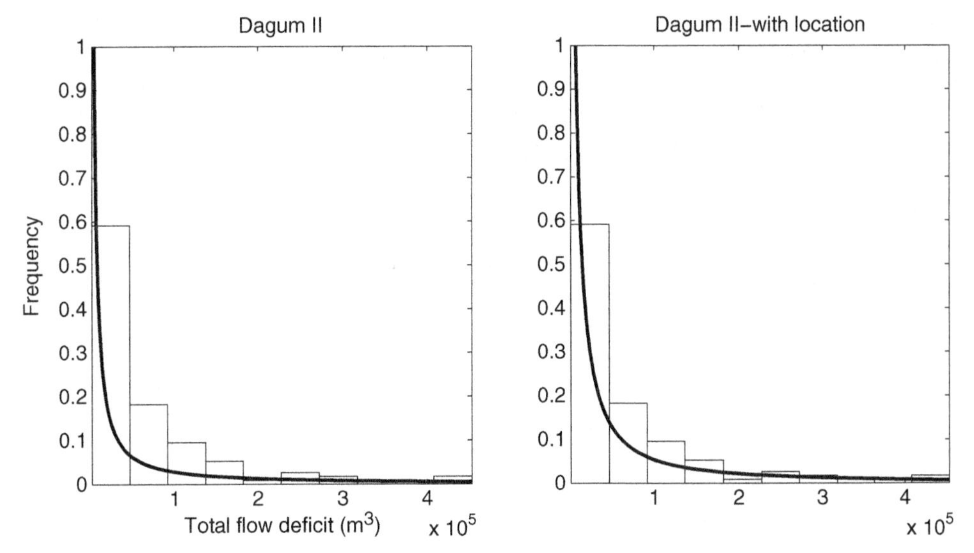

Figure 5.22 Comparison of fitted frequency distributions with empirical frequency distribution for total flow deficit: Dagum II distribution with/without location parameter.

(also called log-logistic), and Singh-Maddala distributions. This provides a new way of deriving frequency distributions. According to the sample applications with the real-world data, one may choose the distributions from other system over the Dagum system for the frequency analysis of water engineering.

References

Arnold, B. C. (1980). Pareto distributions: Pareto and related heavy-tailed distributions. Mimeographed manuscript, University of California at Riverside, California.

Arnold, B. C. (1983). *Pareto Distributions*. Fairland, MD: International Cooperative Publishing House.

Benini, R. (1906). *Principii di Statistica Metodologica*. Torino, Italy: UTET.

Burr, I. W. (1942). Cumulative frequency functions. *Annals of Mathematical Statistics* 13, pp. 215–232.

Champernowne, D. (1952). The graduation of income distribution. *Econometrica* 20, pp. 591–625.

Dagum, C. (1977). A new model of personal income distribution: Specification and estimation. *Economic Appliquee* XXX, no. 3, pp. 413–436.

Dagum, C. (1980a). The generation and distribution of income, the Lorenz curve and Ginni ratio. *Economic Appliquee* 33, pp. 327–367.

Dagum, C. (1980b). Generating systems and properties of income distribution models. *Metron* 38, pp. 3–26.

Dagum, C. (1983). Income distribution models. In *Encyclopedia of Statistical Sciences*, edited by S. Kotz, N. L. Johnson, and C. Read, Vol. 4, pp. 27–34. New York: Wiley.

Dagum, C. (1990). Generation and properties of income distribution functions. In *Income and Wealth Distribution, Inequality and Poverty*, edited by C. Dagum and M. Zenga, pp. 1–17. Berlin: Springer.

Dagum, C. (2006). *Lorenz Curve*. In *Encyclopedia of Statistical Sciences*, edited by S. Kotz and N. Balakrishnan, Vol. 4, pp. 1–5. New York: Wiley-Interscience.

Fisk, P. (1961). The graduation of income distributions. *Econometrica* 29, pp. 171–185.

Pareto, V. (1897). *Cours d'économie politique*. Lausanne: Ed. Rouge.

Pareto, V. (1985). La legge della domanda. *Gionale degli Economisti* 10, 59–68. English translation in Rivista di Politica Economica, Vol. 87 (1997), pp. 691–700.

Singh, S., and Maddala, G. (1976). A function for the size distribution of income. *Econometrica* 44, pp. 963–970.

Sydsaeter, K., and Hammond, P. (1995). *Mathematics for Economic Analysis*, pp. 173–175. Englewood Cliffs, NJ: Prentice Hall.

Weibull, W. (1951). A statistical distribution function of wide applicability. *Journal of Applied Mechanics* 18, pp. 293–297.

6

Stoppa System of Frequency Distributions

6.1 Introduction

Stoppa (1993) employed the elasticity of the cumulative distribution function (CDF) $F(x)$ and then proposed a differential equation that he used to derive a set of distributions that constitute the Stoppa system or family. Examples of these distributions include generalized power distribution, generalized exponential distribution, generalized Pareto distribution, and different Stoppa distributions. This system is revisited and its individual frequency distributions are derived in this chapter.

6.2 Stoppa System of Distributions

The elasticity of a distribution function in econometrics may be expressed as:

$$\eta(x, F(x)) = \frac{d \ln F(x)}{d \ln x} = \frac{F'(x)}{F(x)} x \tag{6.1}$$

Stoppa (1990) proposed a general system (i.e., Stoppa system) for the distributions with the following assumptions: (i) The distribution is nonmodal or unimodal and positively skewed; (ii) the domain of the distribution is (x_0, ∞) where $x_0 > 0$; and (iii) the elasticity of the distribution function $F(x)$, $\eta(x, F(x))$, is a monotonic decreasing function of $F(x)$, depending on the relative development of the survival function $[1-F(x)]$ as $\lim_{x \to x_0} \eta(x, F(x)) \to \infty$ (e.g., diverge) and $\lim_{x \to \infty} \eta(x, F(x)) = 0$ (e.g., converge). According to Stoppa (1990, 1993), the elasticity function is thus written as:

$$\eta(x, F(x)) = \frac{d \ln F(x)}{d \ln x} = \frac{1 - (F(x))^{\frac{1}{\alpha}}}{(F(x))^{\frac{1}{\alpha}}} g(x, F(x)); \alpha > 0, x \in [x_0, \infty), g(x, F(x)) > 0 \tag{6.2}$$

Combining Equations (6.1) and (6.2) and assuming that the generating differential equation for elasticity can be written as

$$\eta(x, F(x)) = \frac{d\ln F(x)}{d\ln x} = \frac{xF'(x)}{F(x)} = \frac{1-(F(x))^{\frac{1}{\alpha}}}{(F(x))^{\frac{1}{\alpha}}} g(x, F(x)) \tag{6.3a}$$

If we simplify the function $g(x, F(x))$ to $g(x)$, Equation (6.3a) can be simplified as:

$$\eta(x, F(x)) = \frac{d\ln F(x)}{d\ln x} = \frac{x}{F(x)}\frac{dF(x)}{dx} = \frac{1-(F(x))^{\frac{1}{\alpha}}}{(F(x))^{\frac{1}{\alpha}}} g(x) \tag{6.3b}$$

Using the simplified case defined in Equation (6.3b), the differential equation may be solved as:

$$\eta(x, F(x)) = \frac{d\ln F(x)}{d\ln x} \Rightarrow \frac{[F(x)]^{\frac{1}{\alpha}-1}}{1-[F(x)]^{\frac{1}{\alpha}}} dF(x) = \frac{g(x)}{x} dx$$

$$\Rightarrow \frac{\alpha}{1-[F(x)]^{\frac{1}{\alpha}}} d[F(x)]^{\frac{1}{\alpha}} = \frac{g(x)}{x} dx \Rightarrow 1-[F(x)]^{\frac{1}{\alpha}} = \exp\left(-\frac{1}{\alpha}\int_{x_0}^{x}\frac{g(x)}{x}dx\right)$$

$$\Rightarrow F(x) = \left(1-\exp\left(-\frac{1}{\alpha}\int_{x_0}^{x}\frac{g(x)}{x}dx\right)\right)^{\alpha} \tag{6.4}$$

Depending on the form of the function $g(x)$, the solution of Equation (6.4) will yield different distribution functions. Interestingly, if $\alpha = 1$, the various distributions of the Burr family are obtained indicating that the Stoppa system is more general than the Burr system. If $a \neq 1$, then type I, a power distribution; type II, a generalized Pareto distribution; and type III, a generalized exponential distribution, are obtained (Kleiber and Kotz, 2003).

Based on Stoppa (1990), Stoppa (1993) introduced the concept of applying linear regression by presenting the elasticity function $\eta(x, F)$ in the logarithmic domain as a linear function of the variables [i.e., independent variables, or regressors] that may be chosen from $\ln F$, $\ln(1-F)$, $\ln x$, $\ln^2 x$, x, x^2, $\ln(1-x)$, $\ln(1+x)$, $\ln(1-x^2)$, $\ln(1+x^2)$, $\ln(e^x-1)$, $\ln(e^x+1)$ etc. Then Stoppa defined the set of distributions with the same number of regressors of different types as one period. For example, he found that 15 distribution families with the elasticity function in the logarithm domain only depended on one regressor, which may be called period 1.

6.3 Derivation of Frequency Distributions

The individual frequency distributions are derived in what follows.

6.3.1 Generalized Power Distribution (Stoppa Type I Distribution)

As discussed in Stoppa (1990) and Kleiber and Kotz (2003), the $g(x)$ and $\eta(x, F(x))$ functions for the generalized power function are given as:

$$g(x) = \frac{abx}{1 - bx}; x \in \left(0, \frac{1}{b}\right), \alpha, b > 0 \tag{6.5a}$$

$$\eta(x, F) = \frac{d \ln F(x)}{d \ln x} = \frac{1 - (F(x))^{\frac{1}{\alpha}}}{(F(x))^{\frac{1}{\alpha}}} \frac{abx}{1 - bx} \tag{6.5b}$$

Substitution of Equation (6.5a) into Equation (6.3b) yields:

$$\frac{F(x)^{\frac{1}{\alpha}-1}}{1 - F(x)^{\frac{1}{\alpha}}} dF = \frac{ab}{1 - bx} dx \tag{6.6}$$

The integration of Equation (6.6) yields

$$F(x) = \left(1 - \exp\left(-\int_0^x \frac{b}{1 - bx} dx\right)\right)^{\alpha} \Rightarrow F(x) = (bx)^{a} \tag{6.7}$$

According to the theory of total probability, x is bounded by $x \in \left[0, \frac{1}{b}\right]$ for $\alpha, b > 0$.

The probability density function (PDF) can then be written as:

$$f(x) = \alpha b^{\alpha} x^{\alpha - 1}; \alpha > 0, b > 0, x \in \left[0, \frac{1}{b}\right] \tag{6.8}$$

Stoppa Type I distribution is a 2-parameter distribution. And from Equation (6.7), Burr I distribution is obtained if $\alpha = 1, b = 1$; and the generalized Burr I distribution is obtained if $\alpha = 1, b \neq 1$. The $g(x), F(x)$, and $f(x)$ functions are plotted in Figures 6.1a, 6.1b, and 6.1c for different parameter values.

6.3.2 Generalized Pareto Type II Distribution

As discussed in Kleiber and Kotz (2003), the generalized Pareto distribution may be considered as the Stoppa type II distribution. Its elasticity function is expressed as:

$$\eta(x, F(x)) = \frac{d \ln F}{d \ln x} = g(x) \frac{1 - [F(x)]^{\frac{1}{\alpha}}}{[F(x)]^{\frac{1}{\alpha}}} = b \frac{1 - [F(x)]^{\frac{1}{\alpha}}}{[F(x)]^{\frac{1}{\alpha}}}; \alpha, b > 0, x \in [x_0, \infty), x_0 > 0 \tag{6.9}$$

where $g(x) = b$.

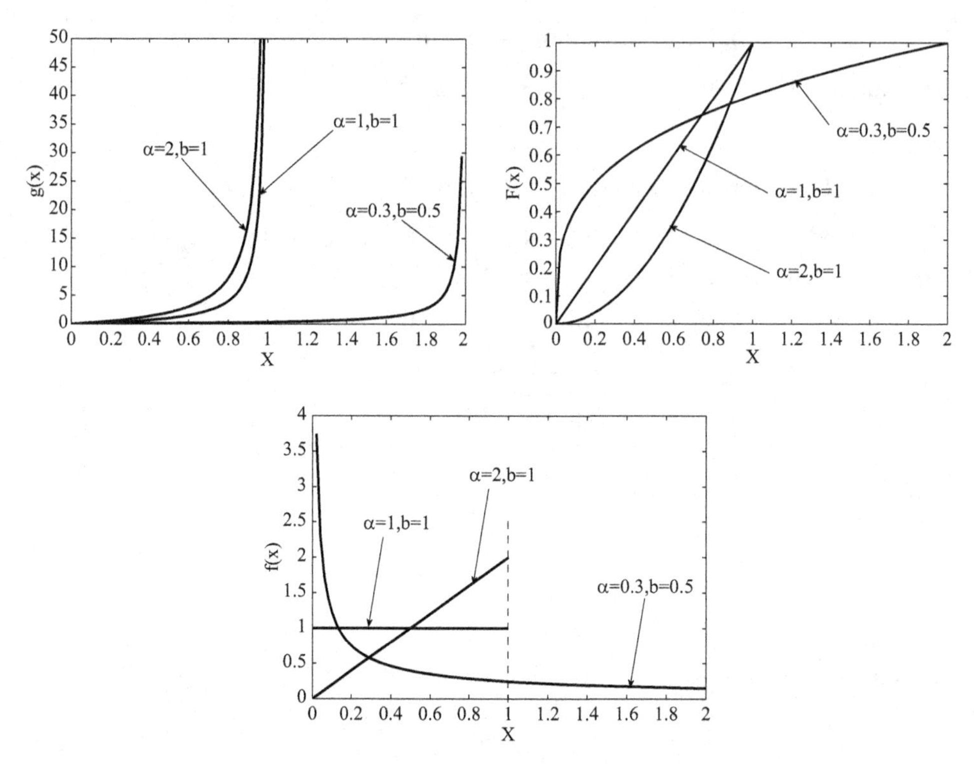

Figure 6.1 (a) Plot of the g(x) function of generalized power (Type I) distribution; (b) plot of the CDF of generalized power (Type I) distribution; and (c) plot of the PDF of generalized power (Type I) distribution.

Integration of Equation (6.9) yields

$$\begin{aligned}\frac{[F(x)]^{\frac{1}{\alpha}-1}}{1-[F(x)]^{\frac{1}{\alpha}}}dF &= \frac{b}{x}dx \Rightarrow F(x) = \left[1-\exp\left(-\tfrac{1}{\alpha}\int_{x_0}^{x}\tfrac{b}{x}dx\right)\right]^{\alpha} \\ &= \left[1-\exp\left(-\tfrac{b}{\alpha}(\ln x-\ln x_0)\right)\right]^{\alpha} \\ &\Rightarrow F(x) = \left(1-\left(\tfrac{x}{x_0}\right)^{-\frac{b}{\alpha}}\right)^{\alpha}; \alpha, b, x_0 > 0, x \in [x_0, \infty)\end{aligned} \tag{6.10}$$

The PDF is then written as:

$$f(x) = bx_0^{\frac{b}{\alpha}}x^{-\left(\frac{b}{\alpha}+1\right)}\left(1-x_0^{\frac{b}{\alpha}}x^{-\frac{b}{\alpha}}\right)^{\alpha-1}; \alpha, b, x_0 > 0, x \in (x_0, \infty) \tag{6.11}$$

The Stoppa type II distribution (the generalized Pareto distribution) is a 3-parameter distribution. The 2-parameter Pareto distribution is obtained if $\alpha = 1$ in Equations (6.10)–(6.11). Figures 6.2a and 6.2b plot the CDF and the PDF of the

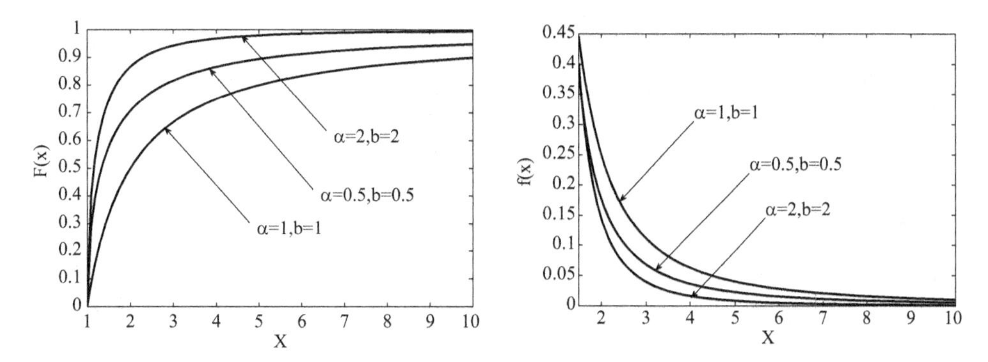

Figure 6.2 (a) Plot of the CDF of 3-parameter generalized Pareto distribution with $x_0 = 1$; and (b) plot of the PDF of 3-parameter generalized Pareto distribution with $x_0 = 1$.

generalized Pareto distribution. The generalized Pareto distribution is positively skewed and L-shaped.

6.3.3 Generalized Exponential Distribution (Type III Distribution)

The generalized exponential distribution is also considered as Stoppa III distribution (Kleiber and Kotz, 2003). Its elasticity function is expressed as:

$$\eta(x, F(x)) = \frac{d \ln F(x)}{d \ln x} = g(x) \frac{1 - [F(x)]^{\frac{1}{\alpha}}}{[F(x)]^{\frac{1}{\alpha}}} = bx \frac{1 - [F(x)]^{\frac{1}{\alpha}}}{[F(x)]^{\frac{1}{\alpha}}}; \alpha, b > 0, x \in (0, \infty) \tag{6.12}$$

where: $g(x) = bx$.

Integration of Equation (6.12) yields:

$$\begin{aligned} \frac{[F(x)]^{\frac{1}{\alpha}-1}}{1 - [F(x)]^{\frac{1}{\alpha}}} dF = \frac{b}{\alpha} dx \Rightarrow F(x) = \left(1 - \exp\left(-\int_0^x \tfrac{b}{\alpha} dx\right)\right)^{\alpha} \\ \Rightarrow F(x) = \left(1 - \exp\left(-\tfrac{b}{a} x\right)\right)^{\alpha} \end{aligned} \tag{6.13}$$

The density function is then given as:

$$f(x) = b \exp\left(-\frac{b}{\alpha} x\right) \left(1 - \exp\left(-\frac{b}{\alpha} x\right)\right)^{\alpha - 1}; \alpha, b > 0, x \in (0, \infty) \tag{6.14}$$

The exponential distribution is obtained if $\alpha = 1$. Figure 6.3 plots the CDF and the PDFs of the generalized exponential distribution with different parameters. The PDF is positively skewed. It is L-shaped if $\alpha \leq 1$ and is bell shaped if $\alpha > 1$.

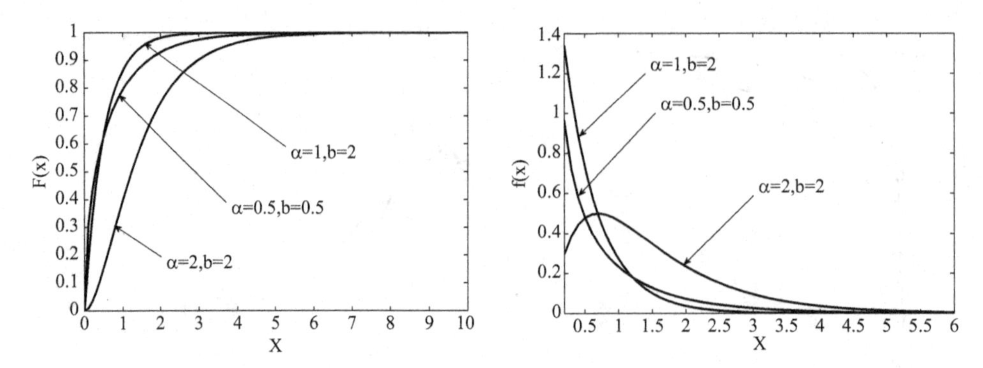

Figure 6.3 (a) Plot of the CDF of the generalized exponential distribution; and (b) plot of the PDF of the generalized exponential distribution.

Figure 6.3a graphs the CDF and Figure 6.3b graphs the PDF of the generalized exponential distribution.

6.3.4 Stoppa Type IV Distribution

The $g(x)$ function of Stoppa IV distribution is given as:

$$g(x) = \frac{1}{b-x}; b > 0, x \in \left(\frac{b}{2}, b\right) \tag{6.15}$$

And the elasticity function is expressed as:

$$\eta(x, F(x)) = \frac{d\ln F}{d\ln x} = \frac{1}{b-x}\frac{1-[F(x)]^{\frac{1}{\alpha}}}{[F(x)]^{\frac{1}{\alpha}}} \tag{6.16}$$

Integration of Equation (6.16) yields:

$$\frac{[F(x)]^{\frac{1}{\alpha}-1}}{1-[F(x)]^{\frac{1}{\alpha}}}dF = \frac{1}{x(b-x)}dx$$

$$\Rightarrow 1-[F(x)]^{\frac{1}{\alpha}} = \exp\left(-\int_{\frac{b}{2}}^{x}\frac{1}{\alpha}\frac{1}{x(b-x)}dx\right) = \left(\frac{b-x}{x}\right)^{\frac{1}{ab}} \tag{6.17}$$

$$\Rightarrow F(x) = \left(1-\left(\frac{b-x}{x}\right)^{\frac{1}{ab}}\right)^{\alpha}; \alpha, b > 0, x \in \left(\frac{b}{2}, b\right)$$

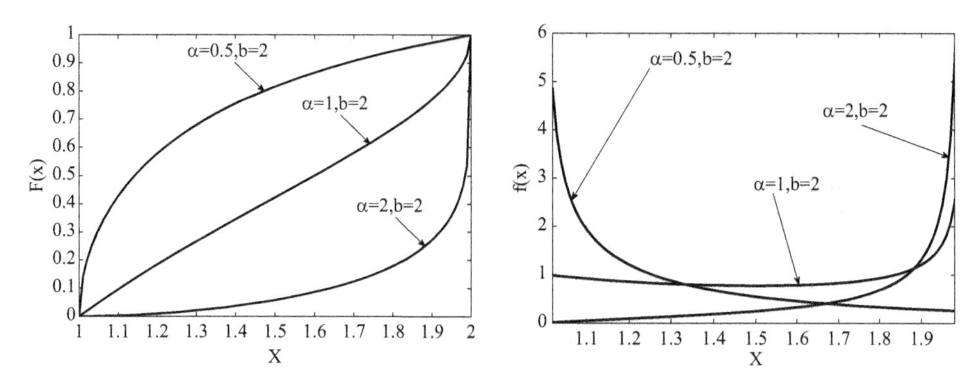

Figure 6.4 (a) Plot of the CDF of Stoppa type IV distribution; and (b) plot of the PDF of Stoppa type IV distribution.

The PDF is then given as:

$$f(x) = x^{-2}\left(\frac{b}{x} - 1\right)^{\frac{1}{ab}-1}\left(1 - \left(\frac{b}{x} - 1\right)^{\frac{1}{ab}}\right)^{\alpha-1}; \alpha, b > 0, x \in \left(\frac{b}{2}, b\right) \tag{6.18}$$

Figures 6.4a and 6.4b plot the CDF and PDF of Stoppa IV distribution. It can be shown that the Stoppa IV distribution is L-shaped if $\alpha < 1$ and is J-shaped if $\alpha \geq 1$.

6.3.5 Stoppa Type V Distribution

The $g(x)$ function of Stoppa V distribution is given as:

$$g(x) = bx\sec^2 x; b > 0, x \in \left[0, \frac{\pi}{2}\right) \tag{6.19}$$

and its elasticity function can be expressed as:

$$\eta(x, F(x)) = \frac{d\ln F}{d\ln x} = bx\sec^2 x\frac{1 - [F(x)]^{\frac{1}{\alpha}}}{[F(x)]^{\frac{1}{\alpha}}} \tag{6.20}$$

Solution of Equation (6.20) yields

$$\begin{aligned}
&\frac{[F(x)]^{\frac{1}{\alpha}-1}}{1 - [F(x)]^{\frac{1}{\alpha}}} dF = b\sec^2 x dx \\
&\Rightarrow \ln\left(1 - [F(x)]^{\frac{1}{\alpha}}\right) = -\frac{b}{\alpha}\int_0^x \sec^2 x dx = -\frac{b}{\alpha}\tan x \\
&\Rightarrow F(x) = \left(1 - \exp\left(-\frac{b}{a}\tan x\right)\right)^{\alpha}; \alpha, b > 0, x \in \left[0, \frac{\pi}{2}\right)
\end{aligned} \tag{6.21}$$

The PDF can then be given as:

$$f(x) = be^{-\frac{b\tan(x)}{\alpha}}\left(1 - e^{-\frac{b\tan(x)}{\alpha}}\right)^{\alpha-1}\sec^2(x); \alpha, b > 0, x \in \left[0, \frac{\pi}{2}\right) \tag{6.22}$$

Figure 6.5a plots the $g(x)$ function, Figure 6.5b plots the CDF, and Figure 6.5c plots the PDF of Stoppa type V distribution.

6.3.6 Four-Parameter Generalized Pareto Distribution

Different from the Stoppa distributions derived in the preceding text, the $g(x, F(x))$ function also depends on $F(x)$ for the 4-parameter generalized Pareto distribution that is given as:

$$g(x, F) = \alpha bx\left(1 - F^{\frac{1}{\alpha}}\right)^{\frac{1}{b}}; \alpha, b > 0; F \neq 0, 1 \tag{6.23}$$

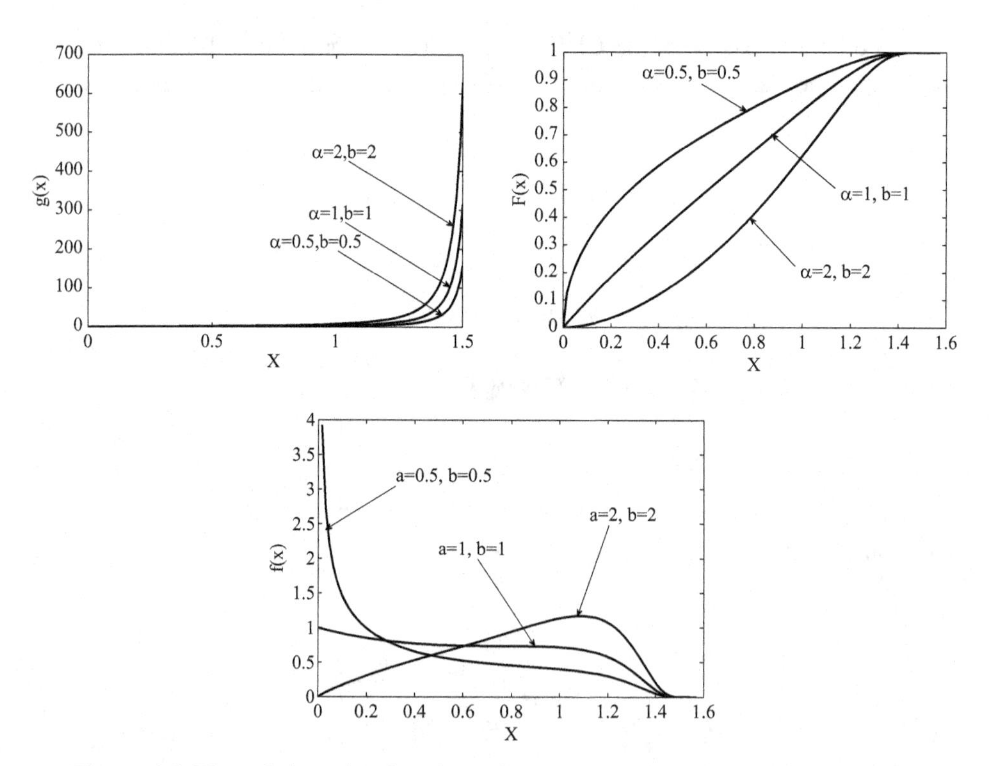

Figure 6.5 Plot of the g(x) function of Stoppa type V distribution; (b) plot of the CDF of Stoppa type V distribution; and (c) plot of the PDF of Stoppa type V distribution.

and its elasticity function can be given as:

$$\eta(x,F) = \frac{d \ln F}{d \ln x} = g(x,F)g(F) = \frac{abx\left(1 - F^{\frac{1}{a}}\right)^{1+\frac{1}{b}}}{F^{\frac{1}{a}}}; F \neq 0, 1 \tag{6.24}$$

Solution of Equation (6.24) yields:

$$F^{\frac{1}{a}-1}\left(1 - F^{\frac{1}{a}}\right)^{-\frac{1}{b}-1} dF = bdx \Rightarrow ad\left(1 - F^{\frac{1}{a}}\right)^{-\frac{1}{b}} = x - c$$
$$\Rightarrow F = \left(1 - x_0^b(x-c)^{-b}\right)^a; x > c \tag{6.25}$$

and the PDF is given as:

$$f(x) = \frac{abx_0^b\left(1 - x_0^b(x-c)^{-b}\right)^{a-1}}{(x-c)^{b+1}} \tag{6.26}$$

In Equations (6.25)–(6.26), c is constant regarded as the integral constant. The 4-parameter Pareto distribution is L-shaped if $a \in (0, 1]$, is bell shaped if $a > 1$, and is positively skewed to the right.

Figure 6.6a plots the $g(x,F)$ function, Figure 6.6b plots the CDF, and Figure 6.6c plots the PDF of the 4-parameter Pareto distribution.

6.4 Relation between Dagum and Stoppa Systems

Both Dagum and Stoppa systems are derived based on the elasticity of the distribution function. The similarity and dissimilarity of these two systems are as following:

(1) There is a slight difference in how the elasticity is being defined. In Dagum system it is defined as the change of $\ln(F(x) - a)$ versus $\ln(x)$ where $a < 1$ [i.e., Equation 5.2 in Chapter 5], while in Stoppa system, it is directly defined as the change of $\ln(F(x))$ versus $\ln(x)$ as shown in Equation (6.2).

(2) The elasticity of the distribution function in both systems are decreasing functions that are formulated as $g(x)g(F)$ for the Dagum system and $\left(\frac{1-F^{\frac{1}{a}}}{F^{\frac{1}{a}}}\right)g(x,F)$ for the Stoppa system.

(3) Most of the distributions in Dagum system may be obtained from the general elasticity function [i.e., Equation 6.2] of the Stoppa system. Table 6.1 lists the distributions in Dagum system that can also be derived from Stoppa system.

Table 6.1 *Distributions of Dagum system that can be derived from Stoppa system*

	Dagum System			Stoppa System	
	a	$g^d(x)$	$G(F)$	$\frac{1-F^{\frac{1}{\alpha}}}{F^{\frac{1}{\alpha}}}$	$g^s(x,F) or\ g^s(x)$
Pareto I	$a=0$	c	$\frac{1-F}{F}$	$\alpha=1$	$g^d(x)$
Pareto II	$a=0$	$\frac{cx}{x-d}$	$\frac{1-F}{F}$	$\alpha=1$	$g^d(x)$
Pareto III	$a=0$	$bx+\frac{cx}{x-d}$	$\frac{1-F}{F}$	$\alpha=1$	$g^d(x)$
Benini	$a=0$	$2c \ln x$	$\frac{1-F}{F}$	$\alpha=1$	$g^d(x)$
Weibull	$a=0$	$bx(x-d)^{c-1}$	$\frac{1-F}{F}$	$\alpha=1$	$g^d(x)$
Fisk	$a=0$	c	$1-F$	$\alpha=1$	$F\cdot g^d(x)$
Singh-Maddala	$a=0$	c	$1-\frac{(1-F)^b}{F(1-F)^{-1}}$	$\alpha=1$	$\left(\frac{F-(1-F)^{b+1}}{1-F}\right)g^d(x)$
Dagum I	$a=0$	c	$1-F^{\frac{1}{b}}$	$\alpha=b$	$F^{\frac{1}{b}}\cdot g^d(x)$

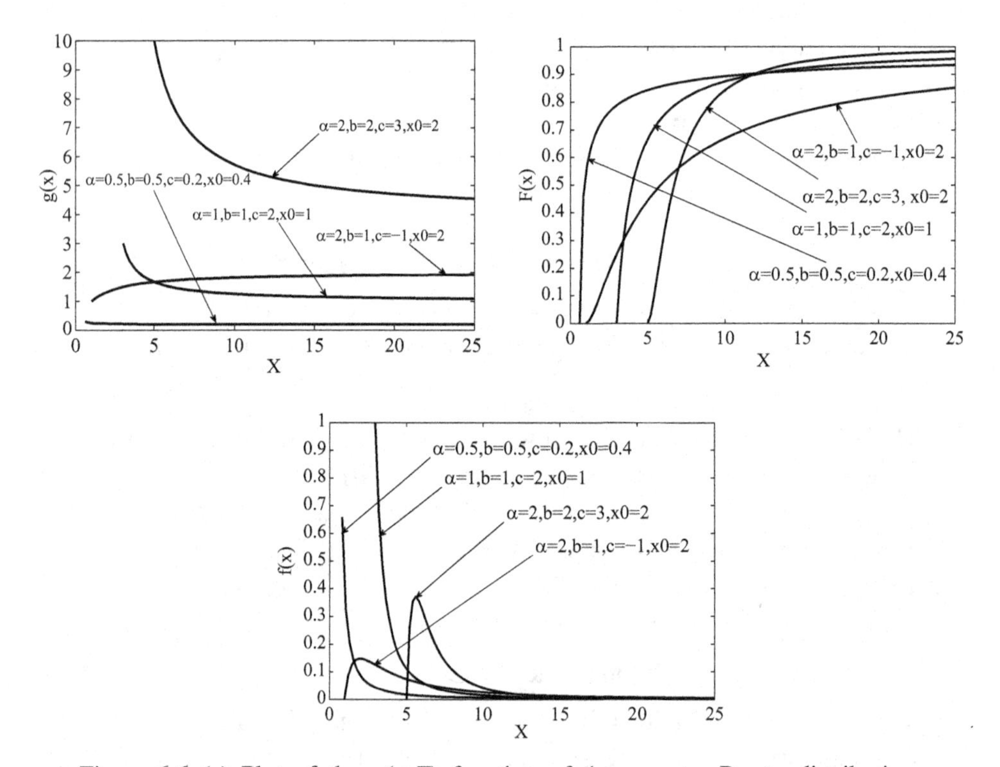

Figure 6.6 (a) Plot of the g(x, F) function of 4-parameter Pareto distribution; (b) plot of the CDF of 4-parameter Pareto distribution; and (c) plot of the PDF of 4-parameter Pareto distribution.

6.5 Relations among Burr distributions and Dagum and Stoppa Systems

As discussed in Chapter 3, the Burr distributions are derived from the differential equation:

$$\frac{dF}{dx} = F(x)(1 - F(x))g(x, F(x)) \tag{6.27}$$

Equation (6.27) indicates the change of distribution function F versus x in the real domain. To associate Burr distributions with Dagum and Stoppa system, Equation (6.28) may be rewritten as:

$$\begin{aligned} &\frac{d \ln F}{d \ln x} = x(1 - F(x))g(x, F(x)) = \frac{1 - F(x)}{F(x)} g^*(x, F(x)); \\ &g^*(x, F(x)) = xF(x)g(x, F(x)) \end{aligned} \tag{6.28}$$

In Equation (6.25) $g^*(x, F(x)) > 0 f$ or $0 \le F(x) \le 1$ with the support of $x > 0$. Then, with the constraints for the support of x, some Burr distributions may be also derived directly from Dagum system with $a = 0$ [i.e., equivalent to Stoppa system], including Burr I, III, IV, X, XI, and XII distributions as shown in Table 6.2.

In what follows, we will show two examples (one from Burr system and one from Dagum system) to illustrate this relationship. First, we will show how to derive Burr XII distribution from Stoppa system. And then we will show how to derive Singh-Maddala distribution from Stoppa system.

Table 6.2 *Burr distributions that can be derived from Dagum/Stoppa system*

	Dagum System (a = 0)/Stoppa System (α = 1)
	$G^*(x, F)$
Burr I	$\frac{F}{1-x}$
Burr III	$\left(\frac{F}{1-F}\right)\left(\frac{kr}{1+x^k}\right)$
Burr IV	$\frac{r\left(\frac{c-x}{x}\right)^{\frac{1}{c}-1}}{x(1-F)F^{-\frac{1}{r}-1}}$
Burr X	$\frac{2re^{-x^2}}{(1-F)F^{\frac{1}{r}-1}}$
Burr XII	$\frac{ckx^c}{1+x^c}$

Burr XII Distribution

Let $g^*(x, F(x)) = \frac{ckx^c}{1+x^c}$, Equation (6.28) may be expressed as:

$$
\begin{aligned}
\frac{d\ln F}{d\ln x} = \frac{ckx^c}{1+x^c}\frac{1-F(x)}{F(x)} &\Rightarrow \frac{1}{(1-F(x))}dF = \frac{ckx^{c-1}}{1+x^c}dx \\
&\Rightarrow \ln(1-F) = -\int_0^x k\, d\ln(1+x^c) \\
&\Rightarrow F = 1-(1+x^c)^{-k}
\end{aligned}
\tag{6.29}
$$

Singh-Maddala Distribution

In case of Singh-Maddala distribution, it may be directly derived from the Stoppa system with the use of $g(x, F(x)) = 1-(1-F)^b$, and the corresponding elasticity function is expressed as:

$$
\eta(x, F(x)) = \frac{d\ln F}{d\ln x} = c\left(1-(1-F)^b\right)\frac{1-F}{F} \tag{6.30}
$$

Integration of Equation (6.30) yields

$$
\begin{aligned}
\frac{1}{(1-F)\left(1-(1-F)^b\right)}dF = c\,d\ln x \Rightarrow \left(\frac{1}{1-F}+\frac{(1-F)^{b-1}}{1-(1-F)^b}\right)dF = c\,d\ln x \\
\Rightarrow F = 1-\left(1+Cx^{bc}\right)^{-\left(\frac{1}{b}\right)}; \\
b, c > 0, x \in [0, \infty)
\end{aligned}
\tag{6.31}
$$

Similar to the discussion in Chapter 5, we can set the integration constant C as 1; and Equation (6.31) may be rewritten as:

$$
F(x) = 1-\left(1+x^{bc}\right)^{-\frac{1}{b}} \tag{6.32}
$$

6.6 Application

In this section, the Stoppa II, III, and V and generalized 4-parameter Pareto distributions are applied for analysis. The maximum likelihood estimation is applied for parameter estimation. In what follows, the likelihood function of each distribution candidate is briefly discussed.

Stoppa II Distribution (Generalized Pareto II Distribution-3 Parameter)

$$
\ln L = n\ln b + n\frac{b}{\alpha}\ln x_0 - \left(\frac{b}{\alpha}+1\right)\sum_{i=1}^{n}\ln x_i + (\alpha-1)\sum_{i=1}^{n}\ln\left(1-\left(\frac{x_0}{x_i}\right)^{\frac{b}{\alpha}}\right) \tag{6.33}
$$

Maximizing the log-likelihood function [Equation 6.33], the parameters may then be estimated.

Stoppa III Distribution (Generalized Exponential Distribution)

$$\ln L = n \ln b - \frac{b}{\alpha} \sum_{i=1}^{n} x_i + (\alpha - 1) \sum_{i=1}^{n} \ln\left(1 - \exp\left(-\frac{b}{\alpha} x_i\right)\right) \quad (6.34)$$

Maximizing Equation (6.34), the parameters may then be estimated.

Stoppa V Distribution

$$\ln L = n \ln b - \frac{b}{\alpha} \sum_{i=1}^{n} \tan(x_i) + (\alpha - 1) \sum_{i=1}^{n} \ln\left(1 - \exp\left(-\frac{b \tan(x_i)}{\alpha}\right)\right) + 2 \sum_{i=1}^{n} \sec(x_i) \quad (6.35)$$

Maximizing Equation (6.35), the parameters may then be estimated.

Four-Parameter Generalized Pareto Distribution

$$\ln L = n \ln \alpha + n \ln b + nb \ln x_0 + (\alpha - 1) \ln\left(1 - \left(\frac{x_0}{x_i - c}\right)^{b}\right) - (b + 1) \ln(x_i - c) \quad (6.36)$$

Maximizing Equation (6.36), the parameters may then be estimated.

6.6.1 Monthly Suspended Sediment

The Stoppa II, III (generalized exponential), and V and generalized 4-parameter Pareto distributions are applied to the monthly sediment data for the month of May at the gaging station USGS4208000. Table 6.3 lists the estimated parameters for the preceding four distributions using the maximum likelihood estimation method. Figure 6.7 compares the fitted frequency distributions with the empirical frequency distribution. It is seen from the figure that the Stoppa III (generalized exponential) and the generalized 4-parameter Pareto distributions may be applied for frequency analysis of monthly sediment data.

Table 6.3 *Parameters estimated for the monthly suspended sediment*

Parameters	Stoppa II	Stoppa III (generalized exponential)	Stoppa V	4-Parameter Pareto
α	13.89	2.03	0.35	24.94
b	15.93	0.029	0.18	2.08
c				–19.34
x_0	6.1			18.25

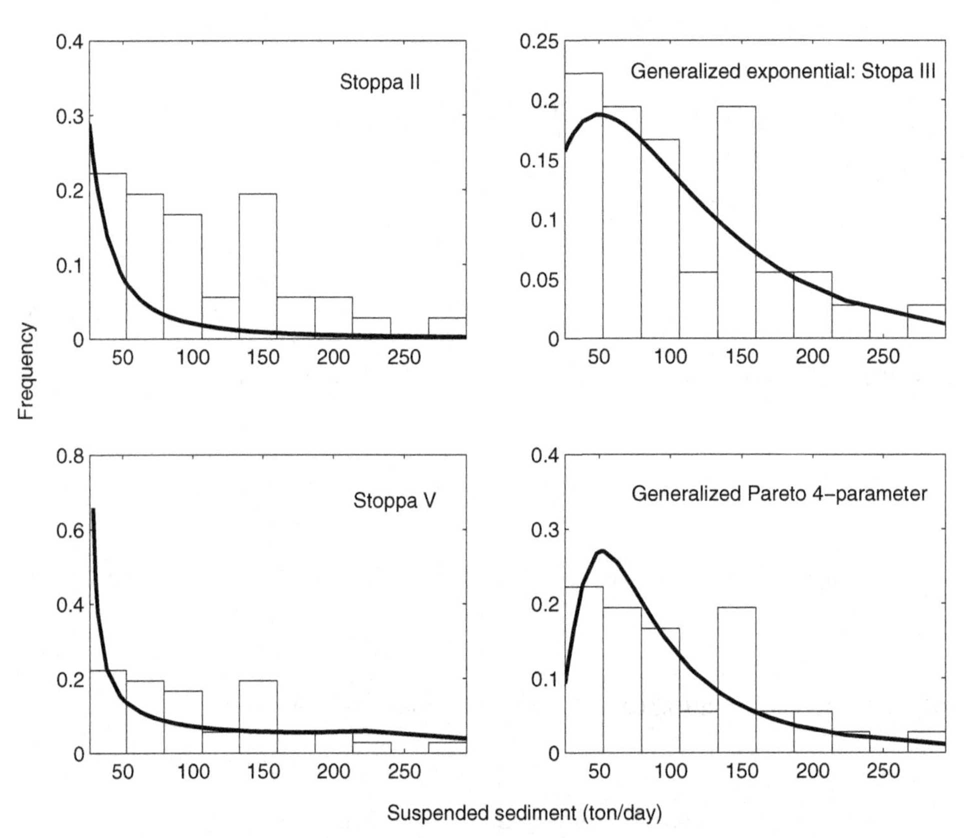

Figure 6.7 Comparison of fitted frequency distribution with the empirical frequency distribution: monthly suspended sediment.

6.6.2 Annual Rainfall Amount

The Stoppa III and 4-parameter generalized Pareto distributions are applied to evaluate the frequency distribution of the annual rainfall amount provided in the appendix. Table 6.4 lists the parameters estimated for the annual rainfall distribution. Figure 6.8 compares the fitted frequency distributions with the empirical

Table 6.4 *Parameters estimated for annual rainfall*

	Parameters			
Distributions	a	b	c	x_0
Stoppa III	3.98	0.008		
4-parameter Pareto	148.42	3.80	178.83	184.96

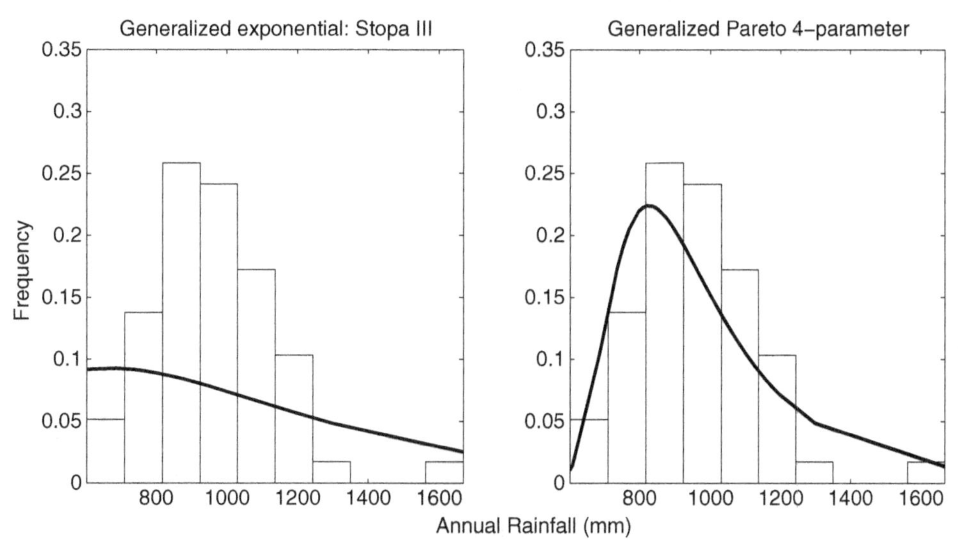

Figure 6.8 Comparison of the fitted frequency distributions with the empirical frequency distribution: annual rainfall.

frequency distribution. Figure 6.8 shows that the 4-parameter generalized Pareto distribution may be applied to model the annual rainfall distribution.

6.6.3 Peak Flow

The Stoppa III and 4-parameter generalized Pareto distributions are applied to evaluate the frequency distribution of peak flow provided in the appendix. Using the maximum likelihood estimation method with the use of genetic algorithm, Table 6.5 lists the parameters estimated for the peak flow. Figure 6.9 compares the fitted frequency with empirical frequency. Comparison in Figure 6.8 indicates that both Stoppa III and the 4-parameter generalized Pareto distributions may be applied to model the peak flow. Furthermore, 4-parameter generalized Pareto distribution outperforms the Stoppa III distribution.

Table 6.5 *Parameters estimated for peak flow*

	Parameters			
Distributions	α	b	c	x_0
Stoppa III	4.90	0.038		
4-parameter Pareto	50.64	2.56	22.35	42.95

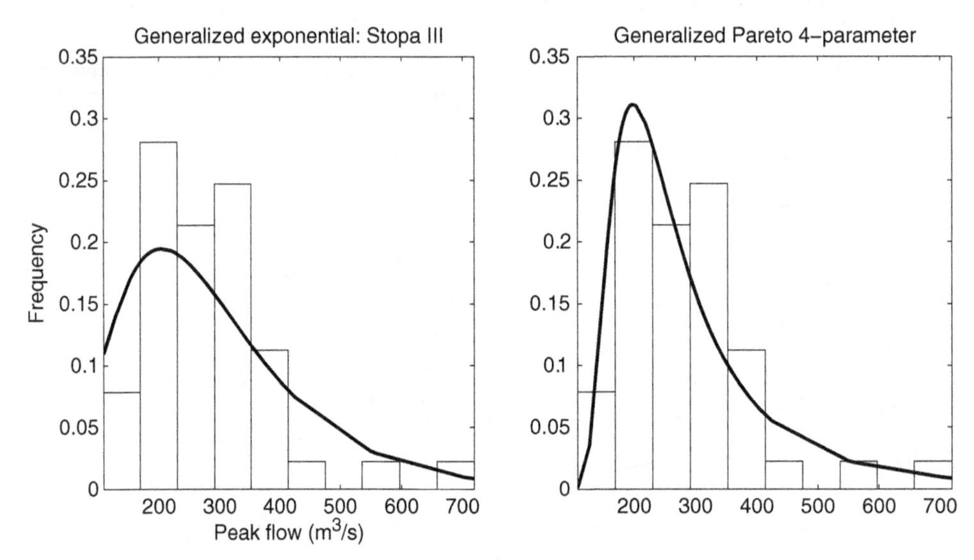

Figure 6.9 Comparison of the fitted frequency distributions with the empirical frequency distribution: peak flow.

6.6.4 Maximum Daily Precipitation

The Stoppa III and 4-parameter generalized Pareto distributions are again applied to evaluate the frequency of maximum daily precipitation provided in the appendix. Using the maximum likelihood estimation method with the use of genetic algorithm, Table 6.6 lists parameters estimated. Figure 6.10 compares the fitted frequency distributions with the empirical frequency distribution. Comparison shows that the 4-parameter generalized Pareto distribution outperforms the Stoppa III distribution, though both Stoppa III and 4-parameter generalized Pareto distributions may be applied to model maximum daily precipitation distribution.

6.6.5 Drought (Total Flow Deficit)

In case of drought application, Stoppa III and 4-parameter generalized Pareto distributions are again applied to evaluate the total flow deficit. The maximum

Table 6.6 *Parameters estimated for maximum daily precipitation*

Distributions	Parameters α	b	c	x_0
Stoppa III	8.76	0.42		
4-parameter Pareto	16.65	2.87	7.93	13.87

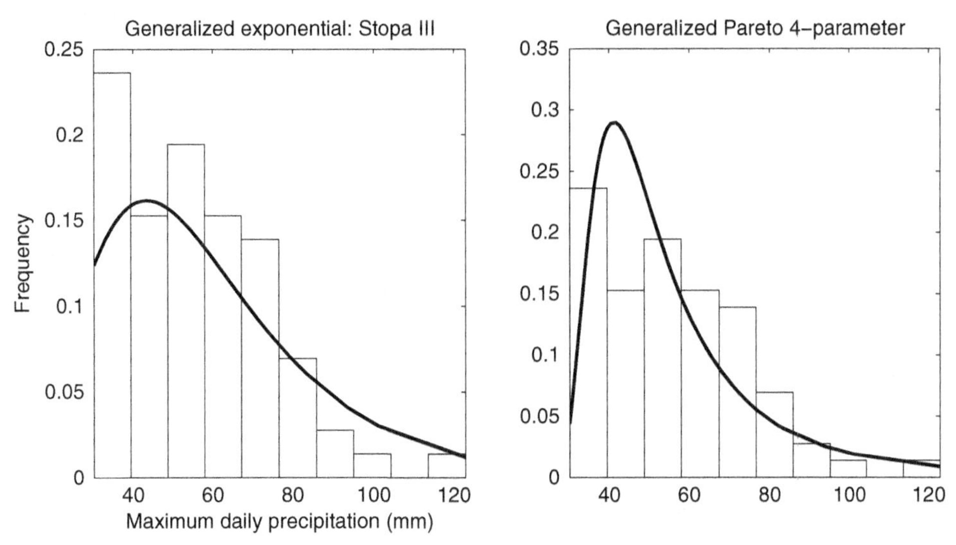

Figure 6.10 Comparison of the fitted frequency distributions with the empirical frequency distribution: maximum daily precipitation.

likelihood estimation method with the use of genetic algorithm is applied. Table 6.7 lists the parameters estimated. Figure 6.11 compares the fitted frequency distributions with the empirical frequency distribution. Comparison shows that the Stoppa III and 4-parameter generalized Pareto distributions yield similar performances. Both distributions may be applied to model the total flow deficit distribution.

6.7 Conclusion

The Stoppa system is a generalized system of distributions. However, the Stoppa system is closely related to the Dagum system. It is also found that several Burr distributions can also be derived from the Stoppa or Dagum system. Thus, there is a close association between these three systems of distributions. Additionally, based on applications, the 4-parameter generalized Pareto distribution may be applied to model frequency distributions of all the datasets.

Table 6.7 *Parameters estimated for total flow deficit*

	Parameters			
Distributions	α	b	c	x_0
Stoppa III	3.39	0.0002		
4-parameter Pareto	18.53	0.46	15.94	26.28

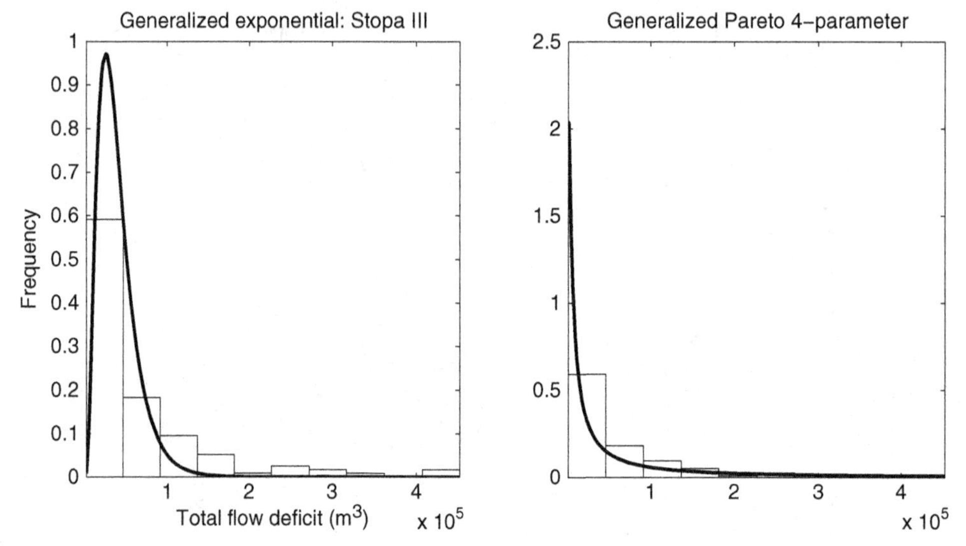

Figure 6.11 Comparison of the fitted frequency distributions with the empirical frequency distribution: total flow deficit.

References

Kleiber, C., and Kotz, S. (2003). *Statistical Size Distributions in Economics and Actuarial Sciences*. Hoboken, NJ: Wiley.

Stoppa, G. (1990). A new generating system of income distribution models. *Quarderni di Statistica e Mathematica Applicata alle Science Economico-Sociali* 12, pp. 47–55.

Stoppa, G. Una tavola per modeli di probabilita. *Metron* 51, pp. 99–117.

7

Esteban System of Frequency Distributions

7.1 Introduction

By defining elasticity in terms of probability density function (PDF), Esteban (1981) derived three generalized probability distributions that were generalized gamma distribution, generalized beta distribution of first kind, and generalized beta distribution of second kind. These generalized distributions include a wide spectrum of frequency distributions used in hydrologic, hydraulic, environmental, and water resources engineering, such as exponential, lognormal, gamma, Weibull, Burr XII, and Pareto distributions. In another study, Esteban (1986) formulated three hypotheses and then derived another set of distributions. He showed that Pareto, gamma, and normal PDFs corresponded to constant, linear, and quadratic elasticities, respectively. The objective of this chapter therefore is to discuss the Esteban system and the derivation of the resulting distributions.

7.2 Esteban System of Distributions

Esteban (1981, 1986) derived the income share elasticity that has a one-to-one relation to the PDF as:

$$\pi(x) = \lim_{h\to 0}\frac{d\ln\left(\frac{1}{\mu}\int_{x}^{x+h} xf(x)dx\right)}{d\ln x} = 1+\frac{xf'(x)}{f(x)} = 1+\frac{x}{f(x)}\frac{df(x)}{dx} \tag{7.1a}$$

in which $\pi(x)$ is the elasticity function of the distribution of X. Furthermore, according to Kleiber and Kotz (2003), the hypotheses that $\pi(x)$ represents the elasticity of given PDFs are:

(i) To satisfy the weak Pareto law:

$$\lim_{x\to\infty} \pi(x) = -\theta;\ \theta > 0 \tag{7.1b}$$

(ii) At least one interior mode exists;

(iii) $\pi(x)$ shows either as a constant or with a constant rate of decline (i.e., $\pi(x)$ is a decreasing function):

$$\pi'(x) = 0 \; or \frac{d\ln(\pi(x))}{d\ln x} = -(1+\epsilon); \epsilon > -1 \tag{7.2}$$

Rearranging Equation (7.1a), we obtain:

$$d\ln f(x) = -\frac{1-\pi(x)}{x}dx \tag{7.3}$$

Integration of Equation (7.3) yields

$$f(x) = C\exp\left\{\int -\frac{1-\pi(x)}{x}dx\right\}; x \in [a_1, a_2] \tag{7.4a}$$

In Equation (7.4a), C is the integration constant obtained by applying the total probability:

$$C = \frac{1}{\exp\left\{\int_{a_1}^{a_2} -\frac{1-\pi(x)}{x}dx\right\}} \tag{7.4b}$$

Depending on the formulation of $\pi(x)$ in terms of x, different distribution functions can be derived.

7.2.1 Three-Parameter Gamma Distribution

If the elasticity of the PDF is given as:

$$\pi(x) = \alpha k - a\left(\frac{x}{b}\right)^a; a, k, b > 0, x \in (0, \infty) \tag{7.5}$$

then substituting Equation (7.5) into Equation (7.3), we obtain:

$$d\ln f(x) = -\frac{1 - ak + a\left(\frac{x}{b}\right)^a}{x}dx = \left[\frac{ak-1}{x} - \frac{a}{b}\left(\frac{x}{b}\right)^{a-1}\right]dx; a, k, b > 0, x \in (0, \infty) \tag{7.6}$$

The general solution of Equation (7.5) can be obtained:

$$f(x) = C\exp\left(\int\left[\frac{ak-1}{x} - \frac{a}{b}\left(\frac{x}{b}\right)^{a-1}\right]dx\right) = C\exp\left((ak-1)\ln x - \left(\frac{x}{b}\right)^a\right)$$
$$\Rightarrow f(x) = Cx^{ak-1}\exp\left(-\left(\frac{x}{b}\right)^a\right) \tag{7.7a}$$

Applying the total probability, the integration constant C can be evaluated as:

$$C = \frac{1}{\displaystyle\int_0^\infty x^{ak-1} \exp\left(-\left(\tfrac{x}{b}\right)^a\right) dx} \tag{7.7b}$$

Let $\left(\frac{x}{b}\right)^a = t$, we have: $x = bt^{\frac{1}{a}}, dt = \frac{ax^{a-1}}{b^a} dx \Rightarrow dx = \frac{b}{a} t^{-\frac{a-1}{a}} dt$, $x^{ak-1} = b^{ak-1} t^{\frac{ak-1}{a}}$. Equation (7.7b) may then be solved as:

$$C = \frac{1}{\displaystyle\int_0^\infty \frac{b^{ak}}{a} t^{k-1} \exp(-t) dt} = \frac{a}{b^{ak}\Gamma(\mathrm{k})} \tag{7.7c}$$

Substituting Equation (7.7c) into Equation (7.7a), we obtain the PDF of 3-parameter generalized gamma distribution as:

$$f(x) = \frac{a}{b^{ak}\Gamma(k)} x^{ak-1} \exp\left(-\left(\frac{x}{b}\right)^a\right) \tag{7.8}$$

In Equation (7.8), b is the scale parameter, and a, b are the shape parameters. Figure 7.1a plots the elasticity and Figure 7.1b plots the PDF of 3-parameter generalized gamma distributions with different parameter values. The PDF of the 3-parameter generalized gamma distribution may be L-shaped or bell-shaped skewed.

7.2.2 Special Cases of Generalized Gamma Distribution

The generalized gamma distribution leads to the lognormal, Weibull, gamma, exponential, normal, and Pareto distributions as special or limiting cases. These distributions are derived here.

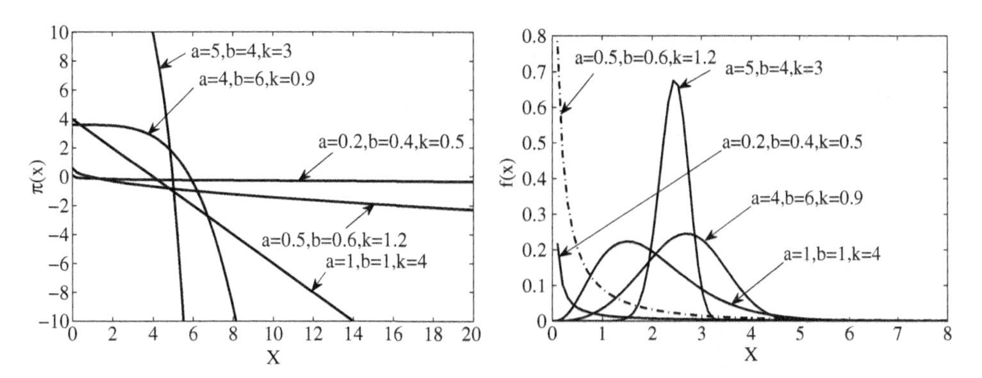

Figure 7.1 (a) Plot of the π(x) function of 3-parameter generalized gamma distribution; and (b) plot of the PDF of 3-parameter generalized gamma distribution.

Exponential Distribution: The exponential distribution may be derived from the 3-parameter generalized gamma distribution when $a = k = 1$ in Equation (7.5). The elasticity of the exponential distribution can be written as:

$$\pi(x) = 1 - \frac{x}{b}; b > 0, x \in (0, \infty) \tag{7.9}$$

Substitution of Equation (7.9) in Equation (7.3) yields

$$d \ln f(x) = -\frac{\frac{x}{b}}{x} dx; b > 0, x \in (0, \infty) \tag{7.10}$$

Equation (7.10) can be solved as

$$f(x) = C \exp\left(\int -\frac{\frac{x}{b}}{x} dx\right) = C \exp\left(-\frac{1}{b}x\right) \tag{7.11a}$$

Applying total probability to Equation (7.11a), the integration constant C can be solved as:

$$\int_0^{\infty} f(x)dx = 1 \Rightarrow C\int_0^{\infty} \exp\left(-\frac{1}{b}x\right) dx = 1 \Rightarrow C = \frac{1}{\int_0^{\infty} \exp\left(-\frac{1}{b}x\right) dx} = \frac{1}{b} \tag{7.11b}$$

Substituting Equation (7.11b) into Equation (7.11a), the PDF can then be given as:

$$f(x) = \frac{1}{b} \exp\left(-\frac{x}{b}\right); x \in (0, \infty) \tag{7.11c}$$

Equation (7.11) is the exponential distribution.

Gamma Distribution: The 3-parameter generalized gamma distribution is the gamma distribution when $a = 1$ in Equation (7.5). The elasticity of gamma distribution can be written as:

$$\pi(x) = k - \frac{x}{b}; k > 0, b > 0, x \in (0, \infty) \tag{7.12}$$

Substituting Equation (7.12) into Equation (7.3) we have:

$$d \ln f(x) = -\frac{1 - k + \frac{x}{b}}{x} dx; k, b > 0, x \in (0, \infty) \tag{7.13}$$

Equation (7.13) can be solved as:

$$f(x) = C\exp\left(\int -\frac{1-k+\frac{x}{b}}{x}dx\right) = C\exp\left((k-1)\ln x - \frac{x}{b}\right) = Cx^{k-1}\exp\left(-\frac{x}{b}\right) \tag{7.14a}$$

Applying total probability, the integration constant C in Equation (7.14a) can be solved for as:

$$\int_0^{\infty} f(x)dx = 1 \Rightarrow C\int_0^{\infty} x^{k-1}\exp\left(-\frac{x}{b}\right)dx = 1$$
$$\Rightarrow C = \frac{1}{\int_0^{\infty} x^{k-1}\exp\left(-\frac{x}{b}\right)dx} = \frac{1}{b^k\int_0^{\infty}\left(\frac{x}{b}\right)^{k-1}\exp\left(-\frac{x}{b}\right)d\left(\frac{x}{b}\right)} = \frac{1}{b^k\Gamma(k)} \tag{7.14b}$$

Substitution of Equation (7.14b) into Equation (7.14a) yield the gamma distribution as:

$$f(x) = \frac{1}{b^k\Gamma(k)}x^{k-1}\exp\left(-\frac{x}{b}\right); k > 0, b > 0, x \in (0,\infty) \tag{7.14c}$$

Weibull Distribution: The Weibull distribution can be deduced from 3-parameter generalized gamma distribution if $k = 1$. The elasticity function of Weibull distribution is given as:

$$\pi(x) = a - a\left(\frac{x}{b}\right)^a; a, b > 0, x \in (0,\infty) \tag{7.15}$$

Substitution of Equation (7.15) into Equation (7.3) yields:

$$d\ln f(x) = -\frac{1-a+a\left(\frac{x}{b}\right)^a}{x}dx; a, b > 0, x \in (0,\infty) \tag{7.16}$$

Equation (7.16) can be solved as:

$$f(x) = C\exp\left(\int -\frac{1-a+a\left(\frac{x}{b}\right)^a}{x}dx\right) = C\exp\left((a-1)\ln x - \left(\frac{x}{b}\right)^a\right)$$
$$= Cx^{a-1}\exp\left(-\left(\frac{x}{b}\right)^a\right) \tag{7.17a}$$

Applying the total probability to Equation (7.17a), the integration constant C can be solved for as:

$$\int_0^\infty f(x)dx = 1 \Rightarrow C\int_0^\infty x^{a-1}\exp\left(-\left(\frac{x}{b}\right)^a\right)dx = 1$$

$$\Rightarrow C\frac{b^a}{a}\int_0^\infty \exp\left(-\left(\frac{x}{b}\right)^a\right)d\left(\frac{x}{b}\right)^a = 1 \Rightarrow C = \frac{a}{b^a\int_0^\infty \exp\left(-\left(\frac{x}{b}\right)^a\right)d\left(\frac{x}{b}\right)^a} = \frac{a}{b^a} \tag{7.17b}$$

Substitution of Equation (7.17b) into Equation (7.17a) yields the density function of Weibull distribution as:

$$f(x) = \frac{a}{b^a}x^{a-1}\exp\left(-\left(\frac{x}{b}\right)^a\right); a, b > 0, x \in (0, \infty) \tag{7.17c}$$

In Equations (7.17a)–(7.17c), parameter a is the shape parameter and b is the scale parameter. The PDF of Weibull distribution is L-shaped if the shape parameter $a \leq 1$, and is bell shaped if the shape parameter $a > 1$. The Weibull distribution is skewed. Figure 7.2a plots the elasticity and Figure 7.2b plots PDF of the Weibull distribution with different parameter values.

Pareto Distribution: The Pareto PDF has a constant elasticity (i.e., $\pi'(x) = 0$) and can be expressed as:

$$\pi(x) = -a;\ a > 0 \tag{7.18}$$

Substituting Equation (7.18) into Equation (7.3), we obtain

$$d\ln f(x) = -\frac{1+a}{x}dx; a > 0, x \geq b > 0 \tag{7.19}$$

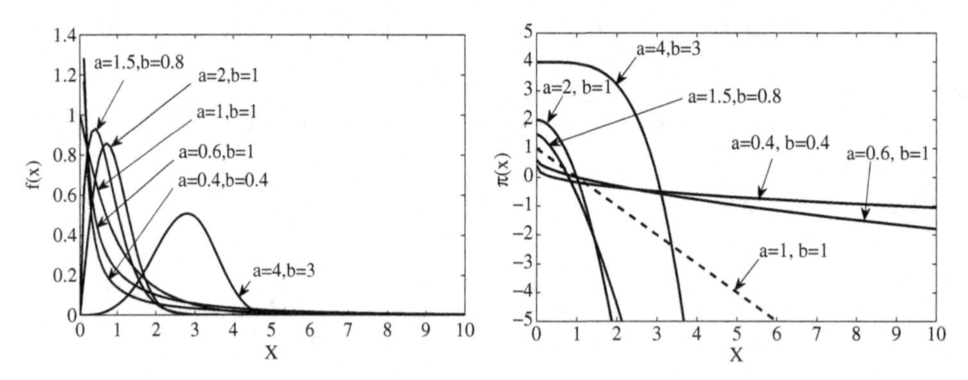

Figure 7.2 (a) Plot of the π(x) function of Weibull distribution; (b) and plot of the PDF of Weibull distribution.

Integration of Equation (7.19) yields

$$f(x) = C\exp\left(\int -\frac{1+a}{x}dx\right) = C\exp\left(-(a+1)\ln x\right) = Cx^{-(a+1)} \tag{7.20a}$$

Applying the total probability, the integration constant C in Equation (7.20a) can be solved for as:

$$\begin{aligned} \int_b^\infty f(x)dx = 1 \Rightarrow C\int_b^\infty x^{-(a+1)}dx = 1 \\ \Rightarrow C = \frac{1}{\int_b^\infty x^{-(a+1)}dx} = ab^a \end{aligned} \tag{7.20b}$$

Substituting Equation (7.20b) into Equation (7.20a), the PDF of Pareto distribution is given as:

$$f(x) = \frac{ab^a}{x^{a+1}}; a, b > 0, x \in [b, \infty) \tag{7.20c}$$

Figure 7.3a plots the elasticity and Figure 7.3b plots the PDF of Pareto distribution with different parameters. The Pareto distribution is L-shaped and skewed.

Lognormal Distribution: The elasticity for the lognormal distribution is given as:

$$\pi(x) = \frac{\mu}{\sigma^2} - \frac{\ln(x)}{\sigma^2}; x > 0, \sigma > 0 \tag{7.21}$$

Substituting Equation (7.21) into Equation (7.3), we obtain:

$$d\ln f(x) = -\frac{1 - \left(\frac{\mu}{\sigma^2} - \frac{\ln(x)}{\sigma^2}\right)}{x}dx; x > 0, \sigma > 0 \tag{7.22}$$

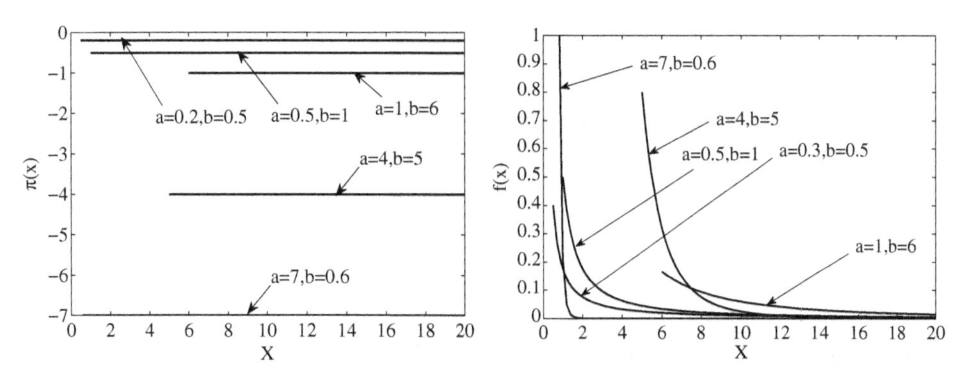

Figure 7.3 (a) Plot of the π(x) function of Pareto distribution; and (b) plot of the PDF of Pareto distribution.

The integration of Equation (7.22) yields

$$f(x) = C\exp\left(\int\left(-\frac{1}{x} - \frac{\ln x - \mu}{x\sigma^2}\right)dx\right) = C\exp\left(-\ln x - \frac{(\ln x - \mu)^2}{2\sigma^2}\right)$$
$$= \frac{C}{x}\exp\left(-\frac{(\ln x - \mu)^2}{2\sigma^2}\right) \tag{7.23a}$$

Applying the total probability, the integration constant C in Equation (7.23a) can be evaluated using the Gaussian integral as:

$$\int_0^{\infty} f(x) = 1 \Rightarrow \int_0^{\infty} \frac{C}{x}\exp\left(-\frac{(\ln x - \mu)^2}{2\sigma^2}\right)dx = 1$$
$$\Rightarrow C = \frac{1}{\int_0^{\infty} \frac{1}{x}\exp\frac{-(\ln x - \mu)^2}{2\sigma^2}dx} = \frac{1}{\int_0^{\infty} \exp\left(-\frac{(\ln x - \mu)^2}{2\sigma^2}\right)d\ln x}$$
$$= \frac{1}{\sqrt{2\sigma^2}\int_0^{\infty} \exp\left(-\left(\frac{\ln x - \mu}{\sqrt{2\sigma^2}}\right)^2\right)d\left(\frac{\ln x - \mu}{\sqrt{2\sigma^2}}\right)} \tag{7.23b}$$
$$= \frac{1}{\sqrt{2\sigma^2}\int_{-\infty}^{+\infty} \exp\left(-t^2\right)dt} = \frac{1}{\sqrt{2\pi\sigma^2}}$$

Substitution of Equation (7.23b) into Equation (7.23a) yields the PDF of lognormal distribution as:

$$f(x) = \frac{1}{x\sqrt{2\pi\sigma^2}}\exp\left(-\frac{(\ln x - \mu)^2}{2\sigma^2}\right) \tag{7.23c}$$

It is well known that the lognormal distribution is bell shaped and skewed. Figure 7.4 plots the elasticity for lognormal distribution.

Normal Distribution: The elasticity for the normal distribution is given as:

$$\pi(x) = 1 - \frac{x(x - \mu)}{\sigma^2}; x, \mu \in (-\infty, \infty), \sigma > 0 \tag{7.24}$$

Substituting Equation (7.24) into Equation (7.3), we obtain:

$$d\ln f(x) = -\frac{x - \mu}{\sigma^2}; x, \mu \in (-\infty, \infty), \sigma > 0 \tag{7.25}$$

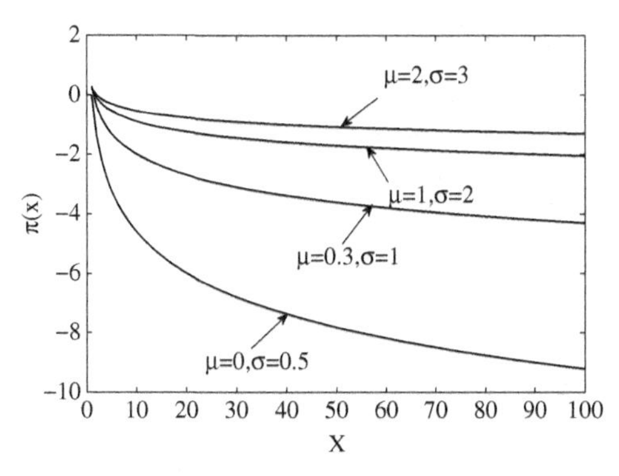

Figure 7.4 Plot of the π(x) function for lognormal distribution.

The integration of Equation (7.25) yields

$$f(x) = C\exp\left(\int -\frac{x-\mu}{\sigma^2}dx\right) = C\exp\left(-\frac{(x-\mu)^2}{2\sigma^2}\right) \tag{7.26a}$$

Applying the total probability, the integration constant C in Equation (7.26a) can be evaluated using the Gaussian integral as:

$$C = \frac{1}{\int_{-\infty}^{\infty}\exp\left(-\frac{(x-\mu)^2}{2\sigma^2}\right)dx} = \frac{1}{\sqrt{2\pi\sigma^2}} \tag{7.26b}$$

Finally, the PDF of normal distribution is given as:

$$f(x) = \frac{1}{\sqrt{2\pi\sigma^2}}\exp\left(-\frac{(x-\mu)^2}{2\sigma^2}\right); x, \mu \in (-\infty, \infty), \sigma > 0 \tag{7.26c}$$

It is well known that the normal distribution is symmetric bell-shaped about $x = \mu$. The skewness of normal distribution is 0. Figure 7.5 plots the elasticity for the normal distribution.

7.2.3 Generalized Beta Distribution of First Kind

This distribution was discussed by Thurow (1970). The elasticity for the 4-parameter generalized beta distribution of first kind is given as:

$$\pi(x) = a(p+q-1) - \frac{a(q-1)}{1-\left(\frac{x}{b}\right)^a} \tag{7.27}$$

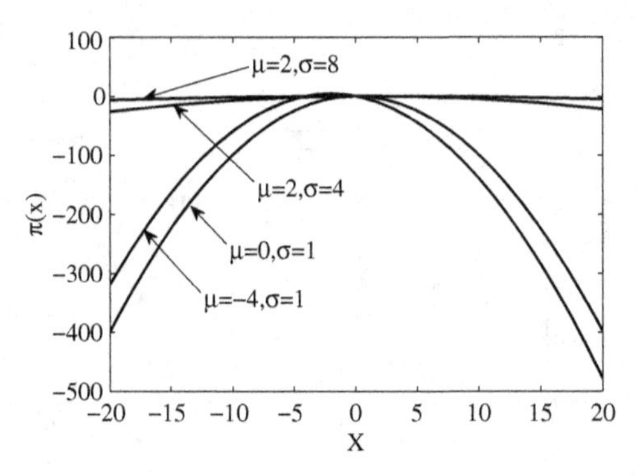

Figure 7.5 Plot of the π(x) function of normal distribution.

Substituting Equation (7.27) into Equation (7.3), we obtain:

$$\begin{aligned}
d\ln f(x) &= -\frac{1 - a(p+q-1) + \dfrac{a(q-1)}{1-\left(\frac{x}{b}\right)^a}}{x}dx \\
&= -\left(\frac{1}{x} - \frac{a(p+q-1)}{x} + \left(\frac{a(q-1)}{x} + \frac{a(q-1)}{b}\frac{\left(\frac{x}{b}\right)^{a-1}}{1-\left(\frac{x}{b}\right)^a}\right)\right)dx \\
&= -\left(\frac{1-ap}{x} + \frac{a(q-1)}{b}\frac{\left(\frac{x}{b}\right)^{a-1}}{1-\left(\frac{x}{b}\right)^a}\right)dx
\end{aligned} \tag{7.28}$$

Integration of Equation (7.28) yields

$$\begin{aligned}
f(x) &= C\exp\left(\int\left(-\frac{1-ap}{x} - \frac{a(q-1)}{b}\frac{\left(\frac{x}{b}\right)^{a-1}}{1-\left(\frac{x}{b}\right)^a}\right)dx\right) \\
&= C\exp\left((ap-1)\ln x + (q-1)\int\frac{1}{1-\left(\frac{x}{b}\right)^a}d\left(1-\left(\frac{x}{b}\right)^a\right)\right) \\
&= C\exp\left((ap-1)\ln x + (q-1)\ln\left(1-\left(\tfrac{x}{b}\right)^a\right)\right) \\
&= Cx^{ap-1}\left(1-\left(\tfrac{x}{b}\right)^a\right)^{q-1}
\end{aligned} \tag{7.29}$$

Applying the total probability, the integration constant C in Equation (7.29) can be evaluated in accordance of $a > 0$ and $a < 0$, respectively, as:

(i) $a > 0$

For $a > 0$, the variable x is bounded as $x \in [0, b]$ where b is the upper bound; and

$$\int_0^b f(x)dx = 1 \Rightarrow C\int_0^b x^{ap-1}\left(1-\left(\frac{x}{b}\right)^a\right)^{q-1}dx = 1$$

$$\Rightarrow C = \frac{1}{\int_0^b x^{ap-1}\left(1-\left(\frac{x}{b}\right)^a\right)^{q-1}dx} = \frac{1}{b^{ap}\int_0^b \left(\frac{x}{b}\right)^{ap-1}\left(1-\left(\frac{x}{b}\right)^a\right)^{q-1}d\left(\frac{x}{b}\right)} \tag{7.30a}$$

Let $y = \left(\frac{x}{b}\right)^a$, Equation (7.30a) can be solved as:

$$C = \frac{1}{\frac{b^{ap}}{a}\int_0^1 y^{p-1}(1-y)^{q-1}dy} = \frac{a}{b^{ap}B(p,q)} \tag{7.30b}$$

(ii) $a < 0$

For $a < 0$, the variable x is bounded at the lower end, i.e., $x \in [b, \infty)$ and

$$\int_b^\infty f(x)dx = 1 \Rightarrow C\int_b^\infty x^{ap-1}\left(1-\left(\frac{x}{b}\right)^a\right)^{q-1}dx = 1$$

$$\Rightarrow C = \frac{1}{\int_b^\infty x^{ap-1}\left(1-\left(\frac{x}{b}\right)^a\right)^{q-1}dx} = \frac{1}{b^{ap}\int_b^\infty \left(\frac{x}{b}\right)^{ap}\left(1-\left(\frac{x}{b}\right)^a\right)^{q-1}d\left(\frac{x}{b}\right)} \tag{7.30c}$$

Again let $y = \left(\frac{x}{b}\right)^a$. Equation (7.30c) can then be solved as:

$$C = \frac{1}{\frac{b^{ap}}{a}\int_1^0 y^{p-1}(1-y)^{q-1}dy} = \frac{1}{\frac{b^{ap}}{-a}\int_0^1 y^{p-1}(1-y)^{q-1}dy} = \frac{-a}{b^{ap}B(p,q)} \tag{7.30d}$$

Combining Equation (7.30b) and (7.30d), we have:

$$C = \frac{|a|}{b^{ap-1}B(p,q)} \tag{7.31}$$

Substituting Equation (7.31) into Equation (7.29), the PDF of 4-parameter generalized beta distribution of the first kind can be given as:

$$f(x) = \frac{|a|}{b^{ap}B(p,q)}x^{ap-1}\left(1-\left(\frac{x}{b}\right)^a\right)^{q-1}; \quad p,q,b > 0, x \in [b,\infty) \text{ if } a < 0, x \in (0,b] \text{ if } a > 0 \tag{7.32}$$

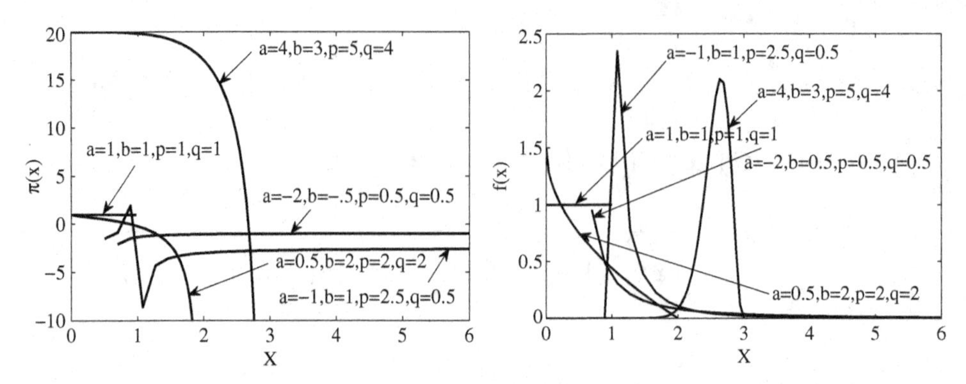

Figure 7.6 (a) Plot of the $\pi(x)$ function of generalized beta distribution of the first kind; and (b) plot of the PDF of generalized beta distribution of the first kind.

Figure 7.6a plots the elasticity and Figure 7.6b plots the PDF of generalized beta distribution of first kind with different parameter values. The GB1 distribution is a uniform distribution in [0, 1] if $a = b = p = q = 1$. The PDF of GB1 distribution may be L-shaped or bell shaped and skewed.

7.2.4 Special Cases of GB1 Distribution

The GB1 distribution includes the generalized gamma, gamma, inverse gamma, Weibull, inverse Weibull, beta distribution of the first kind, beta distribution of the second kind, log-Cauchy, log-normal, IGG, Singh-Maddala, Burr type III, power, exponential, inverse exponential, Raleigh, inverse Raleigh, and uniform distributions as special or limiting cases. These cases are shown in the following text.

Beta Distribution of First Kind: The GB1 distribution reduces to beta distribution of the first kind (B1) if $a = 1$. And its PDF is given as:

$$f(x) = \frac{1}{b^p B(p,q)} x^{p-1}\left(1 - \left(\frac{x}{b}\right)\right)^{q-1}; b, p, q > 0, x \in (0, b] \tag{7.33}$$

The B1 distribution may be also directly derived from Esteban system. The elasticity of B1 distribution can be given as:

$$\pi(x) = (p + q - 1) - \frac{q-1}{1 - \left(\frac{x}{b}\right)} \tag{7.34}$$

Substituting Equation (7.34) into Equation (7.3), we have:

$$d\ln f(x) = -\frac{1-(p+q-1)+\frac{q-1}{1-\left(\frac{x}{b}\right)}}{x} \tag{7.35}$$

Integration of Equation (7.35) yields

$$f(x) = C\exp\left(\int -\frac{1-(p+q-1)+\frac{q-1}{1-\left(\frac{x}{b}\right)}}{x}dx\right)$$

$$= C\exp\left(\int\left[\frac{p-1}{x}-\frac{q-1}{b}\frac{1}{1-\left(\frac{x}{b}\right)}\right]dx\right) = Cx^{p-1}\left(1-\tfrac{x}{b}\right)^{q-1}; x\in(0,b] \tag{7.36a}$$

Applying the total probability, the integration constant C can be evaluated as:

$$\int_0^b f(x)dx = 1 \Rightarrow \int_0^b Cx^{p-1}\left(1-\frac{x}{b}\right)^{q-1}dx = 1$$
$$\Rightarrow C = \frac{1}{\int_0^b x^{p-1}\left(1-\frac{x}{b}\right)^{q-1}dx} \tag{7.36b}$$

Let $y = \frac{x}{b} \Rightarrow dx = bdy$. Then, Equation (7.36b) can be solved as:

$$C = \frac{1}{b^p\int_0^1 y^{p-1}(1-y)^{q-1}dy} = \frac{1}{b^pB(p,q)} \tag{7.36c}$$

Substituting Equation (7.36c) into Equation (7.36a), the PDF of B1 distribution can be given as:

$$f(x) = \frac{1}{b^pB(p,q)}x^{p-1}\left(1-\frac{x}{b}\right)^{q-1}; p,q,b>0, x\in(0,b] \tag{7.37}$$

It is shown that Equation (7.37) is exactly of the same form as Equation (7.33). Figure 7.7a plots the elasticity and Figure 7.7b plots the PDF of B1 distribution with different parameter values using Equation (7.39). The plot of elasticity indicates: (i) constant elasticity if $q = 1$, $\pi(x) = p$; (ii) an increasing function elasticity if $q < 1$; and (iii) a decreasing function of elasticity if $q > 1$. The PDF of B1 distribution is U-shaped if $p < 1, q < 1$; is L-shaped if $p = 1$ *or* $q = 1$; and is bell shaped if $p > 1$, $q > 1$. The B1 distribution is generally skewed.

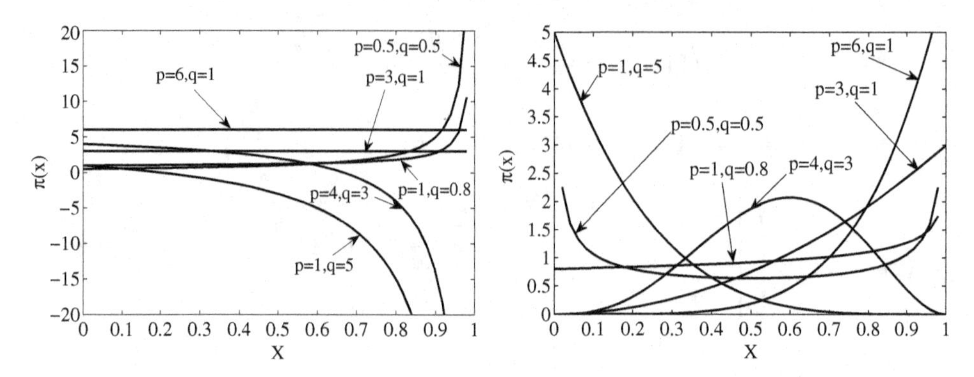

Figure 7.7 (a) Plot of the π(x) function of B1 distribution; and (b) Plot of the PDF of B1 distribution.

Pareto Distribution: Previously, we have shown that the Pareto distribution is a special case of 3-parameter generalized gamma distribution. The Pareto distribution is also a special case of the generalized beta distribution of first kind in which $a = -1$, $q = 1$, $b > 0$, $p > 0$ that is given as:

$$f(x) = \frac{pb^p}{x^{p+1}} \tag{7.38}$$

The Pareto distribution can be directly derived from the Esteban system. Its elasticity function is constant that is given as:

$$\pi(x) = -p \tag{7.39}$$

Substitution of Equation (7.39) into Equation (7.3) yields

$$d\ln f(x) = -\frac{1+p}{x} \tag{7.40}$$

Integration of Equation (7.40) yields

$$f(x) = C\exp\left(\int -\frac{p+1}{x}dx\right) = Cx^{-(p+1)} \tag{7.41a}$$

Applying the total probability, the integration constant C can be evaluated as:

$$\int_b^{\infty} f(x)dx = 1 \Rightarrow C\int_b^{\infty} x^{-(p+1)}dx \Rightarrow C = \frac{1}{\int_b^{\infty} x^{-(p+1)}dx} = pb^p \tag{7.41b}$$

Substituting Equation (7.41b) to Equation (7.41a), we have:

$$f(x) = \frac{pb^p}{(1+x)^{p+1}}; x \in [b,\infty), p, b > 0. \tag{7.41c}$$

It is shown that Equation (7.41c) is exactly the same form as Equation (7.38). Furthermore, $\pi(x) = -p$ is equivalent to the elasticity of Pareto distribution derived from the 3-parameter generalized gamma distribution [i.e., Equation 7.18].

7.2.5 Generalized Beta Distribution of Second Kind

Similar to GB1 distribution, the generalized beta distribution of second kind (GB2) is also a 4-parameter distribution. The elasticity of generalized beta distribution of second kind is given as:

$$\pi(x) = -aq + \frac{a(p+q)}{1+\left(\frac{x}{b}\right)^a}; p, q, b > 0, x \in (0, \infty) \tag{7.42}$$

Substitution of Equation (7.42) into Equation (7.3) yields

$$d\ln f(x) = -\frac{aq + 1 - \frac{a(p+q)}{1+\left(\frac{x}{b}\right)^a}}{x} dx \tag{7.43}$$

Integration of Equation (7.43) yields:

$$\begin{aligned}
f(x) &= C\exp\left(\int -\frac{aq + 1 - \frac{a(p+q)}{1+\left(\frac{x}{b}\right)^a}}{x} dx\right) \\
&= C\exp\left(\int \left(-\frac{aq+1}{x} + \frac{a(p+q)}{x} - \frac{a(p+q)\left(\frac{x}{b}\right)^{a-1}}{b\left(1+\left(\frac{x}{b}\right)^a\right)}\right) dx\right) \\
&= C\exp\left(\int \left(\frac{ap-1}{x} - \frac{a(p+q)\left(\frac{x}{b}\right)^{a-1}}{b\left(1+\left(\frac{x}{b}\right)^a\right)}\right) dx\right) \\
&= C\exp\left((ap-1)\ln x - (p+q)\ln\left(1+\left(\frac{x}{b}\right)^a\right)\right) = Cx^{ap-1}\left(1+\left(\frac{x}{b}\right)^a\right)^{-(p+q)}
\end{aligned} \tag{7.44a}$$

Applying the total probability, the integration constant C can be evaluated as:

$$\begin{aligned}
&\int_0^\infty f(x)dx = 1 \Rightarrow C\int_0^\infty x^{ap-1}\left(1+\left(\frac{x}{b}\right)^a\right)^{-(p+q)} dx = 1 \\
&\Rightarrow C = \frac{1}{\int_0^\infty x^{ap-1}\left(1+\left(\frac{x}{b}\right)^a\right)^{-(p+q)} dx}
\end{aligned} \tag{7.44b}$$

Let $\left(\frac{x}{b}\right)^a = t \Rightarrow x = bt^{\frac{1}{a}} \Rightarrow dx = \frac{b}{a}t^{\frac{1}{a}-1}dt$.

The integral of the constant C can be evaluated based on the value of a as:

(i) $a > 0$

$$\int_0^\infty x^{ap-1}\left(1+\left(\frac{x}{b}\right)^a\right)^{-(p+q)}dx = \frac{b^{ap}}{a}\int_0^\infty t^{p-1}(1+t)^{-(p+q)}dt = \frac{b^{ap}}{a}\mathrm{B}(p,q)$$

$$\Rightarrow C = \frac{1}{\int_0^\infty x^{ap-1}\left(1+\left(\frac{x}{b}\right)^a\right)^{-(p+q)}dx} = \frac{a}{b^{ap}B(p,q)} \tag{7.44c}$$

(ii) $a < 0$

$$\int_0^\infty x^{ap-1}\left(1+\left(\frac{x}{b}\right)^a\right)^{-(p+q)}dx$$

$$= \int_\infty^0 \left(\frac{b}{a}t^{\frac{1}{a}}\right)^{ap-1}(1+t)^{-(p+q)}\left(\frac{b}{a}\right)t^{\frac{1}{a}-1}dt \tag{7.44d}$$

$$= \frac{b^{ap}}{-a}\int_0^\infty t^{p-1}(1+t)^{-(p+q)}dt = \frac{b^{ap}}{-a}B(p,q) \Rightarrow C = -\frac{a}{b^{ap}B(p,q)}$$

Combining Equation (7.44c) and Equation (7.44d), we have:

$$C = \frac{|a|}{b^{ap}B(p,q)} \tag{7.44e}$$

Substituting Equation (7.44e) into Equation (7.44a), the PDF of GB2 distribution is expressed as:

$$f(x) = \frac{|a|}{b^{ap}B(p,q)}x^{ap-1}\left(1+\left(\frac{x}{b}\right)^a\right)^{-(p+q)} \tag{7.44f}$$

The elasticity function of GB2 distribution is a decreasing function, as shown in Figure 7.8a. The PDF of GB2 distribution is L-shaped if $a \le 1$, and is bell shaped if all the parameters are greater than 1, as shown in Figure 7.8b. The GB2 distribution is generally skewed.

7.2.6 Special Cases of GB2 Distribution

The GB2 distribution includes the generalized gamma, gamma, Weibull, beta distribution of the second kind, Singh-Maddala/Burr XII, Burr III, Fisk

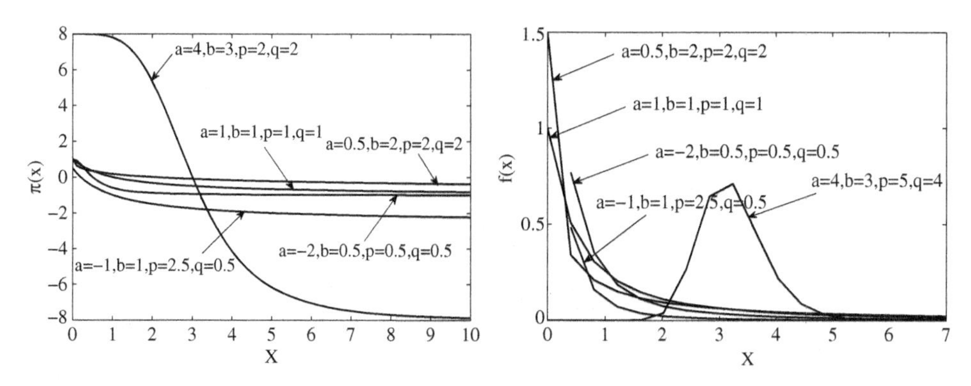

Figure 7.8 (a) Plot of the π(x) function of GB2 distribution; and (b) Plot of the PDF of GB2 distribution.

(log-logistic), Lomax, inverse Lomax, generalized normal, and exponential distribution as special or limiting cases. McDonald (1984) pointed that some of these distributions are known by different names. For example, the GB2 distribution with a nonzero threshold is known as Feller-Pareto distribution (Arnold, 1980); the Singh-Maddala distribution as Beta-P distribution (Johnson and Kotz, 1970; Cronin, 1979); 3-parameter kappa distribution if $q = 1$, or Beta-k or Burr distribution of the third type (Tadikamalla, 1980); and Lomax distribution of $k = a = 2$. The GB2 distribution is also known as generalized F distribution (Kalbfleisch and Prentice, 1980), and transformed beta distribution (Venter, 1984). The special cases are shown in the following text.

Beta Distribution of Second Kind: The Beta distribution of the second kind (B2), also called beta prime or inverse beta distribution, is a special case of the B2 distribution. Its density function is expressed as:

$$f(x) = \frac{1}{B(p,q)} x^{p-1}(1+x)^{-(p+q)}; p,q > 0, x \in (0,\infty) \tag{7.45}$$

The B2 distribution can be derived from the Esteban system. The elasticity function can be expressed as:

$$\pi(x) = -q + \frac{p+q}{1+x} \tag{7.45a}$$

Substituting Equation (7.45a) into Equation (7.3) we have:

$$d\ln f(x) = -\frac{1+q-\frac{p+q}{1+x}}{x}dx \tag{7.46}$$

Integration of Equation (7.46) yields:

$$f(x) = C\exp\left(\int\left[\frac{ak-1}{x} - \frac{a}{b}\left(\frac{x}{b}\right)^{a-1}\right]dx\right) = C\exp\left((ak-1)\ln x - \left(\tfrac{x}{b}\right)^{a}\right)$$
$$\Rightarrow f(x) = Cx^{ak-1}\exp\left(-\left(\tfrac{x}{b}\right)^{a}\right) \tag{7.47a}$$

Applying the total probability, the integration constant C can be evaluated as:

$$\int_0^{\infty} f(x)dx = 1 \Rightarrow C\int_0^{\infty} x^{p-1}(1+x)^{-(p+q)}dx \Rightarrow C = \frac{1}{B(p,q)} \tag{7.47b}$$

Substituting Equation (7.47b) into Equation (7.47a), the PDF of B2 distribution can be written as:

$$f(x) = \frac{1}{B(p,q)}x^{p-1}(1+x)^{-(p+q)}; p,q > 0, x \in (0,\infty) \tag{7.47c}$$

Now we have shown that the density function derived from Esteban system is exactly the same as the B2 distribution. The elasticity of the B2 distribution is an L-shape decreasing function as shown in Figure 7.9a. The PDF of B2 distribution is L-shaped if $p \leq 1$ and is bell shaped if $p > 1$. The B2 distribution is generally skewed as shown in Figure 7.9b.

Generalized Gamma Distribution: As discussed earlier, the generalized distribution may be considered as 3-parameter gamma distribution and was discussed in Section 7.2.1. Furthermore, the generalized gamma distribution is also a limiting

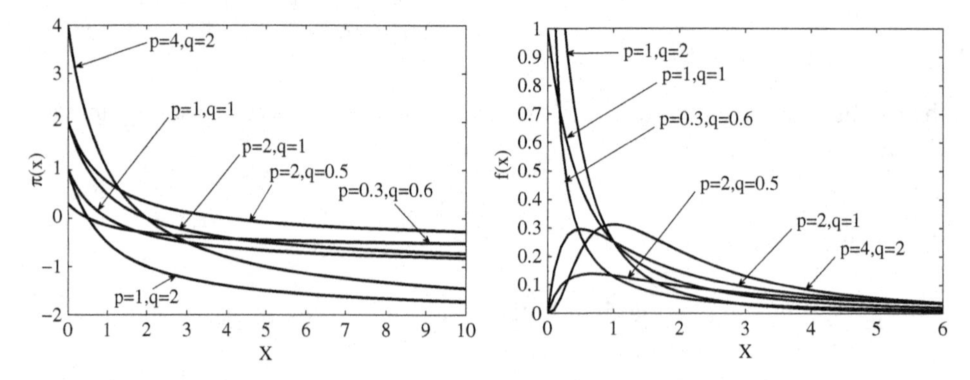

Figure 7.9 (a) Plot of the π(x) function of B2 distribution; and (b) Plot of the PDF of B2 distribution.

case of GB2 distribution in which $q \to \infty, b = kq^{\frac{1}{a}}$, which can be derived from Esteban system directly. The elasticity function is expressed as:

$$\pi(x) = \lim_{q\to\infty}\left(-aq + \frac{a(p+q)}{1+\left(\frac{x}{kq^{\frac{1}{a}}}\right)^{a}}\right); a,p,q,k > 0 \tag{7.48}$$

In Equation (7.48) $\lim_{q\to\infty}\left(\frac{x}{kq^{\frac{1}{a}}}\right)^{a} \to 0$. Application of Taylor series expansion yields

$$\frac{1}{1+\left(\frac{x}{kq^{\frac{1}{a}}}\right)^{a}} = 1 - \left(\frac{x}{kq^{\frac{1}{a}}}\right)^{a} + \left(\frac{x}{kq^{\frac{1}{a}}}\right)^{2a} + \cdots + o(\epsilon) \tag{7.49a}$$

Substitution of Equation (7.49a) into Equation (7.48) yields

$$\begin{aligned}
\pi(x) &= \lim_{q\to\infty}\left(-aq + \frac{a(p+q)}{1+\left(\frac{x}{kq^{\frac{1}{a}}}\right)^{a}}\right)\\
&= \lim_{q\to\infty}\left(-aq + a(p+q)\left(1 - \left(\frac{x}{kq^{\frac{1}{a}}}\right)^{a} + o(\epsilon)\right)\right)\\
&= \lim_{q\to\infty}\left(ap - ap\left(\frac{x}{kq^{\frac{1}{a}}}\right)^{a} - a\left(\frac{x}{k}\right)^{a}\right) = ap - a\left(\frac{x}{k}\right)^{a}, a,p,k > 0
\end{aligned} \tag{7.49b}$$

The derivation of Equation (7.49b) is equivalent to that of Equation (7.5). And the derivation of the density function is in exactly same manner as that discussed in Section 7.2.1.

Gamma Distribution: The gamma distribution is a special case of the Generalized Gamma distribution with $a = 1$. Thus, it is also a limiting case of GB2 distribution. In this case, the elasticity function of the gamma distribution is given as:

$$\begin{aligned}
\pi(x) &= \lim_{q\to\infty}\left(-q + \frac{p+q}{1+\frac{x}{kq}}\right)\\
&= \lim_{q\to\infty}\left(-q + (p+q)\left(1 - \left(\frac{x}{kq}\right) + O(\epsilon)\right)\right)\\
&= \lim_{q\to\infty}\left(p - \frac{px}{kq} - \frac{x}{k}\right) = p - \frac{x}{k}
\end{aligned} \tag{7.50}$$

Equation (7.50) is equivalent to Equation (7.12). And the PDF can then be derived in exactly the same manner as in Section 7.2.2. Furthermore, the elasticity function of exponential distribution is obtained if $p = 0$ (i.e., $\pi(x) = -\frac{x}{k}$); and the elasticity function of Chi-square distribution is deduced from Equation (7.50) if $k = 2, p \in \mathbb{N}^+$ (i.e., $\pi(x) = p - \frac{x}{2}$).

Weibull Distribution: As discussed in Section 7.2.2, Weibull distribution is a special case of the generalized gamma distribution. Thus, it is also a special case of GB2 distribution under the condition of $p = 1, q \to \infty, b = kq^{\frac{1}{a}}$. The elasticity function is expressed as:

$$\begin{aligned}
\pi(x) &= \lim_{q\to\infty}\left(-aq + \frac{a(q+1)}{1+\left(\frac{x}{kq^{\frac{1}{a}}}\right)^a}\right) \\
&= \lim_{q\to\infty}\left(-aq + a(q+1)\left(1-\left(\frac{x}{kq^{\frac{1}{a}}}\right)^a + O(\epsilon)\right)\right) \\
&= \lim_{q\to\infty}\left(a - a\left(\tfrac{x}{k}\right)^a - a\left(\tfrac{x}{k}\right)^a\left(\frac{1}{q}\right)\right) = a - a\left(\tfrac{x}{k}\right)^a; a, k > 0
\end{aligned} \tag{7.51}$$

The elasticity function derived in Equation (7.51) is equivalent to Equation (7.15) discussed in Section 7.2.2.

Rayleigh Distribution: The PDF of Rayleigh distribution is given as:

$$f(x) = \frac{x}{\sigma^2}\exp\left(-\frac{x^2}{2\sigma^2}\right); \sigma > 0, x \in [0, \infty) \tag{7.52}$$

Similar to the Weibull distribution, the Rayleigh distribution is also a special case of GB2 distribution that can be derived from the Esteban system. Its elasticity function can be expressed as:

$$\begin{aligned}
\pi(x) &= \lim_{q\to\infty}\left(-2q + \frac{2(q+1)}{1+\left(\frac{x}{kq^{\frac{1}{a}}}\right)^a}\right) \\
&= \lim_{q\to\infty}\left(-2q + 2(q+1)\left(1-\left(\frac{x}{kq^{\frac{1}{2}}}\right)^2 + O(\epsilon)\right)\right) \\
&= \lim_{q\to\infty}\left(2 - 2\left(\frac{x}{k}\right)^2 - 2\left(\frac{x}{k}\right)^2\left(\frac{1}{q}\right)\right) \Rightarrow \pi(x) = 2 - 2\left(\frac{x}{k}\right)^2
\end{aligned} \tag{7.53}$$

Substituting Equation (7.53) into Equation (7.3), we have:

$$d\ln f(x) = -\frac{2\left(\frac{x}{k}\right)^2 - 1}{x}dx \tag{7.54}$$

Integration of Equation (7.54) yields:

$$f(x) = C\exp\left(\int -\frac{2\left(\frac{x}{k}\right)^2 - 1}{x}dx\right) = C\exp\int\left(-\frac{2x}{k^2} + \frac{1}{x}\right)dx = Cx\exp\left(-\frac{x^2}{k^2}\right) \tag{7.55a}$$

Applying the total probability, the integration constant C in Equation (7.55a) can be evaluated as:

$$\int_0^\infty f(x)dx = 1 \Rightarrow C\int_0^\infty x\exp\left(-\frac{x^2}{k^2}\right)dx = 1 \Rightarrow \frac{Ck^2}{2} = 1 \Rightarrow C = \frac{2}{k^2} \tag{7.55b}$$

Now, Equation (7.55a) can be rewritten as:

$$f(x) = \frac{2}{k^2}x\exp\left(-\frac{x^2}{k^2}\right); k > 0, x \in [0, \infty) \tag{7.56}$$

Letting $\sigma = k/\sqrt{2}$, Equation (7.56) may be written in exactly the same form as Equation (7.52).

The elasticity function of Raleigh distribution is monotonic decreasing and its PDF is bell shaped and skewed. Raleigh distribution is also a special case of Weibull distribution by letting the a = 2 in its elasticity function [i.e., Equation 7.51]. Figure 7.10a plots the $\pi(x)$ function and Figure 7.10b plots the PDF of Raleigh distribution.

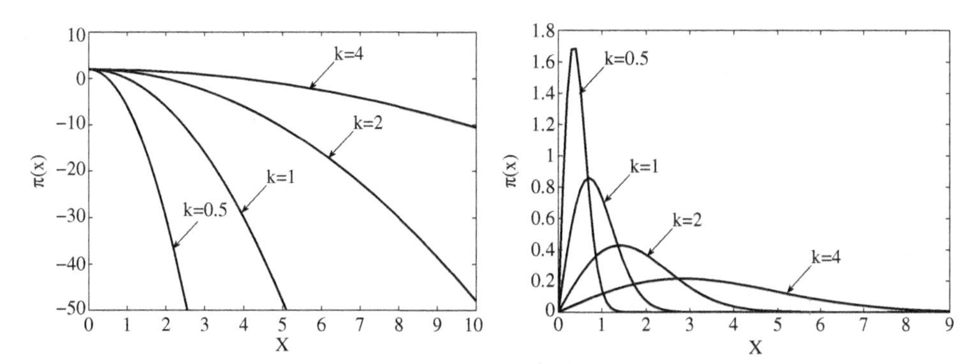

Figure 7.10 (a) Plot of the $\pi(x)$ function of Raleigh distribution; and (b) Plot of the PDF of Raleigh distribution.

Dagum I Distribution: We have discussed Dagum I distribution in Chapter 5. The PDF of Dagum I distribution can be given as:

$$f(x) = \frac{ap}{b}\left(\frac{x}{b}\right)^{ap-1}\left(1+\left(\frac{x}{b}\right)^{a}\right)^{-(p+1)} \tag{7.57}$$

As a special case of the GB2 distribution, the Dagum I distribution may be derived from the Esteban system. Its elasticity function can be given as:

$$\pi(x) = -a + \frac{a(p+1)}{1+\left(\frac{x}{b}\right)^{a}}; a, b, p > 0, x \in (0, \infty) \tag{7.58}$$

Substituting Equation (7.58) into Equation (7.3), we obtain:

$$d\ln f(x) = -\frac{a+1-\frac{a(p+1)}{1+\left(\frac{x}{b}\right)^{a}}}{x}dx \tag{7.59}$$

Integration of Equation (7.59) yields

$$f(x) = C\exp\left(\int -\frac{a+1-\frac{a(p+1)}{1+\left(\frac{x}{b}\right)^{a}}}{x}dx\right)$$

$$= C\int\left(-\frac{a+1}{x}+\frac{a(p+1)}{x}-\frac{a(p+1)}{b}\frac{\left(\frac{x}{b}\right)^{a-1}}{\left(1+\left(\frac{x}{b}\right)^{a}\right)}\right)dx$$

$$= C\int\left(\frac{ap-1}{x}-\frac{a(p+1)}{b}\frac{\left(\frac{x}{b}\right)^{a-1}}{\left(1+\left(\frac{x}{b}\right)^{a}\right)}\right)dx = Cx^{ap-1}\left(1+\left(\frac{x}{b}\right)^{a}\right)^{-(p+1)} \tag{7.60a}$$

Applying the total probability to Equation (7.60a), the integration constant C can be evaluated as:

$$\int_0^\infty f(x) = 1 \Rightarrow C\int_0^\infty x^{ap-1}\left(1+\left(\frac{x}{b}\right)^{a}\right)^{-(p+1)}dx = 1$$

$$\Rightarrow C = \frac{1}{\int_0^\infty x^{ap-1}\left(1+\left(\frac{x}{b}\right)^{a}\right)^{-(p+1)}dx} \tag{7.60b}$$

In Equation (7.60b), let $y = \left(\frac{x}{b}\right)^a \Rightarrow x = by^{\frac{1}{a}} \Rightarrow dx = \frac{b}{a} y^{\frac{1}{a}-1} dy$; and the integral of Equation (7.60b) may be solved for a < 0 as:

$$\int_0^\infty x^{ap-1}\left(1+\left(\frac{x}{b}\right)^a\right)^{-(p+1)} dx = \int_0^\infty \left(by^{\frac{1}{a}}\right)^{ap-1}(1+y)^{-p-1}\left(\frac{b}{a}y^{\frac{1}{a}-1}\right)dy$$
$$= \frac{b^{ap}}{a}\int_0^\infty y^{p-1}(1+y)^{-(p+1)}dy = \frac{b^{ap}}{ap} \Rightarrow C = \frac{ap}{b^{ap}} \tag{7.60c}$$

Substituting Equation (7.60c) into Equation (7.60a), we have:

$$f(x) = \frac{ap}{b^{ap}} x^{ap-1}\left(1+\left(\frac{x}{b}\right)^a\right)^{-(p+1)} = \frac{ap}{b}\left(\frac{x}{b}\right)^{ap-1}\left(1+\left(\frac{x}{b}\right)^a\right)^{-(p+1)} \tag{7.61}$$

It is shown that the Dagum I distribution derived from the Esteban system is of the same form as Equation (7.57).

Figure 7.11a plots the π(x) function and Figure 7.11b plots the PDF of Dagum I distribution.

Singh-Maddala Distribution (Burr XII): In Chapter 3, we have discussed the Singh-Maddala distribution that is also known as Burr XII distribution [i.e., Equation 3.65] as:

$$f(x) = \frac{ckx^{c-1}}{(1+x^c)^{k+1}}; c, k > 0, x \in (0, \infty) \tag{7.62}$$

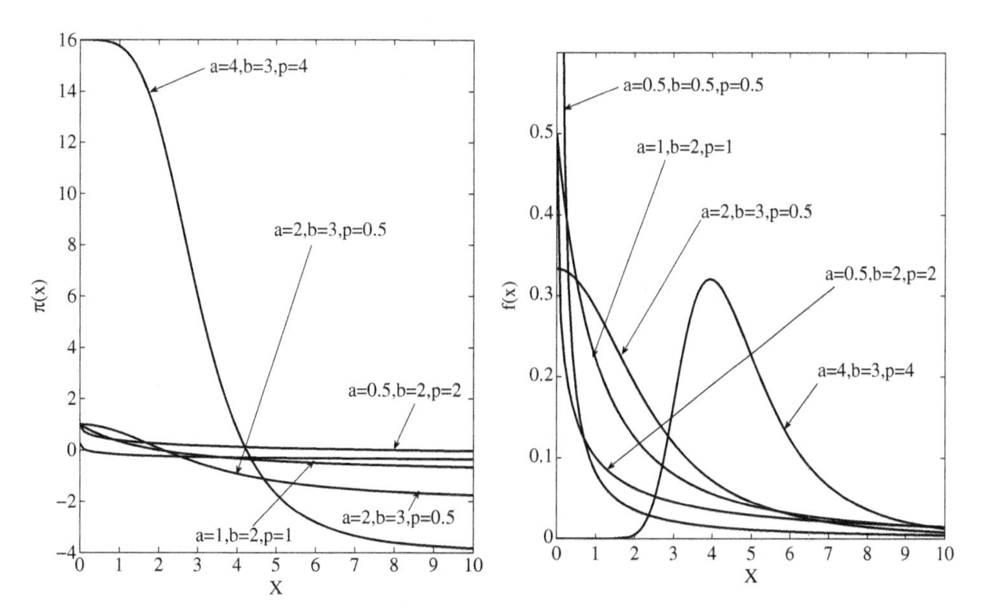

Figure 7.11 (a) Plot of the π(x) function of Dagum I distribution; and (b) Plot of the PDF of Dagum I distribution.

As a special case of the GB2 distribution, the Singh-Maddala distribution may be also derived from the Esteban system. Its elasticity function can be expressed as:

$$\pi(x) = -aq + \frac{a(1+q)}{1+\left(\frac{x}{b}\right)^a}; a, b, q > 0, x > 0 \tag{7.63}$$

Substituting Equation (7.63) into Equation (7.3), we have

$$d\ln f(x) = -\left(\frac{1+aq}{x} - \frac{a(1+q)}{x\left(1+\left(\frac{x}{b}\right)^a\right)}\right)dx \tag{7.64a}$$

Integration of Equation (7.64a) yields

$$\begin{aligned} f(x) &= C\exp\left(\int -\left(\frac{1+aq}{x} - \frac{a(1+q)}{x\left(1+\left(\frac{x}{b}\right)^a\right)}\right)dx\right) \\ &= C\exp\left(\int\left(-\frac{1+aq}{x} + \frac{a(1+q)}{x} - \frac{a(1+q)\left(\frac{x}{b}\right)^{a-1}}{b\left(1+\left(\frac{x}{b}\right)^a\right)}\right)dx\right) \\ &= Cx^{a-1}\left(1+\left(\frac{x}{b}\right)^a\right)^{-(1+q)} \end{aligned} \tag{7.64b}$$

Applying the total probability to Equation (7.64b), the integration constant C may be evaluated as:

$$\begin{aligned} \int_0^\infty f(x)dx = 1 &\Rightarrow C\int_0^\infty x^{a-1}\left(1+\left(\frac{x}{b}\right)^a\right)^{-(1+q)}dx = 1 \\ &\Rightarrow C = \frac{1}{\int_0^\infty x^{a-1}\left(1+\left(\frac{x}{b}\right)^a\right)^{-(1+q)}dx} \end{aligned} \tag{7.64c}$$

Again letting $y = \left(\frac{x}{b}\right)^a$, the integral in Equation (7.64c) may be solved as:

$$\begin{aligned} \int_0^\infty x^{a-1}\left(1+\left(\frac{x}{b}\right)^a\right)^{-(q+1)}dx &= \int_0^\infty \left(by^{\frac{1}{a}}\right)^{a-1}(1+y)^{-(q+1)}\left(\frac{b}{a}y^{\frac{1}{a}-1}\right)dy \\ &= \frac{b^a}{a}\int_0^\infty (1+y)^{-(q+1)}dy = \frac{b^a}{aq} \Rightarrow C = \frac{aq}{b^a} \end{aligned} \tag{7.64d}$$

Substituting Equation (7.64d) into Equation (7.64a), we have:

$$f(x) = \frac{aq}{b^a}x^{a-1}\left(1+\left(\frac{x}{b}\right)^a\right)^{-(q+1)} \tag{7.64e}$$

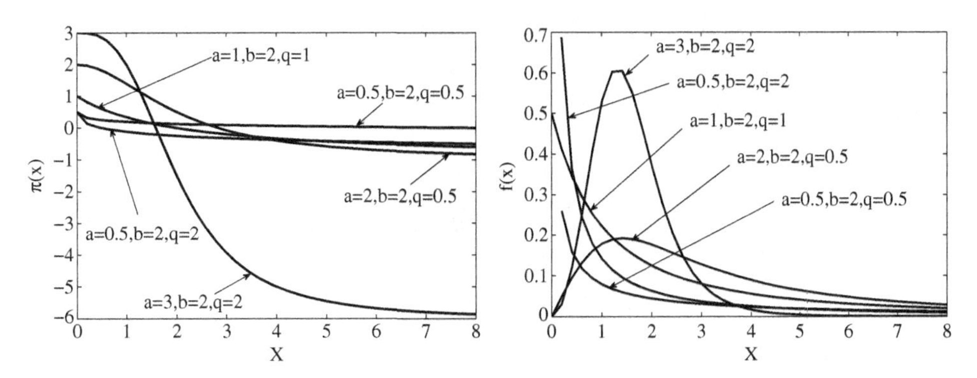

Figure 7.12 (a) Plot of the $\pi(x)$ function of Singh-Maddala distribution; and (b) Plot of the PDF of Singh-Maddala distribution.

Comparing Equation (7.64e) to Equation (7.62) [i.e., Equation 3.65], Equation (7.64e) may be written in the form of Equation (7.62) by setting $c = a, k = q, y = \frac{x}{b}$ with probability density transformation as:

$$\begin{aligned} f(y) &= f_X\left(y^{-1}(x)\right)\left|\frac{dx}{dy}\right| = \frac{q}{b^a}(by)^{a-1}(1+y^a)^{-(q+1)}b \\ &= aqy^{a-1}(1+y^a)^{-(q+1)} = cky^{c-1}(1+y)^{-(k+1)} \end{aligned} \quad (7.65)$$

The elasticity function is L-shaped for the Singh-Maddala distribution, as shown in Figure 7.12a. The PDF is L-shaped if $a \leq 1$, while it is bell shaped if $a > 1$, as shown in Figure 7.12b. The Singh-Maddla distribution is generally skewed.

Fisk Distribution: Fisk distribution is also known as log-logistic distribution. The Fisk distribution is expressed as:

$$f(x) = \left(\frac{\beta}{\alpha}\right)\left(\frac{x}{\alpha}\right)^{\beta-1}\left(1+\left(\frac{x}{\alpha}\right)^{\beta}\right)^{-2}; \alpha, \beta > 0, x \in [0, \infty) \quad (7.66)$$

The Fisk distribution is a special case of Singh-Maddala distribution. Thus, it is also a special case of GB2 distribution that may be derived from Esteban system. Its elasticity function is given as:

$$\pi(x) = \pi(x) = -a + \frac{2a}{1+\left(\frac{x}{b}\right)^a}; a, b > 0, x > 0 \quad (7.67)$$

Substituting Equation (7.67) into Equation (7.3), we have:

$$d\ln f(x) = -\frac{a+1}{x} + \frac{2a}{x\left(1+\left(\frac{x}{b}\right)^a\right)}dx \quad (7.68a)$$

The integration of Equation (7.68a) yields

$$\begin{aligned} f(x) &= C\exp\left(\int\left(-\frac{a+1}{x}+\frac{2a}{x\left(1+\left(\frac{x}{b}\right)^a\right)}\right)dx\right) \\ &= C\exp\left(\int\left(-\frac{a+1}{x}+\frac{2a}{x}-\frac{2a\left(\frac{x}{b}\right)^{a-1}}{1+\left(\frac{x}{b}\right)^a}\right)dx\right) \\ &= C\exp\left(\int\left(\frac{a-1}{x}-\frac{2a\left(\frac{x}{b}\right)^{a-1}}{1+\left(\frac{x}{b}\right)^a}\right)dx\right) = Cx^{a-1}\left(1+\left(\frac{x}{b}\right)^a\right)^{-2} \end{aligned} \tag{7.68b}$$

Applying total probability, the integration constant C may be solved as:

$$\int_0^\infty f(x)dx = 1 \Rightarrow C\int_0^\infty x^{a-1}\left(1+\left(\frac{x}{b}\right)^a\right)^{-2}dx \Rightarrow C = \frac{1}{\int_0^\infty x^{a-1}\left(1+\left(\frac{x}{b}\right)^a\right)^{-2}dx} \tag{7.68c}$$

In Equation (7.68c), let $y = \left(\frac{x}{b}\right)^a \Rightarrow x = by^{\frac{1}{a}} \Rightarrow dx = \frac{b}{a}y^{\frac{1}{a}-1}dy$. And the integral in Equation (7.68c) may be rewritten as:

$$\begin{aligned} \int_0^\infty x^{a-1}\left(1+\left(\frac{x}{b}\right)^a\right)^{-2}dx &= \int_0^\infty \left(by^{\frac{1}{a}}\right)^{a-1}(1+y)^{-2}\left(\frac{b}{a}y^{\frac{1}{a}-1}\right)dy \\ &= \frac{b^a}{a}\int_0^\infty (1+y)^{-2}dy = \frac{b^a}{a} \Rightarrow C = a/b^a \end{aligned} \tag{7.68d}$$

Now the PDF is given as:

$$f(x) = \frac{a}{b^a}x^{a-1}\left(1+\left(\frac{x}{b}\right)^a\right)^{-2} = \frac{a}{b}\left(\frac{x}{b}\right)^{a-1}\left(1+\left(\frac{x}{b}\right)^a\right)^{-2} \tag{7.68e}$$

Equation (7.68e) may be written in the same form as Equation (7.66) by setting $\alpha = b, \beta = a$.

Similar to Singh-Maddala distribution, the elasticity function of Fisk distribution is L-shaped decreasing function. The density function is L-shaped if $a \le 1$, while it is bell shaped if $a > 1$. The Fisk distribution is also skewed. Figure 7.13a plots the $\pi(x)$ function and Figure 7.13b plots the PDF of Fisk distribution.

Lomax Distribution: The Lomax distribution is also known as Pareto II distribution. The PDF of Lomax distribution is expressed as:

$$f(x) = \frac{\alpha}{\lambda}\left(1+\frac{x}{\lambda}\right)^{-(\alpha+1)} \tag{7.69}$$

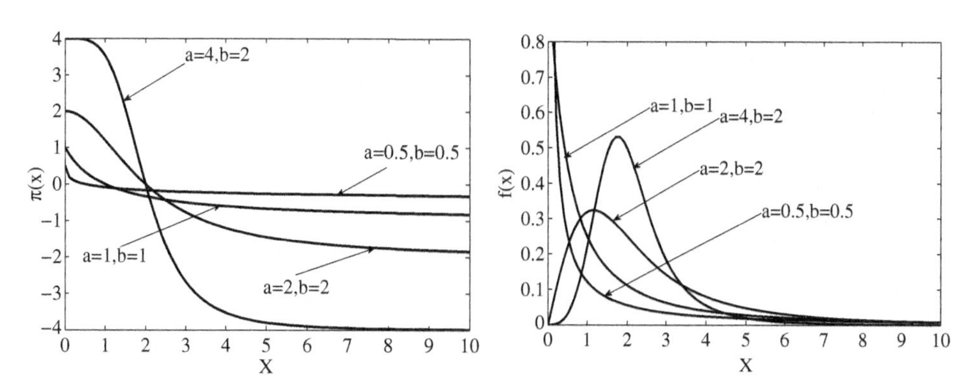

Figure 7.13 (a) Plot of the π(x) function of Fisk distribution; and (b) Plot of the PDF of Fisk distribution.

Similar to the Fisk distribution, the Lomax distribution is also a special case of GB2 that may be derived from the Esteban system. Its elasticity function can be given as:

$$\pi(x) = -q + \frac{q+1}{1+\left(\frac{x}{b}\right)} \tag{7.70}$$

Substitution of Equation (7.70) into Equation (7.3) yields:

$$d\ln f(x) = \left(-\frac{1+q}{x} + \frac{q+1}{x\left(1+\left(\frac{x}{b}\right)\right)} \right) dx \tag{7.71a}$$

Integration of Equation (7.71a) yields:

$$\begin{aligned} f(x) &= C\exp\left(\int \left(-\frac{1+q}{x} + \frac{q+1}{x\left(1+\left(\frac{x}{b}\right)\right)} \right) dx \right) \\ &= C\exp\left(\int \left(-\frac{1+q}{x} + \frac{q+1}{x} - \frac{q+1}{b\left(1+\frac{x}{b}\right)} \right) dx \right) \\ &= C\left(1+\tfrac{x}{b}\right)^{-(q+1)} \end{aligned} \tag{7.71b}$$

Applying the total probability, the integration constant C may be evaluated as:

$$\int_0^\infty f(x)dx = 1 \Rightarrow C\int_0^\infty \left(1+\frac{x}{b}\right)^{-(q+1)} dx = 1 \Rightarrow C = \frac{1}{\int_0^\infty \left(1+\frac{x}{b}\right)^{-(q+1)} dx} = \frac{q}{b} \tag{7.71c}$$

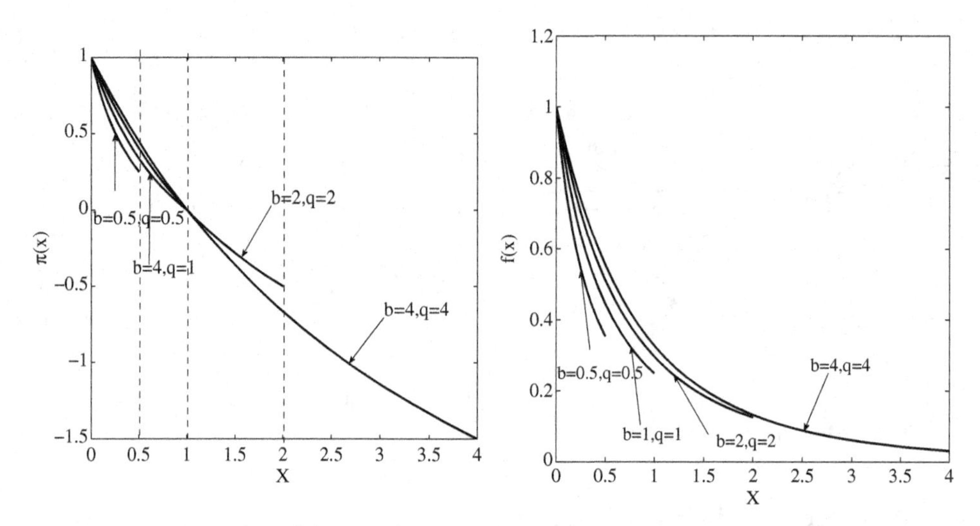

Figure 7.14 (a) Plot of the π(x) function of Lomax distribution; and (b) Plot of the PDF of Lomax distribution.

Substituting Equation (7.71c) into Equation (7.71a), we have:

$$f(x) = \frac{q}{b}\left(1 + \frac{x}{b}\right)^{-(q+1)} \tag{7.72}$$

Equation (7.72) may be written in the same form as Equation (7.69) by setting $\alpha = q$, $\lambda = b$.

The elasticity of Lomax distribution is L-shaped. The PDF of Lomax distribution is L-shaped and skewed. Figure 7.14a plots the π(x) function of Lomax distribution and Figure 7.14b plots the PDF of Lomax distribution.

7.3 Application

In this section, the generalized Gamma distribution (GG3), generalized Beta distribution of first kind (GB1), and generalized Beta distribution of second kind (GB2), and normal distribution are applied for the application. The maximum likelihood estimation method is applied for parameter estimation. The log-likelihood functions of the three generalized distribution previously mentioned can be given as:

Generalized Gamma Distribution (GG3)

$$\ln L = n \ln a - nak \ln b - n\Gamma(k) + (ak - 1)\sum_{i=1}^{n} \ln x_i - \sum_{i=1}^{n} \left(\frac{x_i}{b}\right)^{a} \tag{7.73}$$

Generalized Beta Distribution of First Kind (4-Parameter)

$$\ln L = n \ln \mid a \mid - nap \ln b - n \ln B(p,q) + (ap-1) \sum_{i=1}^{n} \ln x_i + (q-1) \sum_{i=1}^{n} \ln \left(1 - \left(\frac{x_i}{b}\right)^a\right) \tag{7.74}$$

Generalized Beta Distribution of Second Kind (4-Parameter)

$$\ln L = n \ln \mid a \mid - nap \ln b - n \ln B(p,q) + (ap-1) \sum_{i=1}^{n} \ln x_i - (p+q) \sum_{i=1}^{n} \ln \left(1 + \left(\frac{x_i}{b}\right)^a\right) \tag{7.75}$$

Maximizing Equations (7.73)–(7.75), we can then estimate the parameters for each generalized distribution candidate.

7.3.1 TPN

Here, we will illustrate the application of distributions of the Esteban system to water quality (TPN) as examples. Due to the periodicity existing in the monthly TPN data, the dataset is deseasonalized using:

$$X_{i,j}^{*} = \frac{\left(X_{i,j} - \hat{\mu}_i\right)}{\hat{\sigma}_i}, i = 1, 2, \ldots, s \tag{7.76}$$

In Equation (7.76), X, X^* represent the seasonal observations and the corresponding deseasonalized sequence; $\hat{\mu}_i$, $\hat{\sigma}_i$ represent the sample mean and sample standard deviation for i-th season, respectively; and s is the total number of periods. Applying the normal distribution, the parameters are estimated as: $\hat{\mu} = 0$, $\hat{\sigma} = 0.98$. Figure 7.15 compares the fitted frequency distributions to the empirical

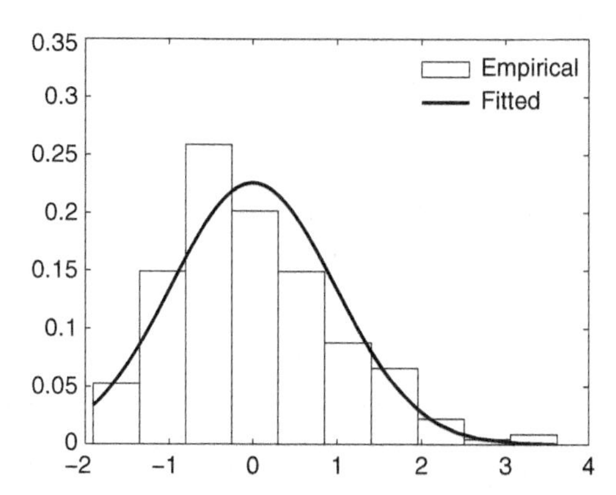

Figure 7.15 Comparison of fitted frequency distributions to empirical frequency distribution for deseasonalized TPN data.

frequency distribution of the deseasonalized monthly TPN. Comparison indicates that the normal distribution may be applied to model the deseasonalized monthly TPN.

7.3.2 Peak Flow

According to the characteristics of peak flow, the GG3, GB1, and GB2 are applied for analysis. Table 7.1 lists the parameters estimated with the use of genetic algorithm. Figure 7.16 compares the fitted frequency with the empirical frequency. Comparisons in Figure 7.16 indicate that all three generalized distributions may be applied to model peak flow. Furthermore, the GG3 and GB2 distributions perform better than does the GB1 distribution.

Table 7.1 *Parameters estimated for peak flow*

	Distributions		
Parameters	GG3	GB1	GB2
a	0.65	–0.41	1.08
b	3.56	30.20	42.66
k	16.66		
p		40.21	55.46
q		50.35	8.13

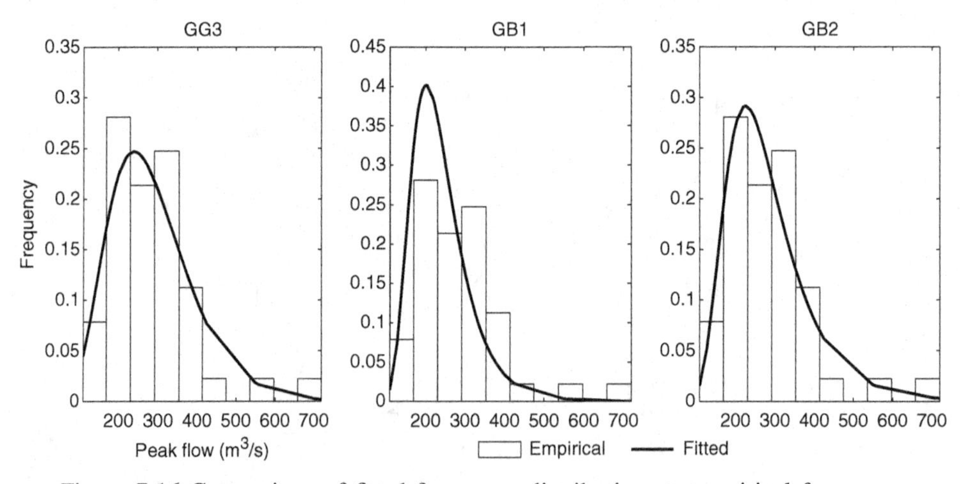

Figure 7.16 Comparison of fitted frequency distributions to empirical frequency distribution for peak flow.

7.3.3 Drought (Total Flow Deficit)

For the right-skewed drought variable-total flow deficit, all three generalized distributions are applied to model total flow deficit. Table 7.2 lists the parameters estimated using the maximum likelihood estimation method with genetic algorithm. Figure 7.17 compares the fitted frequency distributions with the empirical frequency distribution. Comparisons in Figure 7.17 indicate that GG3 and GB2 distributions may be applied to model the total flow deficit. Additionally, GG3 and GB2 distributions yield similar performances.

7.3.4 Annual Rainfall

Besides the three generalized distributions, the normal distribution is also applied to model annual rainfall. Applying the maximum likelihood estimation method, Table 7.3

Table 7.2 *Parameters estimated for total flow deficit*

	Distributions		
Parameters	GG3	GB1	GB2
a	0.21	–0.27	0.30
b	0.91	30.20	21.83
k	9.47		
p		24.33	32.22
q		47.99	4.25

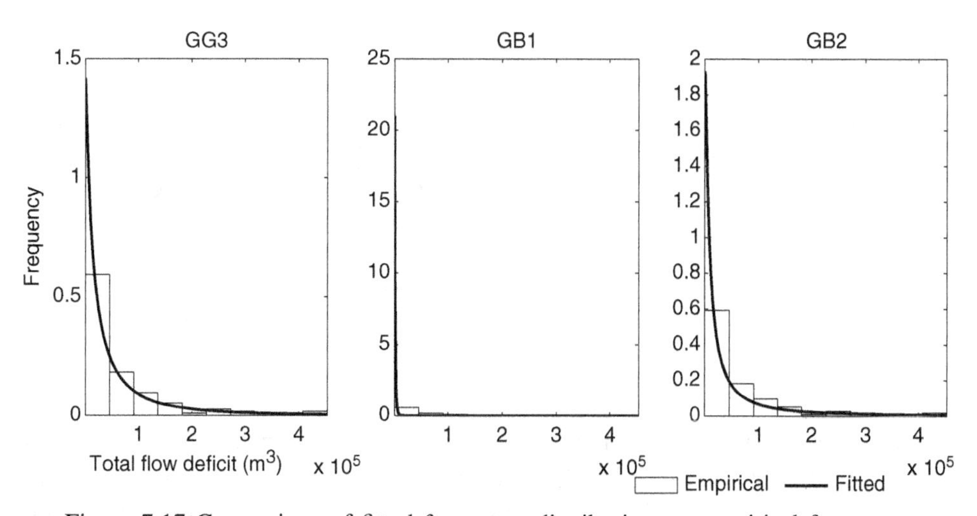

Figure 7.17 Comparison of fitted frequency distributions to empirical frequency distribution for total flow deficit.

Table 7.3 *Parameters estimated for annual rainfall*

	Distributions			
Parameters	GG3	GB1	GB2	Normal (μ, σ)
a	0.57	–0.32	0.79	$\hat{\mu} = 957.51$
b	2.92	30.20	95.77	$\hat{\sigma} = 174.87$
k	27.09			
p		29.52	110.96	
q		58.11	18.89	

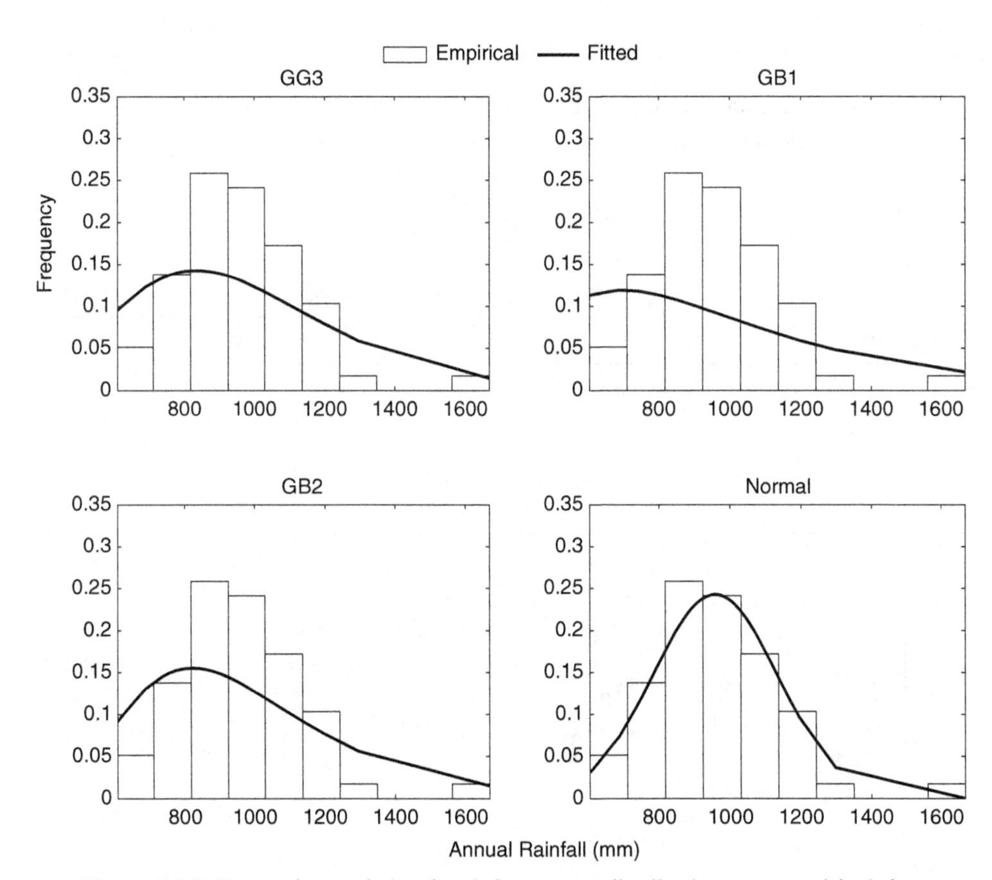

Figure 7.18 Comparison of the fitted frequency distributions to empirical frequency distribution to annual rainfall.

lists the estimated parameters. Figure 7.18 compares the fitted frequency distributions with the empirical frequency distribution. Comparisons in Figure 7.18 indicate that the normal distribution outperforms other distributions and may be applied to model annual rainfall. This is consistent with the statistics of annual rainfall data.

7.4 Conclusion

The Esteban system of frequency distributions is a general system that yields as special cases several generalized distributions that lead to a number of frequency distributions as special cases that are commonly used in hydrology, hydraulics, and water resources engineering. The sample applications prove that the distributions of Esteban system can be applied in water engineering.

References

Arnold, B. C. (1980). *Pareto distributions: Pareto and related heavy-tailed distributions*. Mimeographed manuscript, University of California at Riverside.

Cronin, D. C. (1979). A function for the size distribution of income: A further comment. *Econemetrica* 47, pp. 773–774.

Esteban, J. (1981). *Income-share elasticity, density function and the size distribution of income*. Mimeographed manuscript, University of Barcelona.

Esteban, J. (1986). Income-share elasticity and the size distribution of income. *International Economic Review* 27, pp. 439–444.

Johnson, N. L., and Kotz, S. (1970). *Continuous Univariate Distributions (1)*. New York: John Wiley & Sons.

Kalbfleisch, J. D., and Prentice, R. L. (1980). *The Statistical Analysis of Failure Time Data*. New York: John Wiley & Sons.

Kleiber, C., and Kotz, S. (2003). *Statistical Size Distributions in Economics and Actuarial Sciences*. Hoboken, NJ: John Wiley & Sons.

McDonald, J. B. (1984). Some generalized functions for the size distribution of income. *Econometrics* 52, no. 3, pp. 647– 663.

Tadikamalla, P. R. (1980). A look at the Burr and related distributions. *International Statistical Review* 48, pp. 337–344.

Thurow, L. C. (1970). Analyzing the American income distribution. *American Economic Review* 48, pp. 261–269.

Venter, G. (1984). Transformed beta and gamma distributions and aggregate losses. Proceedings of the Casualty Actuarial Society, pp. 156–193.

8

Singh System of Frequency Distributions

8.1 Introduction

A wide spectrum of frequency distributions is used in hydrologic, hydraulic, environmental, and water resources engineering. Many of these distributions have closed-form cumulative distribution functions (CDFs) but others do not. In practical applications, CDFs need to be often computed. If a CDF cannot be expressed in closed form, it is then computed numerically, for example, for lognormal, gamma, Pearson, and log-Pearson type III distributions. Based on empirical data from hydrology and hydraulics, it is possible to hypothesize a relation between the probability density function (PDF) and CDF. Using this hypothesis, it is then possible to derive CDFs for a number of frequency distributions. The objective of this chapter therefore is to derive a number of frequency distributions, based on empirical evidence from hydrology, that are commonly employed in environmental and water engineering.

8.2 Singh System of Distributions

It is considered that the CDF should be zero as the value of the random variable goes to lower limit and approaches unity as the value of the random variable tends to upper limit. Furthermore, the distribution should be unimodal. These considerations must be satisfied by any distribution.

The PDF, $f(x)$, and the CDF, $F(x)$, of a continuous random variable X are related as

$$\frac{dF(x)}{dx} = f(x) \tag{8.1}$$

Equation (8.1) contains two unknowns, $f(x)$ and $F(x)$, and therefore its solution requires a relation between $f(x)$ and $F(x)$. Based on empirical observations, it is hypothesized that

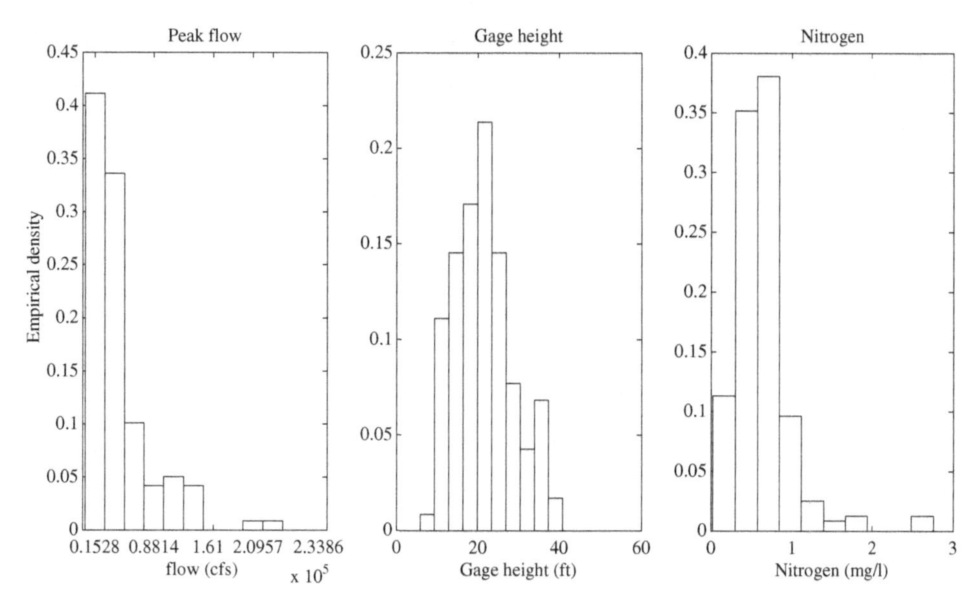

Figure 8.1 Empirical frequencies for peak flow, gage height, and nitrogen datasets.

$$f(x) \propto (F(x))^m \tag{8.2}$$

It can also be hypothesized that

$$f(x) \propto (1 - F(x))^n \tag{8.3}$$

where m and n are exponents. Combining Equations (8.2) and (8.3), it is postulated that

$$f(x) = g(x)(F(x))^m(1 - F(x))^n \tag{8.4}$$

where $g(x)$ is some function of x but can also be a constant.

Using the peak flow and gage height at USGS08096500 (Brazos River at Waco, Texas), and nitrogen close to USGS 08015500 (Calcasieu River near Kinder, Louisiana), Figure (8.1) plots the empirical frequency for the three datasets. It shows that peak flow (skewness = 2.20, kurtosis = 9.16) and nitrogen (skewness = 2.48, kurtosis = 13.6) datasets are skewed to the right with heavy tail compared with the gage height (skewness = 0.54, kurtosis = 2.83) dataset. Figures (8.2)–(8.4) plot the relations of $f(x)$ versus $F(x)$, $f(x)$ versus $1 - F(x)$, and $f(x)$ versus $(F(x))^m(1 - F(x))^n$ in the logarithm domain for peak flow, gage height, and nitrogen datasets.

Equation (8.4) can be expressed in a more general way as

$$f(x) = J(x, F) \tag{8.5}$$

where $J(x, F)$ is some function of x and F. However, for simplicity it is assumed that this function can be partitioned into two independent functions as: $J(x, F) = g(x)A(F)$. Therefore, Equation (8.4) can be written as

$$\frac{dF(x)}{dx} = g(x)A(F), A(F) = (F(x))^m(1 - F(x))^n \tag{8.6}$$

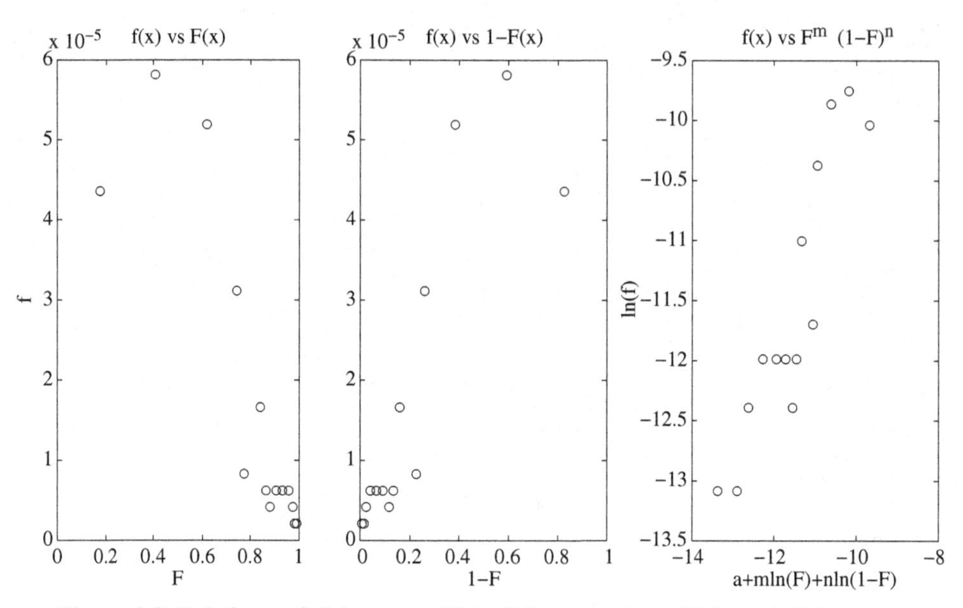

Figure 8.2 Relations of $f(x)$ versus $F(x)$, $f(x)$ versus $1 - F(x)$, and $f(x)$ versus $(F(x))^m(1 - F(x))^n$ in the logarithm domain for peak flow.

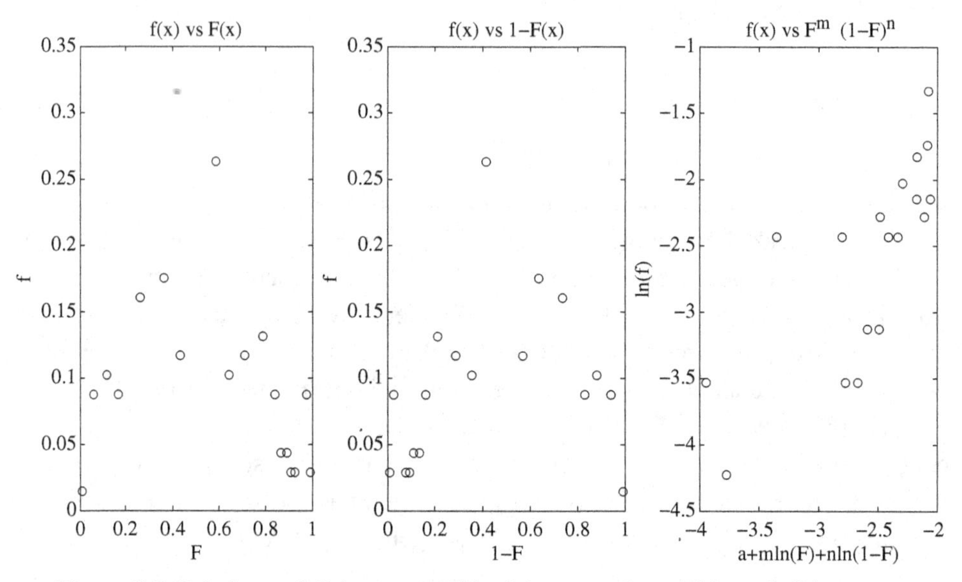

Figure 8.3 Relations of $f(x)$ versus $F(x)$, $f(x)$ versus $1 - F(x)$, and $f(x)$ *versus* $(F(x))^m(1 - F(x))^n$ in the logarithm domain for gage height.

Equation (8.6) can be solved for $F(x)$ for given values of m and n and a given form of $g(x)$. Hence, a large number of frequency distributions can be derived by solving this equation for different values of m, n, and different forms of $g(x)$, and this framework can be regarded as a generalized framework. One immediate advantage

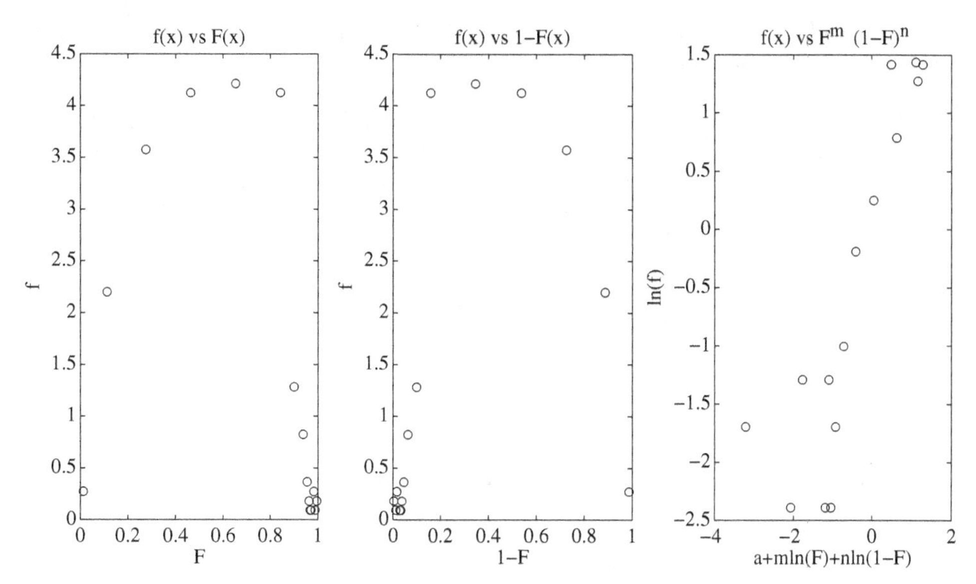

Figure 8.4 Relations of $f(x)$ versus $F(x)$, $f(x)$ versus $1 - F(x)$, and $f(x)$ versus $(F(x))^m(1 - F(x))^n$ in the logarithm domain for nitrogen.

of this framework is that it yields CDF, which can be differentiated to produce PDF but the reverse is more difficult, as is the case with the Pearson system. It can be shown that Equation (8.6) satisfies the first two considerations and the consideration depending on the form of $g(x)$. In what follows, we will use several examples to illustrate the concept.

Example 8.1 Determine function $g(x)$ and exponents m, n for uniform distribution

Solution: For the uniform probability distribution, its PDF $f(x)$ and CDF $F(x)$ are written, respectively, as:

$$f(x) = \frac{1}{b-a} \tag{8.7}$$

$$F(x) = \frac{x-a}{b-a} \tag{8.8}$$

Equation (8.7) can be written as

$$f(x) = \frac{1}{x-a}\frac{x-a}{b-a} = g(x)F(x), g(x) = \frac{1}{x-a} \tag{8.9}$$

Equation (8.9) shows that $m = 1$ and $n = 0$. Differentiating Equation (8.7) one obtains,

$$\frac{df(x)}{dx} = 0 \tag{8.10}$$

Equation (8.10) shows that Equation (8.7) does not have any mode.

Example 8.2 Determine function $g(x)$ and exponents m, n for the exponential distribution

Solution: For the exponential distribution, its PDF and CDF can be written, respectively, as:

$$f(x) = \frac{1}{k} \exp\left(-\frac{x}{k}\right); x > 0 \tag{8.11}$$

$$F(x) = 1 - \exp\left(-\frac{x}{k}\right); x > 0 \tag{8.12}$$

Equation (8.11) can be expressed as

$$f(x) = \frac{1}{k}(1 - F(x)) = g(x)(1 - F(x)), g(x) = \frac{1}{k} \tag{8.13}$$

Equation (8.13) shows that $m = 0$ and $n = 1$. Differentiating Equation (8.11) we have:

$$\frac{df}{dx} = -\frac{1}{k^2} \exp\left(-\frac{x}{k}\right) < 0 \tag{8.14}$$

Equation (8.14) shows that density function of exponential distribution is a decreasing function with single mode at 0.

Example 8.3 Determine function $g(x)$ and exponents m, n for the Extreme Value Type I (EV I) distribution

Solution: For EV I distribution (may also be called as Gumbel distribution), its PDF and CDF can be written, respectively, as:

$$\begin{aligned} f(x) &= a \exp(-a(x-b) - \exp(-a(x-b))); a > 0, b \in \mathbb{R} \\ &= a \exp(-a(x-b)) \exp(-\exp(-a(x-b))) \end{aligned} \tag{8.15}$$

$$F(x) = \exp(-\exp(-a(x - b))) \tag{8.16}$$

Based on Equation (8.6), Equation (8.15) can be rewritten as:

$$f(x) = a \exp(-a(x-b)) F(x)(1 - F(x))^0 = g(x)F(x) \tag{8.17}$$

Equation (8.17) shows that for EV I distribution, $g(x) = a \exp(-a(x-b))$, $m = 1, n = 0$. Differentiating Equation (8.15) we have:

$$\frac{df}{dx} = -a \exp(-a(x-b) - \exp(-a(x-b)))(a - a \exp(-a(x-b))) \tag{8.18}$$

In Equation (8.18), $\frac{df}{dx} > 0, \textit{if } x \leq b$ (i.e., $f(x)$ is an increasing function if $x \leq b$), and $\frac{df}{dx} < 0, \textit{if } x > b$ (i.e., $f(x)$ is a decreasing function if $x > b$). Thus, there also exists a single mode for EV I distribution at $x = b$.

Example 8.4 Determine function $g(x)$ and exponents m, n for the logistic distribution

Solution: For the logistic distribution, its PDF and CDF can be written, respectively, as:

$$f(x) = \frac{\exp\left(-\frac{(x-b)}{a}\right)}{a\left(1+\exp\left(-\frac{x-b}{a}\right)\right)^2}; a > 0, b \in \mathbb{R} \tag{8.19}$$

$$F(x) = \frac{1}{1+\exp\left(-\frac{x-b}{a}\right)} \tag{8.20}$$

According Equation (8.6), Equation (8.19) can be rewritten as:

$$\begin{aligned} f(x) &= \frac{\exp\left(-\frac{x-b}{a}\right)}{a}\left(\frac{1}{1+\exp\left(-\frac{x-b}{a}\right)}\right)^2 \\ &= \frac{\exp\left(-\frac{x-b}{a}\right)}{a}[F(x)]^2[1-F(x)]^0 \end{aligned} \tag{8.21}$$

In Equation (8.21), $g(x) = \frac{\exp\left(-\frac{x-b}{a}\right)}{a}$, $m = 2, n = 0$. Differentiating Equation (8.19) we have:

$$\frac{df}{dx} = \frac{\exp\left(-\frac{x-b}{a}\right)\left(\exp\left(-\frac{x-b}{a}\right)-1\right)}{a^2\left(1+\exp\left(-\frac{x-b}{a}\right)\right)^3} \tag{8.22}$$

From Equation (8.22) we have that $f(x)$ is an increasing function if $x \leq b$, i.e., $\frac{df}{dx} \geq 0, if\ x \leq b$, and $f(x)$ is a decreasing function if $x > b$, i.e., $\frac{df}{dx} < 0, if\ x > b$. Thus, there exists a single mode for logistic distribution at $x = b$.

In this manner, a number of distributions can be derived as shown in Tables 8.1–8.3. Tables 8.2–8.3 list the derived frequency distributions for Burr (1942) and Stoppa (1990, 1993) systems. Comparing the hypotheses of the Singh system with those of Burr and Stoppa systems, it is seen that all three systems require the following: (i) CDF has a closed form; and (ii) PDF may be written as a function of CDF, its survival function, and some function of x. The Burr and Stoppa systems may also be considered as special cases of the Singh system. More specifically

Table 8.1 *Distributions derived using the proposed system (BMS = Burr-Singh-Moddala; MJ = Marczewski-Jaronice)*

Distribution	PDF	CDF	*m, n*	*g*(*x*)
Uniform	$f(x) = \frac{1}{b-a}$	$F(x) = \frac{x-a}{b-a}$	$m = 1$ $n = 0$	$g(x) = \frac{1}{x-a}$
Exponential	$f(x) = \frac{1}{k}\exp\left(-\frac{x}{k}\right)$	$F(x) = 1 - \exp\left(-\frac{x}{k}\right)$	$m = 0$ $n = 1$	$g(x) = \frac{1}{k}$
EV I	$f(x) = a\exp(-a(x-b) - \exp(-a(x-b))$	$F(x) = \exp(-\exp(-a(x-b)))$	$m = 1$ $n = 0$	$g(x) = a\exp(-a(x-b))$
Log-EV I	$f(x) = \frac{a}{x}\exp(-a(\ln x - b) - \exp(-a(\ln x - b)))$	$F(x) = \exp(-\exp(-a(\ln x - b)))$	$m = 1$ $n = 0$	$g(x) = \frac{a}{x}\exp(-a(\ln x - b))$
EV III	$f(x) = \frac{a}{b-c}\left(\frac{x-c}{b-c}\right)^{a-1}\exp\left(-\left(\frac{x-c}{b-c}\right)^{a}\right)$	$F(x) = \exp\left(-\left(\frac{x-c}{b-c}\right)^{a}\right)$	$m = -1$ $n = 0$	$g(x) = \frac{a}{b-c}\left(\frac{x-c}{b-c}\right)^{a-1}$
GEV	$f(x) = \frac{1}{a}\left(1 - \frac{b}{a}(x-c)\right)^{\frac{1-b}{b}}\exp\left(-\left(1 - \frac{b}{a}(x-c)\right)^{\frac{1}{b}}\right)$	$F(x) = \exp\left(-\left(1 - \frac{b}{a}(x-c)\right)^{\frac{1}{b}}\right)$	$m = 1$ $n = 0$	$g(x) = \frac{1}{a}\left(1 - \frac{b}{a}(x-c)\right)^{\frac{1-b}{b}}$
Weibull	$f(x) = \frac{a}{b}\left(\frac{x}{b}\right)^{a-1}\exp\left(-\left(\frac{x}{b}\right)^{a}\right)$	$F(x) = \exp\left(-\left(\frac{x}{b}\right)^{a}\right)$	$m = 1$ $n = 0$	$g(x) = \frac{a}{b}\left(\frac{x}{b}\right)^{a-1}$
Generalized Weibull	$f(x) = \frac{a}{b}x^{a-1}\left(1 - \frac{1-b}{2-b}\left(\frac{x^a}{b}\right)\right)^{\frac{1}{1-b}}$	$F(x) = 1 - \left(1 - \frac{1-b}{2-b}\frac{x^a}{b}\right)^{\frac{2-b}{1-b}}$	$m = 1$ $n = 0$	$g(x) = \frac{a}{b}x^{a-1}\left(1 - \frac{1-b}{2-b}\left(\frac{x^a}{b}\right)\right)^{-1}$
Logistic	$f(x) = \frac{\exp\left(-\frac{(x-b)}{a}\right)}{a\left(1+\exp\left(-\frac{x-b}{a}\right)\right)^2}$	$F(x) = \frac{1}{1+\exp\left(-\frac{x-b}{a}\right)}$	$m = 2$ $n = 0$	$g(x) = \frac{1}{a}\exp\left(-\frac{x-b}{a}\right)$
Log-logistic	$f(x) = \frac{a}{x}\frac{\left(\frac{x}{b}\right)^a}{\left(1+\left(\frac{x}{b}\right)^a\right)^2}$	$F(x) = 1 - \left(1 + \left(\frac{x}{b}\right)^a\right)^{-1}$	$m = 0$ $n = 1$	$g(x) = \frac{a}{x}\left(\frac{x}{b}\right)^a\left(1 + \left(\frac{x}{b}\right)^a\right)^{-1}$

2-parameter Pareto	$f(x) = ba^b x^{-1-b}$	$F(x) = 1 - \left(\frac{x}{a}\right)^{-b}$	$m = 0$ $n = 1$	$g(x) = \frac{b}{x}$
Generalized Pareto	$f(x) = \begin{cases} \frac{1}{a}\left(1 - \frac{b}{a}x\right)^{\frac{1-b}{b}}; b \neq 0 \\ \frac{1}{a}\exp\left(-\frac{x}{a}\right); b = 0 \end{cases}$	$F(x) = \begin{cases} 1 - \left(1 - \frac{b}{a}x\right)^{\frac{1}{b}}; b \neq 0 \\ 1 - \exp\left(-\frac{x}{a}\right); b = 0 \end{cases}$	$m = 0$ $n = 1$	$g(x) = \begin{cases} \frac{1}{a}\left(1 - \frac{b}{a}x\right)^{-1}; b \neq 0 \\ \frac{1}{a}; \quad b = 0 \end{cases}$
3-Parameter Pareto	$f(x) = \begin{cases} \frac{1}{a}\left(1 - \frac{b}{a}(x-c)\right)^{\frac{1-b}{b}}; b \neq 0 \\ \frac{1}{a}\exp\left(-\frac{x-c}{a}\right); b = 0 \end{cases}$	$F(x) = \begin{cases} 1 - \left(1 - \frac{b}{a}(x-c)\right)^{b}; b \neq 0 \\ 1 - \exp\left(-\frac{x-c}{a}\right); b = 0 \end{cases}$	$m = 0$ $n = 1$	$g(x) = \begin{cases} \frac{1}{a}\left(1 - \frac{b}{a}x\right)^{-1}; b \neq 0 \\ \frac{1}{a}; \quad b = 0 \end{cases}$
Lomax	$f(x) = \frac{k}{a}\left(1 + \frac{x}{a}\right)^{-k-1}$	$F(x) = 1 - \left(1 + \frac{x}{a}\right)^{-k}$	$m = 0$ $n = 1$	$g(x) = \frac{k}{a}\left(1 + \frac{x}{a}\right)^{-1}$
Langmuir I	$f(x) = \frac{1}{b+x} - \frac{x}{(b+x)^2}$	$F(x) = \frac{x}{b+x}$	$m = 0$ $n = 1$	$g(x) = \frac{1}{b+x}$
Langmuir II	$f(x) = \frac{a}{b}\left(\frac{x}{b}\right)^{a-1}\left(1 + \left(\frac{x}{b}\right)^a\right)^{-2}$	$F(x) = 1 - \left(1 + \left(\frac{x}{b}\right)^a\right)^{-1}$	$m = 0$ $n = 1$	$g(x) = \frac{a}{b}\left(\frac{x}{b}\right)^{a-1} \left(1 + \left(\frac{x}{b}\right)^a\right)^{-1}$
Dagum I	$f(x) = \frac{a}{b}\left(\frac{x}{b}\right)^{-c-1} \left(1 + \left(\frac{x}{b}\right)^{-c}\right)^{-\left(\frac{a+c}{c}\right)}$	$F(x) = \left(1 + \left(\frac{x}{b}\right)^{-c}\right)^{-\frac{a}{c}}$	$m = 1$ $n = 0$	$g(x) = \frac{1}{x}\left(\frac{b}{x}\right)^c \left(1 + \left(\frac{x}{b}\right)^{-c}\right)^{-1}$
Dagum II	$f(x) = \frac{cp}{b}(1-a) \left(1 + \left(\frac{x}{b}\right)^c\right)^{-p-1}\left(\frac{x}{b}\right)^{-c-1}$	$F(x) = a + (1-a)\left(1 + \left(\frac{x}{b}\right)^{-c}\right)^{-p}$; a<0	$m = 0$ $n = 1$	$g(x) = \frac{cp}{b}\left(1 + \left(\frac{x}{b}\right)^{-c}\right)^{-1} \left(\frac{x}{b}\right)^{-c-1}$
Dagum III	$f(x) = \frac{cp}{b}(1-a) \left(1 + \left(\frac{x}{b}\right)^{-c}\right)^{-p-1}\left(\frac{x}{b}\right)^{-c-1}$	$F(x) = a + (1-a)\left(1 + \left(\frac{x}{b}\right)^{-c}\right)^{-p}$; a>0	$m = 0$ $n = 1$	$g(x) = \frac{cp}{b}\left(1 + \left(\frac{x}{b}\right)^{-c}\right)^{-1} \left(\frac{x}{b}\right)^{-c-1}$

Table 8.1 (*cont.*)

Distribution	PDF	CDF	m, n	$g(x)$
Fisk	$f(x) = \frac{a}{b}\left(\frac{x}{b}\right)^{a-1}\left(1+\left(\frac{x}{b}\right)^{a}\right)^{-2}$	$F(x) = 1-\left(1+\left(\frac{x}{b}\right)^{a}\right)^{-1}$	$m = 0$ $n = 1$	$g(x) = \frac{a}{b}\left(\frac{x}{b}\right)^{a-1}\left(1+\left(\frac{x}{b}\right)^{a}\right)^{-1}$
BMS	$f(x) = \frac{a}{b}\left(\frac{x}{b}\right)^{a-1}\left(1+c\left(\frac{x}{b}\right)^{a}\right)^{-\frac{1}{c}-1}$	$F(x) = 1-\left(1+c\left(\frac{x}{b}\right)^{a}\right)^{-\frac{1}{c}}$	$m = 0$ $n = 1$	$g(x) = \frac{a}{b}\left(\frac{x}{b}\right)^{a-1}$ $\left(1+c\left(\frac{x}{b}\right)^{a}\right)^{-1}$
GB XII	$f(x) = \frac{acp}{b}\left(\frac{x}{b}\right)^{a-1}\left(1+\left(\frac{x}{b}\right)^{a}\right)^{-c-1}$ $\left(1-\left(1+\left(\frac{x}{b}\right)^{a}\right)^{c}\right)^{p-1}$	$F(x) = \left(1-\left(1+\left(\frac{x}{b}\right)^{a}\right)^{-c}\right)^{p}$	$m = 1$ $n = 0$	$g(x) = \frac{acp}{b}\left(\frac{x}{b}\right)^{a-1}$ $\left(1+\left(\frac{x}{b}\right)^{a}\right)^{-c-1}$ $\left(1-\left(1+\left(\frac{x}{b}\right)^{a}\right)^{c}\right)^{-1}$
2-parameter kappa	$f(x) = \frac{a}{b}\frac{\left(\frac{x}{b}\right)^{a-1}}{\left(a+\left(\frac{x}{b}\right)^{a}\right)^{\frac{1}{a}}}$ $\left(1-\frac{\left(\frac{x}{b}\right)^{a}}{b\left(a+\left(\frac{x}{b}\right)^{a}\right)^{\frac{1}{a}}}\right)$	$F(x) = \frac{\left(\frac{x}{b}\right)^{a}}{\left(a+\left(\frac{x}{b}\right)^{a}\right)^{\frac{1}{a}}}$	$m = 1$ $n = 0$	$g(x) = \frac{1}{b}\left(a\left(\frac{x}{b}\right)^{-1} -\right.$ $\left.\left(\frac{x}{b}\right)^{a-1}\left(1+\left(\frac{x}{b}\right)^{a}\right)^{-1}\right)$
3-parameter kappa	$f(x) = \frac{\left(\frac{x}{b}\right)^{ac}}{\left(a+\left(\frac{x}{b}\right)^{ac}\right)}\frac{ac}{x}$ $\left(1-\left(\frac{x}{b}\right)^{ac}\left(a+\left(\frac{x}{b}\right)^{ac}\right)^{-1}\right)$	$F(x) = \frac{\left(\frac{x}{b}\right)^{ac}}{a+\left(\frac{x}{b}\right)^{ac}}$	$m = 1$ $n = 1$	$g(x) = \frac{ac}{b}\left(\frac{x}{b}\right)^{-1}$
Beta-kappa	$f(x) = \frac{c\left(\frac{x}{b}\right)^{a-1}}{b\left(1+\left(\frac{x}{b}\right)^{c}\right)}$ $\left(1-\left(\frac{x}{b}\right)^{c}\left(1+\left(\frac{x}{b}\right)^{c}\right)^{-1}\right)$	$F(x) = \frac{\left(\frac{x}{b}\right)^{c}}{\left(1+\left(\frac{x}{b}\right)\right)^{a}}$	$m = 1$ $n = 1$	$g(x) = \frac{c}{b}\left(\frac{x}{b}\right)^{-1}$
MJ	$f(x) = cb^{a}x^{-a-1}$ $\left(1+\left(\frac{x}{b}\right)^{-a}\right)^{-\frac{c}{a}-1}$	$F(x) = \left(1+\left(\frac{x}{b}\right)^{-a}\right)^{-\frac{c}{a}}$	$m = 1$ $n = 0$	$g(x) = cb^{a}x^{-a-1}$ $\left(1+\left(\frac{x}{b}\right)^{-a}\right)^{-1}$

2-parameter Benin	$f(x) = \frac{2b}{x} \exp \left(-b \left(\log \left(\frac{x}{x_0} \right) \right)^2 \right) \log \left(\frac{x}{x_0} \right)$	$F(x) = 1 - \exp \left(-b \left(\log \left(\frac{x}{x_0} \right) \right)^2 \right)$ $x > x_0$	$m = 0$ $n = 1$	$g(x) = \frac{2b}{x} \log \left(\frac{x}{x_0} \right)$
3-parameter Benin	$f(x) = \exp \left(-a \log \left(\frac{x}{x_0} \right) - b \left(\log \left(\frac{x}{x_0} \right) \right)^2 \right) \left(\frac{a}{x} + \frac{2b}{x} \log \left(\frac{x}{x_0} \right) \right)$	$F(x) = 1 - \exp \left(-a \log \left(\frac{x}{x_0} \right) - b \left(\log \left(\frac{x}{x_0} \right) \right)^2 \right)$ $x > x_0$	$m = 0$ $n = 1$	$g(x) = \frac{a}{x} + \frac{2b}{x} \log \left(\frac{x}{x_0} \right)$
Verhulst	$f(x) = \frac{\exp(-x)}{(1+\exp(-x))^2}$	$F(x) = \frac{1}{1+\exp(-x)}$	$m = 2$ $n = 0$	$g(x) = \exp(-x)$
General Conic Dist.	$f(x) = \frac{1}{x} \exp \left(-a\sqrt{1 + (\ln x - b)^2} + c(\ln x - b) + p \right) \left(\frac{a(\ln x - b)}{\sqrt{1+(\ln x - b)^2}} - c \right)$	$F(x) = 1 - \exp \left(-a\sqrt{1 + (\ln x - b)^2} + c(\ln x - b) + p \right)$ $0 < x \leq x_0;\ a > 0,$ $-\infty < c < a$ $p = a\sqrt{1 + (\ln x - b)^2} - c(\ln x_0 - b)\ \frac{a(\ln x_0 - b)}{\sqrt{1+(\ln x - b)^2}} - c \geq 0$	$m = 0$ $n = 1$	$g(x) = \frac{1}{x} \left(\frac{a(\ln x - b)}{\sqrt{1+(\ln x - b)^2}} - c \right)$

Table 8.2 *The Burr system of distributions derived using the proposed system [*x $\in$ *(−∞, +∞), unless specified otherwise*

Distribution	$f(x)$	$F(x)$	m, n	$g(x)$
I	$f(x) = 1$	$F(x) = x,\ x \in [0, 1]$	$m = 1$ $n = 1$	$g(x) = 1$
II	$f(x) = r\exp(-x)(\exp(-x) + 1)^{-r-1}$	$F(x) = (\exp(-x) + 1)^{-r}$	$m = 1$ $n = 1$	$g(x) = r\exp(-x)(\exp(-x) + 1)^{-1}$
III	$f(x) = rkx^{-k-1}(1 + x^{-k})^{-r-1}$	$F(x) = (1 + x^{-k})^{-r},\ x \in (0, \infty)$	$m = 1$ $n = 1$	$g(x) = rkx^{-k-1}(1 + x^{-k})^{-1}$
IV	$f(x) = \frac{r}{c}\left(1 + \left(\frac{c-x}{x}\right)^{\frac{1}{c}}\right)^{-r-1} \times \left(\frac{c-x}{x}\right)^{\frac{1}{c}-1}\left(\frac{c}{x} - \frac{c-x}{x^2}\right)$	$F(x) = \left(1 + \left(\frac{c-x}{x}\right)^{\frac{1}{c}}\right)^{-r},\ x \in (0, c)$	$m = 1$ $n = 1$	$g(x) = \frac{r}{c}\left(1 + \left(\frac{c-x}{x}\right)^{\frac{1}{c}}\right)^{-1}\left(\frac{c-x}{x}\right)^{\frac{1}{c}-1}\left(\frac{c}{x} + \frac{c-x}{x^2}\right)$
V	$f(x) = rk(1 + k\exp(-\tan x))^{-r-1} \times \exp(-\tan x)\sec^2 x$	$F(x) = (1 + k\exp(-\tan x))^{-r}$ $x \in \left(-\frac{\pi}{2}, \frac{\pi}{2}\right)$	$m = 1$ $n = 1$	$g(x) = rk(1 + k\exp(-\tan x))^{-1} \times \exp(-\tan x)\sec^2 x$
VI	$f(x) = kc\cosh x\exp(-c\sinh x) \times (1 + k\exp(-c\sinh x))^{-r-1}$	$F(x) = (k\exp(-c\sinh x) + 1)^{-r}$	$m = 1$ $n = 1$	$g(x) = kc\cosh x\exp(-c\sinh x) \times (1 + k\exp(-c\sinh x))^{-1}$
VII	$f(x) = 2^{-r}r(1 + \tanh x)^{r-1}\operatorname{sech}^2 x$	$F(x) = 2^{-r}(1 + \tanh x)^{r}$	$m = 1$ $n = 1$	$g(x) = 2^{-r}r(1 + \tanh x)^{-1}\sinh^2 x$
VIII	$f(x) = \frac{2r}{\pi}\left(\frac{2}{\pi}\tan^{-1}(\exp(x))\right)^{r-1} \times \frac{\exp(x)}{1+\exp(2x)}$	$F(x) = \left(\frac{2}{\pi}\tan^{-1}(\exp(x))\right)^{r}$	$m = 1$ $n = 1$	$g(x) = \frac{2r}{\pi}\left(\frac{2}{\pi}\tan^{-1}(\exp(x))\right)^{-1}\frac{\exp(x)}{1+\exp(2x)}$
IX	$f(x) = \frac{2kr(1+\exp(x))^{r-1}\exp(x)}{(k((1+\exp(x))^{r}-1)+2)^2}$	$F(x) = 1 - \frac{2}{k((1+\exp(x))^{r}-1)+2}$	$m = 1$ $n = 1$	$g(x) = \frac{kr\exp(x)(1+\exp(x))^{r-1}}{k((1+\exp(x))^{r}-1)+2}$
X	$f(x) = 2rx(1 - \exp(-x^2))^{r-1} \times \exp(-x^2)$	$F(x) = (1 - \exp(-x^2))^{r}\ x \in (0, \infty)$	$m = 1$ $n = 1$	$g(x) = 2rx(1 - \exp(-x^2))^{-1}\exp(-x^2)$
XI	$f(x) = r\left(x - \frac{\sin 2\pi x}{2\pi}\right)^{r-1} \times (1 - \cos 2\pi x)$	$F(x) = \left(x - \frac{\sin 2\pi x}{2\pi}\right)^{r},\ x \in [0, 1]$	$m = 1$ $n = 1$	$g(x) = r\left(x - \frac{\sin 2\pi x}{2\pi}\right)^{-1}(1 - \cos 2\pi x)$
XII	$f(x) = kcx^{c-1}(1 + x^{c})^{-k-1}$	$F(x) = 1 - (1 + x^{c})^{-k},\ 0 \le x < \infty$ $F(x) = 1 - (1 + x^{c})^{-k},\ x \in [0, \infty)$	$m = 1$ $n = 1$	$g(x) = kcx^{c-1}(1 + x^{c})^{-1}$

Table 8.3 *Stoppa distributions derived using the proposed system*

Distribution	$f(x)$	$F(x)$	m, n	$g(x)$
I	$f(x) = ab(bx)^{a-1}$	$F(x) = (bx)^a,\ x \in \left(0, \frac{1}{b}\right)$	$m = 1$ $n = 0$	$g(x) = \frac{a}{x}$
II	$f(x) = abx^{-b-1}(1 - x^{-b})^{a-1}$	$F(x) = (1 - x^{-b})^a\ \ x \in (0, \infty)$	$m = 1$ $n = 0$	$g(x) = \frac{ab}{x^{1+b}(1-x^{-b})}$
III	$f(x) = ab \exp(-bx) \times (1 - \exp(-bx))^{a-1}$	$F(x) = (1 - \exp(-bx))^a$	$m = 1$ $n = 0$	$g(x) = \frac{ab \exp(-bx)}{1-\exp(-bx)}$
IV	$f(x) = x^{-2}\left(1 - (bx^{-1} - 1)^{\frac{1}{ab}}\right)^{a-1} \times (bx^{-1} - 1)^{\frac{1}{ab}-1}$	$F(x) = \left(1 - (bx^{-1} - 1)^{\frac{1}{ab}}\right)^a$	$m = 1$ $n = 0$	$g(x) = x^{-2}\left(1 - (bx^{-1} - 1)^{\frac{1}{ab}}\right)^{-1} \times (bx^{-1} - 1)^{\frac{1}{ab}-1}$

$m = n = 1$ of the differential equation for the Singh system [i.e., Equation 8.6] yields the Burr system.

8.3 Conclusion

The following conclusions can be drawn from this study: (1) Empirical data from hydrology show that the PDF is a function of the survival function (1-CDF) and CDF; (2) the functional relation between PDF and CDF depends on the data and the resulting distribution; and (3) the Singh system may be considered as the generalized Burr and Stoppa system.

References

Burr, I. W. (1942). Cumulative frequency functions. *Annals of Mathematical Statistics* 13, pp. 215–232.

Singh, V. P. (2018). System of frequency distributions for water and environmental engineering, *Physica A* 506, pp. 50–74. doi: 10.1016/j.physa.2018.03.038.

Stoppa, G. (1990). A new generating system of income distribution models. *Quarderni di Statistica e Mathematica Applicata alle Science Economico-Sociali* 12, pp. 47–55.

Stoppa, G. (1993). Una tavola per modeli di probabilita. *Metron* 51, pp. 99–117.

9

Systems of Frequency Distributions Using Bessel Functions and Cumulants

9.1 Introduction

The preceding chapters have discussed several approaches for deriving different systems of frequency distributions. Indeed, these approaches together lead to virtually all the distributions used in environmental and water engineering. The objective of this chapter is to discuss the methods of Bessel functions and expansions in terms of cumulants or moments and resulting systems of distributions. Some of the distributions can be useful in hydrology, environmental, and water engineering.

9.2 Bessel Function Distributions

McKay (1932) proposed a family of Bessel function distributions for fitting data for which $(\beta_1 - 3)/\beta_2 > 1.5$ falls below the Pearson type III curve, where β_1 represents the third moment and β_2 represents the fourth moment. At least one distribution would correspond to one pair of (β_1, β_2). He provided two types of distributions that can be expressed in general as a product of three functions as functions of random variable X as:

$$f(x) = f_0 f_1(x) f_2(x) \tag{9.1}$$

where

$$f_0 = \frac{\left|1 - c^2\right|^{m+\frac{1}{2}}}{\pi^{\frac{1}{2}} 2^m b^{m+1} \Gamma\left(m + \frac{1}{2}\right)} \tag{9.2}$$

$$f_1(x) = \exp\left(-\frac{cx}{b}\right)|x|^m \tag{9.3}$$

and

$$f_2(x) = \begin{cases} \pi I_m \left|\frac{x}{b}\right| \\ or \\ K_m \left|\frac{x}{b}\right| \end{cases} \tag{9.4}$$

where $f(x)$ is the probability density function (PDF), b is a parameter, c is a parameter, m is an exponent and order of the Bessel function, I_m is the modified Bessel function of the first kind of order m defined as

$$I_m\left(\frac{x}{b}\right) = \left(\frac{x}{2b}\right)^m \sum_{r=0}^{\infty} \frac{\left(\frac{x}{2b}\right)^{m+2r}}{\Gamma(r+1)\Gamma(m+r+1)}, \tag{9.5}$$

and K_m is the modified Bessel function of the second kind of order m defined as

$$K_m\left(\frac{x}{b}\right) = \frac{\Gamma\left(m+\frac{1}{2}\right)\left(\frac{x}{2b}\right)^m}{\sqrt{\pi}} \int_0^{\infty} \frac{\cos t}{\left(t^2 + \frac{x^2}{b^2 t^2}\right)^{m+\frac{1}{2}}} dt \tag{9.6}$$

If $|c| > 1$, the modified Bessel function of the first kind is applied for the PDF with $x \in (0, \infty)$ or $c > 1$ and $x \in (-\infty, 0)$ or $c < -1$. However, if $|c| < 1$, the modified Bessel function of the second kind is applied for the PDF with $x \in (-\infty, \infty)$. The parameters b and m satisfy $b > 0$ and $m > -\frac{1}{2}$. Johnson et al. (1994) stated that the distribution with modified Bessel function of the second kind was more useful than that with the first kind. Figure 9.1 plots the density functions for Bessel functions of the first and second kind.

9.2.1 Moments of Bessel Function Distributions

Following McKay (1932), the first four moments of the Bessel function distribution may be expressed, respectively, as:

$$\mu = \frac{bc(2m+1)}{c^2 - 1} \tag{9.7}$$

$$\sigma_x^2 = \frac{b^2(c^2+1)(2m+1)}{(c^2-1)^2} \tag{9.8}$$

$$\beta_1 = \frac{4c^2(c^2+3)^2}{(2m+1)(c^2+1)^3} \tag{9.9}$$

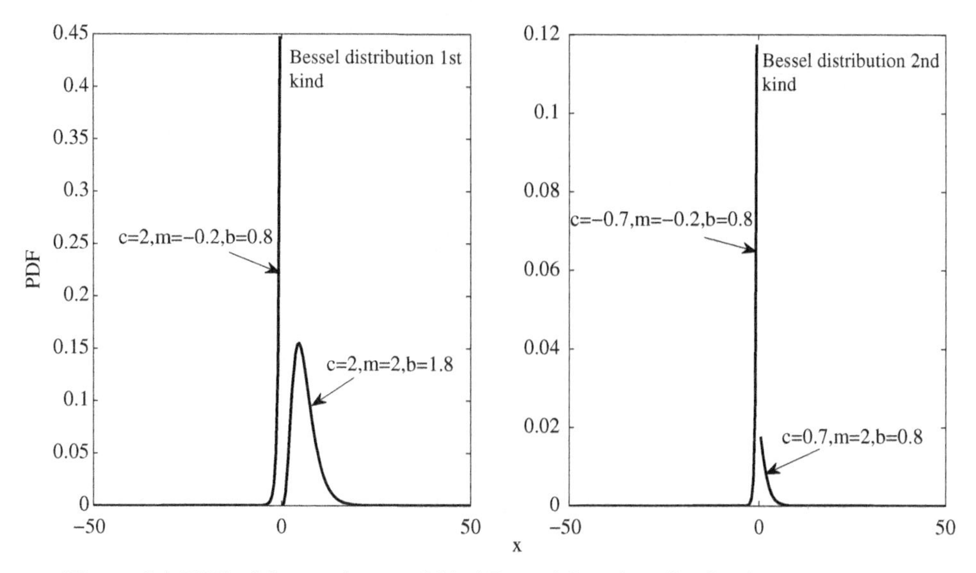

Figure 9.1 PDF of first and second kind Bessel function distributions.

$$\beta_2 = 3 + \frac{6(c^4 + 6c^2 + 1)}{(2m+1)(c^2+1)^2} \tag{9.10}$$

From Equations (9.10) and (9.11), one can write

$$2c^2\left(c^2+3\right)^2\left(\frac{\beta_2 - 3}{\beta_1}\right) - 3\left(c^2+1\right)\left(c^4+6c^2+1\right) = 0 \tag{9.11}$$

Equation (9.11) can be considered as a cubic equation in c^2. It can be shown that this equation has three distinctive positive roots of c^2 (one root less than 1 and two roots greater than 1) with the following constraint:

$$1.5\beta_1 + 3 < \beta_2 < 1.57735\beta_1 + 3 \tag{9.12}$$

Within this range, both Bessel function (first kind) and (second kind) distributions may be obtained.

9.2.2 Bessel Function Line

The Bessel function line (McKay, 1932) is defined as:

$$\beta_2 - 1.57735\beta_1 - 3 = 0 \tag{9.13}$$

The Bessel function lines are shown in Figure 9.2.

Along the Bessel function line, the roots of c^2 are evaluated as: $c^2_{(1)} = c^2_{(2)} = 3 + 2\sqrt{3}, c^2_{(3)} = 2\sqrt{3} - 3$. This leads to a unique Bessel distribution.

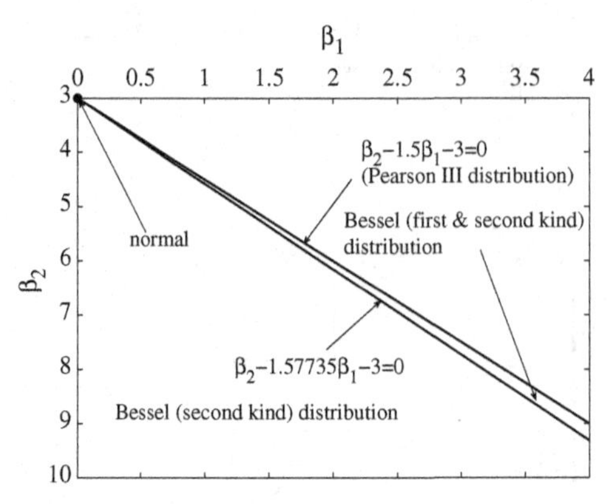

Figure 9.2 Graph of Bessel function lines.

It is interesting to note that the Pearson type III distribution is obtained on the line

$$\beta_2 - 1.5\beta_1 - 3 = 0 \tag{9.14}$$

Along the line for Pearson type III distribution, one has: $c_{(1)}^2 = c_{(2)}^2 = 1, c_{(3)}^2 = \infty$. Furthermore, if $\beta_1 = 0$ in Equation (9.14), the normal curve is obtained.

Finally, there only exists one positive root, which is less than 1 if

$$\beta_2 - 1.57735\beta_1 - 3 > 0 \tag{9.15}$$

In this region, the Bessel function distribution (second kind) is obtained.

9.2.3 Inverse Gaussian Distribution

Barndorff-Nielsen (1997) proposed a normal inverse Gaussian distribution, which is expressed through the modified Bessel function of the third kind with order 1 (also called modified Bessel function of the second kind with order 1). The normal inverse Gaussian distribution (shown in Figure 9.3) is a 4-parameter distribution that may be expressed as:

$$f(x;\alpha,\beta,c,\delta) = \frac{\alpha}{\pi}p(x)(q(x))^{-1}K_1(\alpha\delta q(x)); x \in \mathbb{R}, c \in \mathbb{R}, |\beta| \le \alpha, \delta > 0 \tag{9.16}$$

In Equation (9.16), $p(x)$ and $q(x)$ are expressed as:

$$p(x) = \exp\left(\delta\sqrt{\alpha^2 - \beta^2} - \beta c + \beta x\right) \tag{9.16a}$$

$$q(x) = \frac{1}{\delta}\sqrt{\delta^2 + (x-c)^2} \tag{9.16b}$$

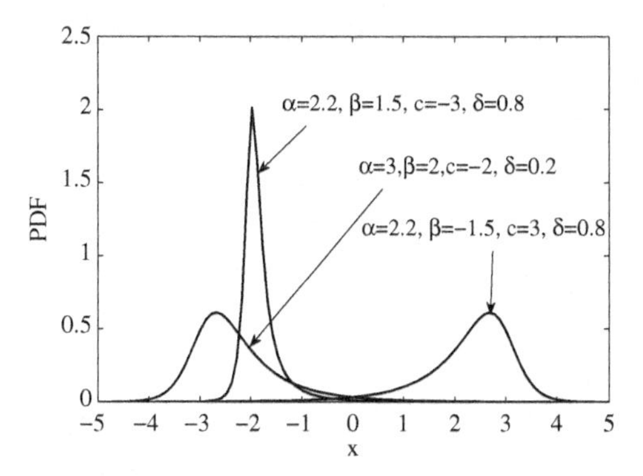

Figure 9.3 PDF plots for 4-parameter inverse Gaussian distribution.

Similar to the Bessel function distributions of the first kind and the second kind, the moments of the normal inverse Gaussian distribution may be explicitly expressed as:

$$\mu = c + \frac{\delta\beta}{\left(\alpha^2 - \beta^2\right)^{\frac{1}{2}}} \tag{9.16c}$$

$$\sigma^2 = \frac{\alpha^2\delta}{\left(\alpha^2 - \beta^2\right)^{3/2}} \tag{9.16d}$$

$$\beta_1 = 3\beta\left(\alpha^2 - \beta^2\right)^{-\frac{1}{4}}\alpha^{-1}\delta^{-\frac{1}{2}} \tag{9.16e}$$

$$\beta_2 = 3 + \frac{3}{\delta\sqrt{\alpha^2 - \beta^2}} + \frac{12\beta^2}{\alpha^2\delta\sqrt{\alpha^2 - \beta^2}} \tag{9.16f}$$

Furthermore, 2-parameter inverse Gaussian distribution is defined as:

$$f(x; c, \delta) = \sqrt{\frac{\delta}{2\pi x^3}}\exp\left(-\frac{\delta}{2c^2 x}(x - c)^2\right); c > 0, \delta > 0, x \in (0, \infty) \tag{9.17}$$

9.2.4 Other Distributions

Up to now, we have discussed three types of Bessel function distributions derived from the first and second (third) kind Bessel functions. Nadarajah (2008) proposed the distribution derived from the multiplication of two Bessel functions of first kind, two Bessel functions of second kind, and one Bessel function of first kind and one Bessel function of second kind. According to Nadarajah (2008), these PDFs are expressed as follows:

Product of Two Bessel Functions of the First Kind (I_m, I_n)

$$f(x) = Cx^{m+n}\exp(-\gamma x)I_m\left(\frac{x}{\alpha}\right)I_n\left(\frac{x}{\beta}\right); x>0, \alpha>0, \beta>0, \gamma>\frac{1}{\alpha}+\frac{1}{\beta}, m>1, n>1 \tag{9.18}$$

In Equation (9.18), C is the normalizing constant given as:

$$C = \frac{\gamma^{2m+2n+1}\Gamma(m+1)\Gamma(n+1)(2\alpha)^m(2\beta)^n}{\Gamma(2m+2n+1)F_4\left(m+n+\frac{1}{2}, m+n+1; m+1, n+1; \frac{1}{\gamma^2\alpha^2}, \frac{1}{\gamma^2\beta^2}\right)} \tag{9.19a}$$

and F_4 is the Appell function (Appell, 1925) given as:

$$F_4(a,b;c,d;x,y) = \sum_{k=0}^{\infty}\sum_{l=0}^{\infty}\frac{(a)_{k+l}(b)_{k+l}x^k y^l}{(c)_k(d)_l k!l!} \tag{9.19b}$$

According to Nadarajah (2008), if $\alpha = \beta$, Equation (9.19a) may be simplified as:

$$C = \frac{\gamma^{2m+2n+1}(2\beta)^{m+n}\Gamma(m+1)\Gamma(n+1)}{\Gamma(2m+2n+1)\,{}_3F_2\left(\frac{m+n+1}{2}, \frac{m+n}{2}+1, m+n+\frac{1}{2}; m+1, n+1; \frac{4}{\beta^2\gamma^2}\right)} \tag{9.20a}$$

$${}_3F_2(a,b,c;d,e;x) = \sum_{k=0}^{\infty}\frac{(a)_k(b)_k(c)_k}{(d)_k(e)_k}\frac{x^k}{k!} \tag{9.20b}$$

Furthermore, if $\alpha = \beta$ and $m = n$; Equation (9.19a) may be further simplified as:

$$C = \frac{\pi\beta^{2m}\gamma^{4m+1}\Gamma(m+1)}{2^{4m}\Gamma\left(m+\frac{1}{2}\right)\Gamma\left(2m+\frac{1}{2}\right){}_2F_1\left(m+\frac{1}{2}, 2m+\frac{1}{2}; m+1; \frac{4}{\beta^2\gamma^2}\right)} \tag{9.21a}$$

$${}_2F_1(a,b;c;x) = \sum_{k=0}^{\infty}\frac{(a)_k(b)_k}{(c)_k}\frac{x^k}{k!} \tag{9.21b}$$

Figure 9.4 plots the PDF of the density function composed with the product of two Bessel functions of the first kind with $\alpha = 1.2$, $\beta = 0.7$, $\gamma = 4.0$ and different m, n values.

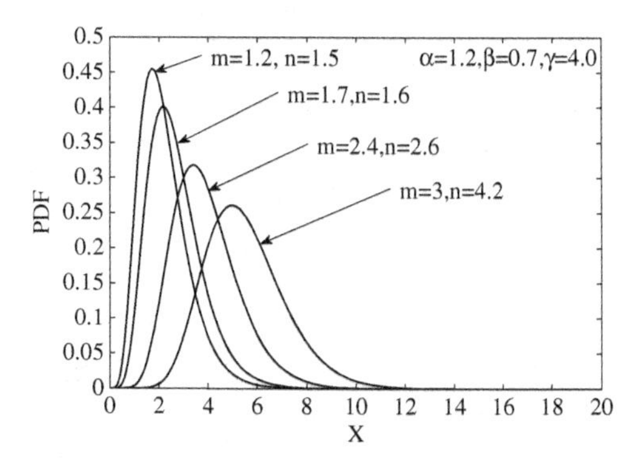

Figure 9.4 PDF plot for the product of two first kind Bessel functions.

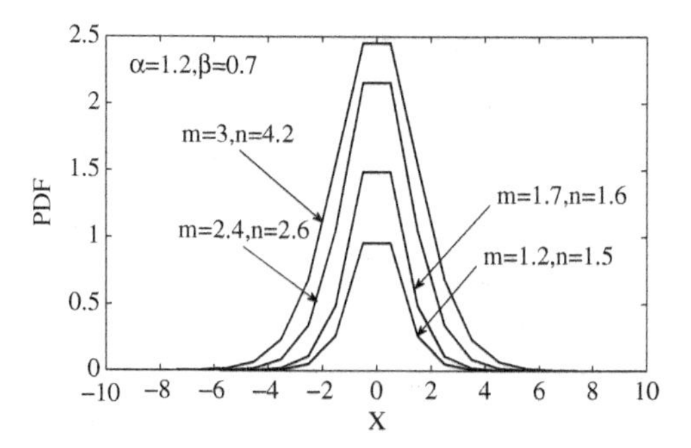

Figure 9.5 PDF plot for the product of two Bessel functions of the second kind.

Product of the Two Bessel Functions of the Second Kind (K_m, K_n)

$$f(x) = C|x|^{m+n} K_m\left(\left|\frac{x}{\alpha}\right|\right) K_n\left(\left|\frac{x}{\beta}\right|\right); x \in \mathbb{R}, \alpha > 0, \beta > 0, m > 1, n > 1 \quad (9.22a)$$

where C is the normalizing constant given as:

$$C = \left\{ \sqrt{\pi} 2^{m+n-1} \alpha^{-m} \beta^{2m+n+1} \Gamma\left(m+n+\frac{1}{2}\right) B\left(m+\frac{1}{2}, n+\frac{1}{2}\right) \right. \\ \left. {}_2F_1\left(m+n+\frac{1}{2}, n+\frac{1}{2}; m+n+1; 1-\frac{\beta^2}{\alpha^2}\right) \right\}^{-1} \quad (9.22b)$$

Figure 9.5 plots the PDF of the density function composed with the product of two Bessel functions of the second kind with $\alpha = 1.2$, $\beta = 0.7$ and different m, n values. It is shown that this PDF is symmetric about $x = 0$.

Product of the Bessel Functions of the First Kind (I_m) *and the Second Kind* (K_n)

The PDF can be written as

$$f(x) = Cx^{m+n} I_m\left(\frac{x}{\alpha}\right) K_n\left(\frac{x}{\beta}\right); x > 0, \beta > 0, \alpha > \beta, m > 1, n > 1 \quad (9.23a)$$

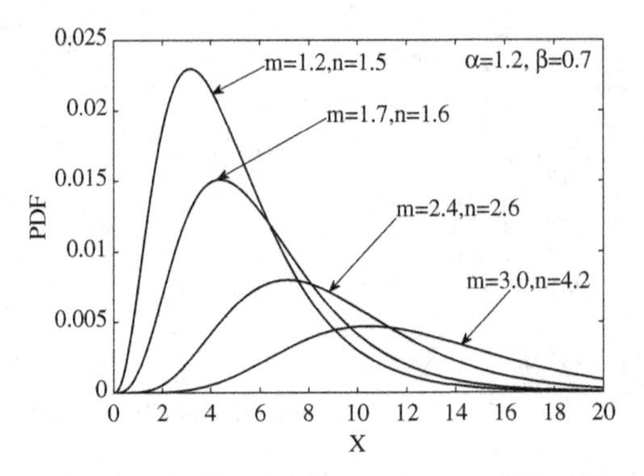

Figure 9.6 PDF plot for the product of first and second kind Bessel functions.

where C is The normalizing constant given as:

$$C = \left\{ \frac{2^{m+n+1}\beta^{2m+n+1}}{b^m \Gamma(m+1)} \Gamma\left(m+n+\frac{1}{2}\right)\Gamma\left(m+\frac{1}{2}\right) \right. \\ \left. {}_2F_1\left(m+n+\frac{1}{2}, m+\frac{1}{2}; m+1; \frac{\beta^2}{\alpha^2}\right) \right\}^{-1} \tag{9.23b}$$

Figure 9.6 plots the PDF of the density function composed with the product of the Bessel functions of the first and second kind with $\alpha = 1.2$, $\beta = 0.7$ and different m, n values.

9.3 Frequency Distributions by Series Approximation

In this section, we will focus on introducing the concept of a frequency distribution by series approximation with Gaussian distribution as the baseline density function, i.e., Gram-Charlier type A series and Edgeworth type A series. Before we present these two widely applied series, we will first introduce the Chebyshev (also called Probabilists')-Hermite polynomials and cumulants.

9.3.1 Chebyshev (Probabilists')-Hermite Polynomials

The Chebyshev-Hermite polynomial is derived based on the standard Gaussian distribution:

$$\phi(x) = \frac{1}{\sqrt{2\pi}} \exp\left(-\frac{x^2}{2}\right) \tag{9.24}$$

As shown in Kendall and Stuart (1977), the r-th order Chebyshev-Hermite polynomial (H_r) of normally distributed random variable may be expressed with the use of differential equation as:

$$H_r(x)\phi(x) = (-1)^r \frac{d^r\phi(x)}{dx^r} \tag{9.25}$$

Substituting Equation (9.24) into Equation (9.25), we can express H_r as:

$$\begin{cases} H_1 = 1 \\ H_2 = x \\ H_3 = x^2 - 1 \\ H_4 = x^4 - 6x^2 - 3 \\ H_5 = x^5 - 10x^3 + 15x \\ H_6 = x^6 - 15x^4 + 45x^2 - 15 \\ \quad \cdots \end{cases} \tag{9.26}$$

Two properties in the following text are the interesting properties held by the Chebyshev-Hermite polynomials:

(1) Orthogonality

$$\int_{-\infty}^{+\infty} H_m(x)H_n(x)\phi(x)dx = n!\delta_{mn} \tag{9.27a}$$

In Equation (9.27a), δ_{mn} is Kronecker delta. Equation (9.27a) may also be expressed as:

$$\int_{-\infty}^{+\infty} H_m(x)H_n(x)\phi(x)dx = \begin{cases} 0, & m \neq n \\ n!, & m = n \end{cases} \tag{9.27b}$$

(2) Recursive relation

The recursive relation of the Chebyshev-Hermite polynomial may be expressed as:

$$\frac{d}{dx}H_r(x) = rH_{r-1}(x), r \geq 1 \tag{9.28a}$$

Equation (9.28a) may also be written as:

$$H_r(x) = xH_{r-1}(x) - (r-1)H_{r-2}(x),\ r \geq 2 \tag{9.28b}$$

9.3.2 Cumulants

Similar to ordinary moments that may be derived from the moment generating function $[M_X(t)]$ or characteristic function $[\varphi_X(t)]$, cumulants may also be derived from the cumulant generating function $[K_X(t)]$ with the use of $M_X(t)$ and $\varphi_X(t)$.

If the moment generating function exists (i.e., $M_X(t) < \infty$), the cumulant generating function $[K_X(t)]$ may be written as:

$$K_X(t) = \ln(M_X(t)) \text{ or equivalantly } M_X(t) = \exp(K_X(t)) \tag{9.29a}$$

Applying Taylor series expansion, Equation (9.29a) may be rewritten as:

$$1 + \mu_1' t + \frac{\mu_2' t^2}{2!} + \cdots + \frac{\mu_r' t^r}{r!} + \cdots = \exp\left(k_1 t + \frac{k_2 t^2}{2!} + \cdots + \frac{k_r t^r}{r!} + \cdots\right) \tag{9.29b}$$

In Equation (9.29b), μ_r' represents the r-th moment about the origin and k_r represents the r-th cumulant.

If the moment generating function does not exist, the characteristic function defined as $\varphi_X(t) = E(\exp(itX))$ may be applied to substitute for $M_X(t)$ in Equations (9.29a)–(9.29b), which yields:

$$K_X(t) = \ln(\varphi_X(t)) \tag{9.30a}$$

Equation (9.30a) may also be rewritten using Taylor series expansion as:

$$1 + \mu_1'(it) + \frac{\mu_2'(it)^2}{2!} + \cdots + \frac{\mu_r'(it)^r}{r!} + \cdots = \exp\left(k_1(it) + \frac{k_2(it)^2}{2!} + \cdots + \frac{k_r(it)^r}{r!} + \cdots\right) \tag{9.30b}$$

The cumulants expressed in Equations (9.29b) and (9.30b) may be evaluated as:

$$k_r = \left.\frac{d(K_X(t))}{dt}\right|_{t=0} \tag{9.31}$$

From Equations (9.24)–(9.26), the relations between cumulants (k_i) and moments about origin (μ_i') for random variable X may be expressed as:

$$\begin{cases} k_1 = \mu_1' \\ k_2 = \mu_2' - \mu_1' 2 \\ k_3 = \mu_3' - 3\mu_1'\mu_2' + 2\mu_1'^3 \\ k_4 = \mu_4' - 4\mu_1'\mu_3' - 3\mu_2'^2 + 12\mu_2'\mu_1'^2 - 6\mu_1'^4 \\ \quad\cdots \end{cases} \tag{9.32a}$$

$$\begin{cases} \mu_1' = k_1 \\ \mu_2' = k_2 + k_1^2 \\ \mu_3' = k_1^3 + 3k_1k_2 + k_3 \\ \mu_4' = k_1^4 + 6k_1^2k_2 + 3k_2^2 + 4k_1k_3 + k_4 \\ \quad\cdots \end{cases} \tag{9.32b}$$

Similarly, we can express the relation of cumulants (k_i) and moments about mean (μ_i) for random variable X as:

$$\begin{cases} k_2 = \sigma^2 \\ k_3 = \mu_3 = \beta_1 \sigma^3 \\ k_4 = \mu_4 - 3\mu_2^2 = \beta_2 \sigma^4 \\ \quad \ldots \end{cases} \tag{9.33a}$$

$$\begin{cases} \mu_2 = k_2 \\ \mu_3 = k_3 \\ \mu_4 = k_4 + 3k_2^2 \\ \quad \ldots \end{cases} \tag{9.33b}$$

In Equation (9.33a) β_1 and β_2 represent the skewness and excess kurtosis of the random variable.

9.3.3 Basic Concept of Approximating Frequency Distribution with Series Approximation

The basic premise of the series approximation method is that an approximate frequency distribution [f_X] of target random variable X may be expressed as a function of a baseline density function [g] (Kolassa, 2006). The conditions for expansion methods are: (1) random variables X (target) and Y (baseline) follow the density function f_X and g_Y, respectively; (2) random variables X and Y share the common probability space with $Y^* = X - Y$, which is independent of Y. Thus, for $Y^* = y^*$, the PDF of X can be expressed through the expansion of the baseline density function of Y as follows:

$$f_X(x) = g_Y(x) \sum_{r=0}^{\infty} \frac{h_r(x) \mu_r(Y^*)}{r!} = g_Y(x) \sum_{r=0}^{\infty} C_r h_r(x); \qquad C_r = \frac{\mu_r(Y^*)}{r!} \tag{9.34a}$$

$$h_r(x) = \frac{(-1)^r g_Y^{(r)}(x)}{g_Y(x)} = \frac{(-1)^r}{g_Y(x)} \frac{d^r g_Y(x)}{dx^r} \tag{9.34b}$$

The probability density approximation of Equation (9.34a) indicates that $f_X(x)$ depends on the moments of the difference variable between random variable X and its baseline variable Y.

9.3.4 Gram-Charlier Type A Series

Let the standard Gaussian density be the baseline density. Then $h_r(x)$ in Equation (9.34b) is the Chebyshev-Hermite polynomials expressed as Equation (9.26). The density function of random variable X is given as:

$$f(x) = \frac{1}{\sqrt{2\pi}} \exp\left(-\frac{x^2}{2}\right) \int_{r=0}^{\infty} C_r H_r(x) \tag{9.35}$$

Following Kendall and Stuart (1977), the coefficient C_r is given as:

$$\begin{cases} C_0 = 1 \\ C_1 = 0 \\ C_2 = \frac{1}{2}(\mu_2 - 1) \\ C_3 = \frac{1}{6}\mu_3 \\ C_4 = \frac{1}{24}(\mu_4 - 6\mu_2 + 3) \\ \dots \end{cases} \tag{9.36}$$

Substituting Equation (9.36), Equation (9.35) may be rewritten as:

$$f(x) = \frac{1}{\sqrt{2\pi}} \exp\left(-\frac{x^2}{2}\right)\left[1 + \frac{1}{2}(\mu_2 - 1)H_2 + \frac{1}{6}\mu_3 H_3 + \frac{1}{24}(\mu_4 - 6\mu_2 + 3)H_4 + \cdots\right] \tag{9.36a}$$

If random variable X is standardized, i.e., $\mu_1 = 0$, $\mu_2 = 1$, Equation (9.36a) is rewritten as:

$$f(x) = \frac{1}{\sqrt{2\pi}} \exp\left(-\frac{x^2}{2}\right)\left[1 + \frac{1}{6}\mu_3 H_3 + \frac{1}{24}(\mu_4 - 6\mu_2 + 3)H_4 + \cdots\right] \tag{9.36b}$$

Equation (9.36a) is called the Gram-Charlier type A series.

By definition, the Chi-squared distribution with a v degree of freedom is the summation of v squared independent random variables (Y) following the standard Gaussian distribution as:

$$f(\chi^2) = \frac{(\chi^2)^{\frac{v}{2}-1} \exp\left(-\frac{\chi^2}{2}\right)}{2^{\frac{v}{2}}\Gamma\left(\frac{v}{2}\right)}; \chi^2 = \sum_{i=1}^{v} Y_i^2, Y_i \sim N(0,1) \tag{9.37}$$

Standardizing χ^2 using $\mu = v$, $\sigma^2 = 2v$, we have the standardized Chi-squared variable as: $x = \frac{\chi^2 - v}{\sqrt{2v}}$. Then the density function of Equation (9.37) is rewritten as:

$$f(x) = \frac{\sqrt{2v}}{2^{\frac{v}{2}}\Gamma\left(\frac{v}{2}\right)}\left(x\sqrt{2v} + v\right)^{\frac{v}{2}-1} \exp\left(-\frac{x\sqrt{2v} + v}{2}\right); x > -\sqrt{\frac{v}{2}} \tag{9.38}$$

Using the chi-squared distribution with six degrees of freedom (d.f.), Figures 9.7–9.8 plot standardized chi-squared distribution (d.f. = 5, 10) and the corresponding density approximation with Gram-Charlier type A series. Figure 9.7 shows that the approximation becomes inaccurate as the term of the expansion increases for the low degree of freedom. Figure 9.8 shows that the approximation performs better for the high degree of freedom. In other words, the chi-squared distribution is closer to Gaussian distribution with the increase of the degree of freedom.

9.3.5 Edgeworth Series with Baseline Gaussian Distribution

As stated in Kolassa (2006) and Cavalcante et al. (2004), the Edgeworth series is consistent with Gram-Charlie series. In the case of the Edgeworth series, the density function of target random variable is expressed based on the Fourier transformation of $\phi(x)H_r(x)$ where $\phi(x)$ represents the density function of standard normal distribution. Following Kendall and Stuart (1977), the PDF of standardized target random variable may be also expressed with the use of the Chebyshev-Hermite polynomials and the baseline standard normal distribution with same general equation [i.e., Equation 9.34a]. The coefficient C_r is expressed through the cumulants as:

$$\begin{cases} C_0 = 1, \quad C_1 = C_2 = 1 \\ C_3 = \dfrac{k_3}{3!} \\ C_4 = \dfrac{k_4}{4!} \\ \ldots \end{cases} \tag{9.39}$$

In Equation (9.39) the r-th order cumulant is computed from Equation (9.31) or (9.32a). Then the Edgeworth series expansion is given as:

$$f(x) = \frac{1}{\sqrt{2\pi}} \exp\left(-\frac{x^2}{2}\right)\left[1 + \frac{k_3}{3!}H_3 + \frac{k_4}{4!}H_4 + \frac{k_5}{5!}H_5 + \frac{k_6 + 10k_3^2}{6!}H_6\right] \tag{9.40}$$

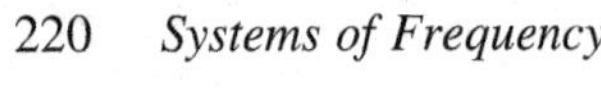

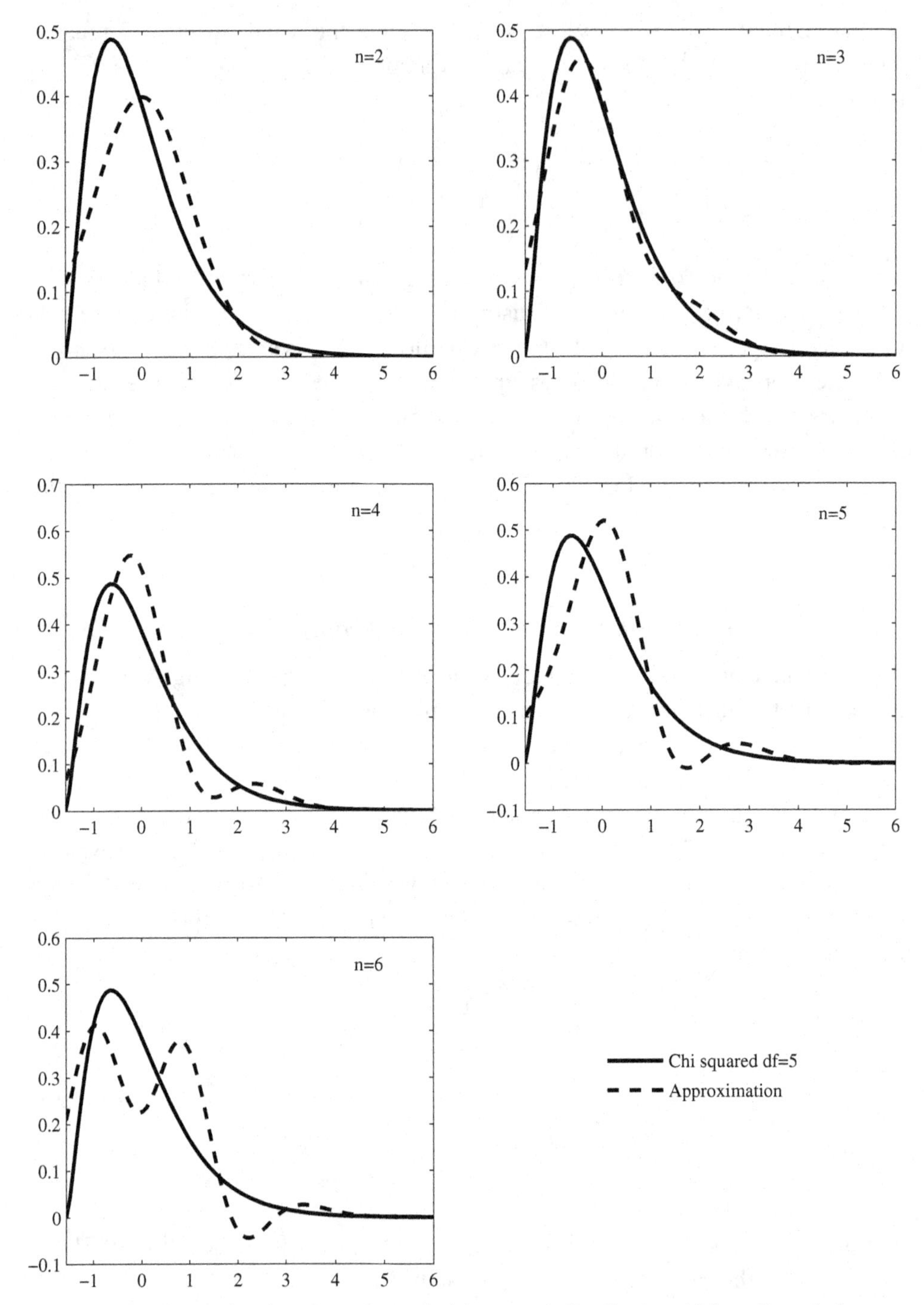

Figure 9.7 Density function plots of chi-squared distribution (d.f. = 5) and the density approximation with Gram-Charlier type A series.

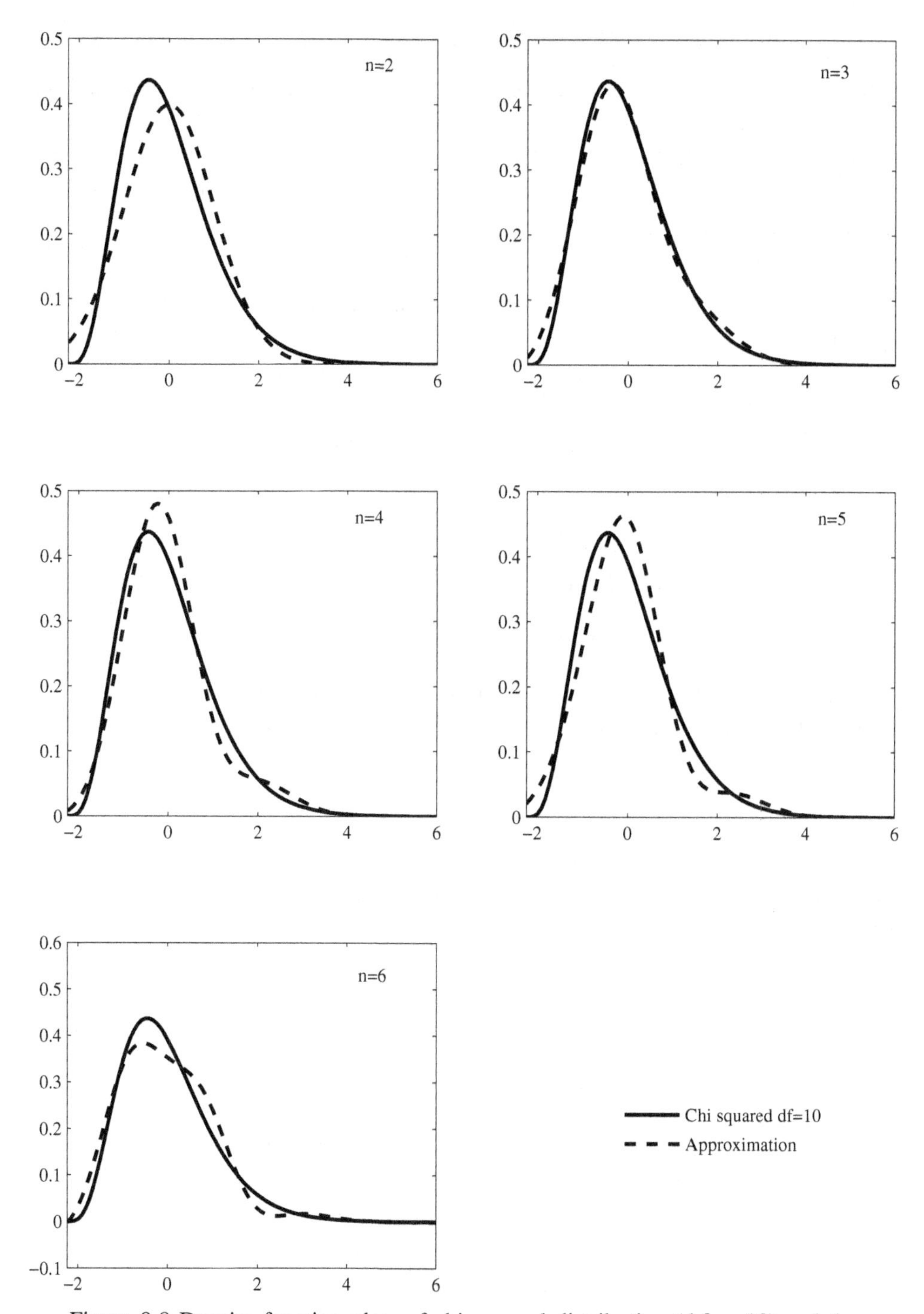

Figure 9.8 Density function plots of chi-squared distribution (d.f. = 10) and the density approximation with Gram-Charlier type A series.

9.3.6 Gram-Charlier/Edgeworth Series with Non-Gaussian Distribution

In this section, we will introduce two approximations of a nonnormal distribution with gamma and beta distributions as the baseline distribution. The baseline density of gamma distribution leads to the expansion in terms of the generalized Laguerre polynomials. The baseline PDF of the beta distribution leads to the expansion in terms of Jacobi polynomials.

Gamma Distribution as Baseline PDF

The PDF, characteristic function, cumulant generating function, and cumulant of the gamma distribution (scale parameter $b = 1$) can be given as:

$$\begin{cases} g_Y(x) = \dfrac{1}{\Gamma(a)} x^{a-1} \exp(-x); a > 0, x \in (0, \infty) \\ \varphi(t) = \dfrac{1}{(1-it)^a} \\ K(t) = -a \ln(1-t) \\ k_r = a\Gamma(r); r = 0, 1, \ldots \end{cases} \tag{9.41}$$

Let $L_n^{(a)}$ denote the n-th order Laguerre polynomial of $x^{a-1} \exp(-x)$ (i.e. $\mu = \sigma^2 = a$) we can write the Laguerre polynomial in its generalized form following Bowers (1966) as:

$$L_n^{(a)} = (-1)^n x^{1-a} e^x \frac{d^n}{dx^n} \left(x^{n+a-1} e^{-x}\right), n = 0, 1, 2, \ldots \tag{9.42}$$

Following Bowers (1966), The first several order Laguerre polynomials are evaluated as:

$$\begin{cases} L_0^{(a)} = 1 \\ L_1^{(a)} = x - a \\ L_2^{(a)} = x^2 - 2(a+1)x + (a+1)a \\ L_3^{(a)} = x^3 - 3(a+2)x^2 + 3(a+1)(a+2)x - a(a+1)(a+2) \\ L_4^{(a)} = \begin{matrix} x^4 - 4(a+3)x^3 + 6(a+3)(a+2)x^2 - 4(a+3)(a+2)(a+1)x \\ +a(a+1)(a+2)(a+3) \end{matrix} \\ L_5^{(a)} = \begin{matrix} x^5 - 5(a+4)x^4 + 10(a+3)(a+4)x^3 - 10(a+2)(a+3)(a+4)x^2 \\ +5(a+1)(a+2)(a+3)(a+4)x - a(a+1)(a+2)(a+3)(a+4) \end{matrix} \\ \cdots\cdots \end{cases} \tag{9.43}$$

It is known that the Laguerre polynomials are orthogonal to the gamma PDF given in Equation (9.41) in the domain of $x \in (0, \infty)$ as:

$$\begin{cases} \int_0^\infty \frac{x^{a-1}\exp(-x)}{\Gamma(a)} L_n^{(a)}(x) L_m^{(a)}(x)dx = 0, \ for\ m \neq n \\ \int_0^\infty \frac{x^{a-1}\exp(-x)}{\Gamma(a)} \left[L_n^{(a)}(x)\right]^2 dx \\ = \frac{1}{\Gamma(a)} \int_0^\infty x^{a-1}e^{-x}\left[L_n^{(a-1)}(x)\right]^2 dx = \frac{n!\Gamma(n+a)}{\Gamma(a)} \end{cases} \tag{9.44}$$

Applying the expansion, we may approximate the PDF for random variable X with the gamma PDF as the baseline distribution as:

$$f(x) = \frac{1}{\Gamma(a)} x^{a-1}e^{-x} \sum_{i=0}^{\infty} C_i L_i^{(a)} \tag{9.45}$$

The coefficient C_i may be evaluated by applying the orthogonal property of the Laguerre polynomials as:

$$\begin{aligned} \int_0^\infty f(x) L_i^{(a)}(x)dx &= \int_0^\infty \frac{x^{a-1}e^{-x}}{\Gamma(a)} \left(C_0 L_0^{(a)}(x) + C_1 L_1^{(a)}(x) + \cdots\right) L_i^{(a)}(x)dx \\ &= \frac{1}{\Gamma(a)} \int_0^\infty C_i x^{a-1}e^{-x}\left[L_i^{(a)}(x)\right]^2 dx = C_i \frac{i!\Gamma(i+a)}{\Gamma(a)} \end{aligned} \tag{9.46}$$

From Equation (9.46), we obtain:

$$C_i = \frac{\Gamma(a)}{i!\Gamma(a+i)} \int_0^\infty f(x) L_i^{(a)}(x)dx \tag{9.47}$$

$$
\begin{cases}
C_0 = \int_0^\infty f(x)\, L_0^{(a)}(x)dx = \int_0^\infty f(x)dx = 1 \\
C_1 = \frac{1}{a}\int_0^\infty f(x)L_1^{(a)}dx = \frac{1}{a}[E(X) - a] = 0;\;\; E(X) = a \\
C_2 = \frac{\Gamma(a)}{2!\Gamma(a+2)}\int_0^\infty f(x)L_2^{(a)}dx = 0 \\
\quad = \frac{\Gamma(a)}{2!\Gamma(a)}\left[E(X^2) - 2(a+1)E(X) + a(a+1)\right] \\
\quad = \frac{\Gamma(a)}{2!\Gamma(a+2)}\left((a+a^2) - 2a(a+1) + a(a+1)\right) = 0; Var(X) = a \\
C_3 = \frac{\Gamma(a)}{3!\Gamma(a+3)}\int_0^\infty f(x)L_3^{(a)}dx \\
\quad = \frac{\Gamma(a)}{3!\Gamma(a+3)}\left[E(X^3) - 3(a+2)E(X^2) + 3(a+1)(a+2)E(X)\right. \\
\qquad\qquad \left. - a(a+1)(a+2)\right. \\
\quad = \frac{\Gamma(a)}{3!\Gamma(a+3)}\left[\mu_3' - 3(a+2)(a+a^2) + 3a(a+1)(a+2) - a(a+1)(a+2)\right] \\
\quad = \frac{\Gamma(a)}{3!\Gamma(a+3)}(\mu_3 - 2a) \\
\text{Similarly,} \\
C_4 = \frac{\Gamma(a)}{4!\Gamma(a+4)}\int_0^\infty f(x)\, L_4^{(a)}dx \\
\quad = \frac{\Gamma(a)}{4!\Gamma(a+4)}\left(\mu_4 - 12\mu_3 - 3a^2 + 18a\right) \\
C_5 = \frac{\Gamma(a)}{5!\Gamma(a+5)}\int_0^\infty f(x)L_5^{(a)}dx \\
\quad = \frac{\Gamma(a)}{5!\Gamma(a+5)}\left(\mu_5 - 20\mu_4 - (10a - 120)\mu_3 + 60a^2 - 144a\right)
\end{cases}
\tag{9.48}
$$

Beta Distribution as Baseline PDF

The PDF and moment generating function of beta distribution can be given as:

$$
\begin{cases}
g(x, \alpha, \beta) = \frac{x^{\alpha-1}(1-x)^{\beta-1}}{B(\alpha, \beta)}; x \in [0, 1] \\
\mathcal{M}_X(t) = {}_1F_1(\alpha; \alpha+\beta; t)
\end{cases}. \tag{9.49}
$$

where ${}_1F_1$ is the confluent hypergeometric function.

Jacobi polynomials $P_n^{(a,b)}$ may be written using hypergeometric function as:

$$P_n^{(a,b)}(x) = \frac{(a+1)_n}{n!}\ {}_2F_1\left(-n, a+b+1+n; a+1; \frac{1}{2}(1-x)\right);$$
$$x \in [-1, 1], a, b > -1 \tag{9.50}$$

It may also be written using the differential equation as:

$$P_n^{(a,b)} = \frac{(-1)^n}{2^n n!}(1-x)^{-a}(1+x)^{-b}\frac{d^n}{dx^n}\left((1-x)^{a+n}(1+x)^{b+n}\right) \tag{9.51a}$$

Here, we will show the first several formulations for Jacobi polynomials as:

$$\begin{cases} P_0^{(a,b)} &= 1 \\ P_1^{(a,b)} &= \frac{a-b}{2} + \frac{(a+b+2)}{2}x \\ P_2^{(a,b)} &= \frac{1}{8}\left[(a2+b^2+7a+7b+2ab+12)x^2 + \right. \\ & \left. (2a^2-2b^2+6a-6b)x + (a^2+b^2-(a+b)-2ab-4)\right] \\ P_3^{(a,b)} &= \frac{1}{48}\left[(a+1)(a+2)(a+3)(1+x)^3 - (b+1)(b+2)(b+3)(1-x)^3 - \right. \\ & 3(a+2)(a+3)(b+3)(1-x)(1+x)^2 + \\ & \left. 3(a+3)(b+2)(b+3)(1-x)^2(1+x)\right] \\ \ldots \end{cases} \tag{9.51b}$$

Additionally, the Jacobi polynomials satisfy the orthogonality condition as:

$$\int_{-1}^{1}(1-x)^a(1+x)^b P_m^{(a,b)}(x), P_n^{(a,b)}(x)dx = \frac{2^{a+b+1}}{a+b+1+2n} \times \frac{\Gamma(a+1+n)\Gamma(b+1+n)}{n!\Gamma(a+b+1+n)}\delta_{mn} \tag{9.52}$$

In Equation (9.52), let $x = 1 - 2u$, Equation (9.52) may be rewritten as:

$$\begin{aligned} &\int_{-1}^{1}(1-x)^a(1+x)^b P_m^{(a,b)}(x)P_n^{(a,b)}(x)dx \\ &= 2^{a+b+1}\int_0^1 u^a(1-u)^b P_m^{(a,b)}(1-2u)P_n^{(a,b)}(1-2u)du \end{aligned} \tag{9.53}$$

The right-hand side of Equation (9.53) clearly shows that the Jacobi polynomials is orthogonal to the beta distribution with parameters of $\alpha = a + 1$; $\beta = b + 1$. Equation (9.53) may be rewritten as:

$$\begin{aligned} &\int_0^1 x^a(1-x)^b P_m^{(a,b)}(1-2x)P_n^{(a,b)}(1-2x)dx \\ &= \frac{1}{a+b+1+2n}\frac{\Gamma(a+1+n)\Gamma(b+1+n)}{n!\Gamma(a+1+b+n)}\delta_{mn} \end{aligned} \tag{9.54}$$

Similar to the density approximation with Hermite and Laguerre polynomials, the PDF may be also approximated with beta baseline distribution as:

$$\begin{aligned} f(x) &= g(x)\sum\nolimits_{i=0}^{\infty} C_i P_i^{(\alpha-1,\beta-1)}(1-2x) \\ &= x^{\alpha-1}(1-x)^{\beta-1}\sum\nolimits_{i=0}^{\infty} C_i P_i^{(\alpha-1,\beta-1)}(1-2x) \end{aligned} \tag{9.55}$$

The coefficients are then computed from the orthogonal property of the Jacobi polynomials as:

$$\begin{aligned} &\int_0^{\infty} f(x)P_i^{(\alpha-1,\beta-1)}(1-2x)dx \\ &= \int_0^1 x^{\alpha-1}(1-x)^{\beta-1}\left(C_0P_0^{(\alpha-1,\beta-1)} + C_1P_1^{(\alpha-1,\beta-1)} + \cdots\right)P_i^{(\alpha-1,\beta-1)}dx \\ &= C_i\left(\frac{1}{\alpha+\beta-1+2i}\frac{\Gamma(\alpha+i)\Gamma(\beta+i)}{i!\Gamma(\alpha+\beta-1+i)}\right) \end{aligned} \tag{9.56}$$

Then, We can solve for C_i using:

$$C_i = \frac{i!(\alpha+\beta-1+2i)\Gamma(\alpha+\beta-1+i)}{\Gamma(\alpha+i)\Gamma(\beta+i)}\int_0^1 f(x)P_i^{(\alpha-1,\beta-1)}(1-2x)dx \tag{9.57}$$

For the random variable X with $E(X) = \frac{\alpha}{\alpha+\beta}$, $Var(X) = \frac{\alpha\beta}{(\alpha+\beta)^2(\alpha+\beta+1)}$, we have:

$$\begin{aligned} C_0 &= \frac{(\alpha+\beta-1)\Gamma(\alpha+\beta-1)}{\Gamma(\alpha)\Gamma(\beta)}\int_0^1 f(x)dx = \frac{\Gamma(\alpha+\beta)}{\Gamma(\alpha)\Gamma(\beta)} = \frac{1}{B(\alpha,\beta)} \\ C_1 &= \frac{(\alpha+\beta+1)\Gamma(\alpha+\beta)}{\Gamma(\alpha+1)\Gamma(\beta+1)}\int_0^1 f(x)P_1^{(\alpha-1,\beta-1)}(1-2x)dx \\ &= \frac{(\alpha+\beta+1)\Gamma(\alpha+\beta)}{\Gamma(\alpha+1)\Gamma(\beta+1)}\int_0^1 f(x)[(\alpha-\beta)-(\alpha+\beta)x]dx \\ &= \frac{(\alpha+\beta+1)\Gamma(\alpha+\beta)}{\Gamma(\alpha+1)\Gamma(\beta+1)}[\alpha-\beta-\alpha] = -\frac{\beta(\alpha+\beta+1)\Gamma(\alpha+\beta)}{\Gamma(\alpha+1)\Gamma(\beta+1)} \\ &\cdots \end{aligned} \tag{9.58}$$

From Equation (9.58), it is seen that the coefficient C_0, C_1 estimated based on the beta baseline distribution are no longer equal to 1 and 0, respectively. In addition, compared with the density approximation based on normal (Hermite polynomial) and gamma distributions (Laguerre polynomials), the approximation with the beta distribution is not as widely applied.

If $g(x)$ is chosen as a standard normal distribution, then the polynomial is the Hermite polynomial and the Gram-Charlier expansions are obtained. If $g(x)$ is a standard gamma density function, then the expansion is in terms of Laguerre polynomials. If $g(x)$ is a standard beta distribution, then the expansion is in terms of Jacobi polynomials.

In the case of standardized sums of independent and identically distributed random variables, $k_l^n = k_l^1 n^{1-\frac{l}{2}}$. These series expansion methods are also often used in the case of sums of nonindependent and identically distributed random variables if $k_l^n = O\left(n^{1-\frac{l}{2}}\right)$. Note specially that k_1^n increases rather than decreasing as $n \to \infty$.

9.4 Applications

In this section, the inverse Gaussian distribution is applied to the real-world data. Based on the statistical characteristics of the inverse Gaussian distribution, the inverse Gaussian distribution is applied to evaluate peak flow, total flow deficit, maximum daily precipitation, annual rainfall, and monthly suspended sediment. Applying the 2-parameter inverse Gaussian distribution, i.e., commonly applied in statistical analysis, the parameters are estimated with the use of maximum likelihood estimation method as listed in Table 9.1. Figures 9.9–9.13 compare the fitted frequencies of 2-parameter inverse Gaussian distribution with the empirical frequencies. Comparison indicates that the inverse Gaussian distribution (2-parameter) can be applied to model the frequency distribution of all the datasets applied here.

Table 9.1 *Parameters estimated for Inverse Gaussian distribution (2-parameter)*

	Parameters	
Variable	c	δ
Peak flow	283.27	2.22×10^3
Total flow deficit	6.32×10^4	1.11×10^4
Maximum daily precipitation	56.24	548.03
Annual rainfall	9.58×10^2	3.02×10^4
Monthly suspended sediment	105.85	237.19

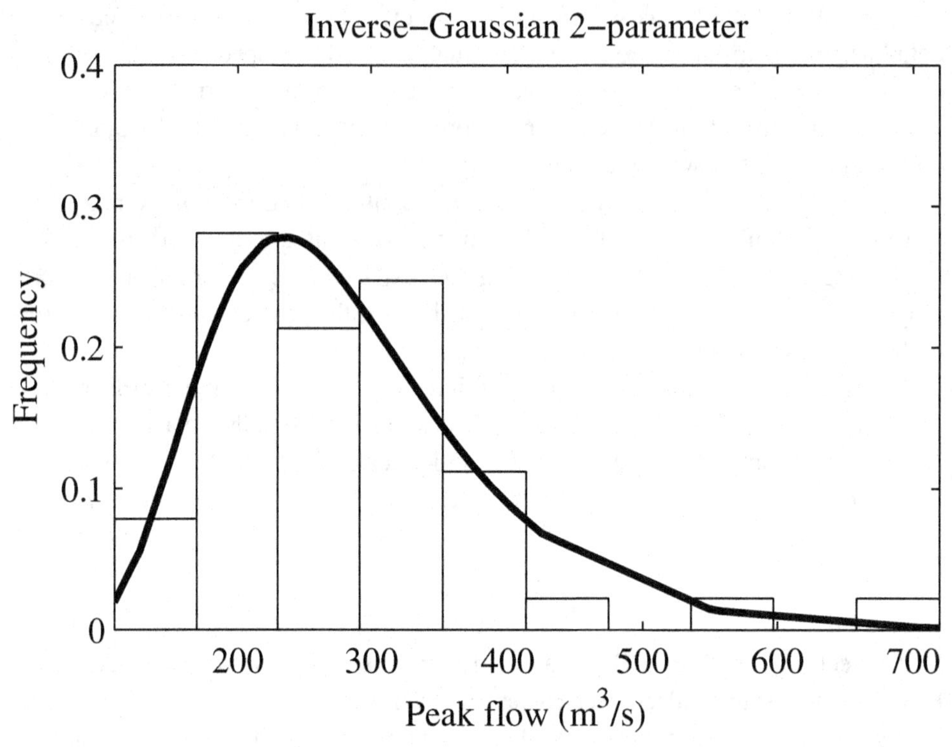

Figure 9.9 Comparison of fitted frequency distribution to empirical frequency distribution for peak flow.

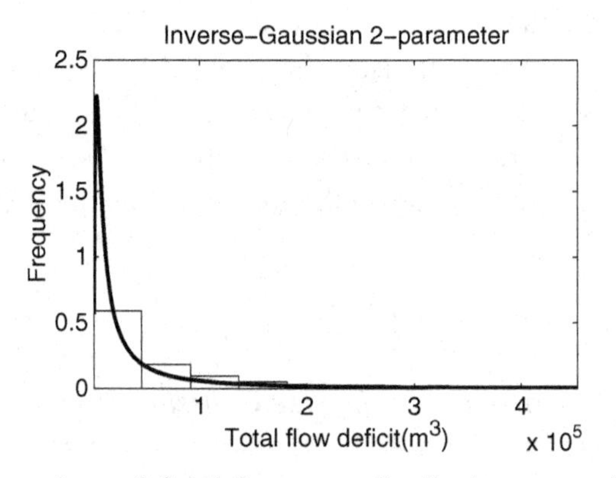

Figure 9.10 Comparison of fitted frequency distribution to empirical frequency distribution for total flow deficit.

9.5 Conclusion

Several probability distributions can be derived with the use of Bessel functions and the method of expansions in terms of cumulants or moments. These distributions do not seem to have been used in hydrologic, hydraulic, environmental, and

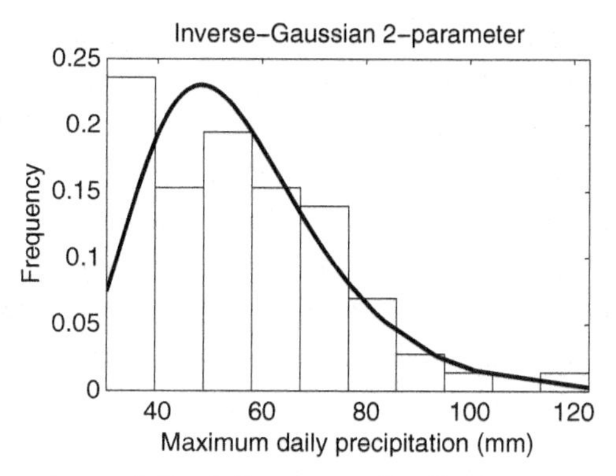

Figure 9.11 Comparison of fitted frequency distribution to empirical frequency distribution for maximum daily precipitation.

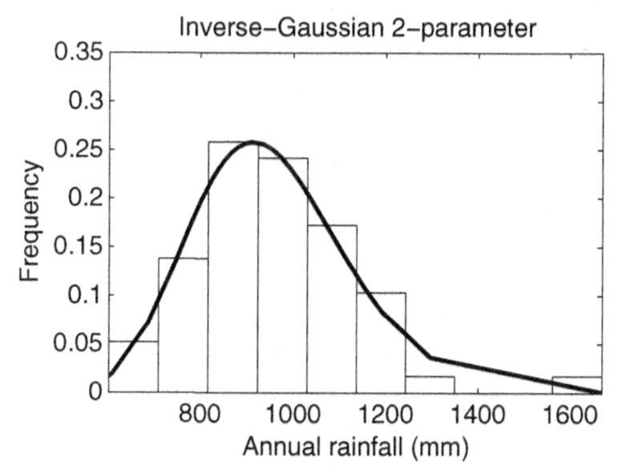

Figure 9.12 Comparison of fitted frequency distribution to empirical frequency distribution for annual rainfall.

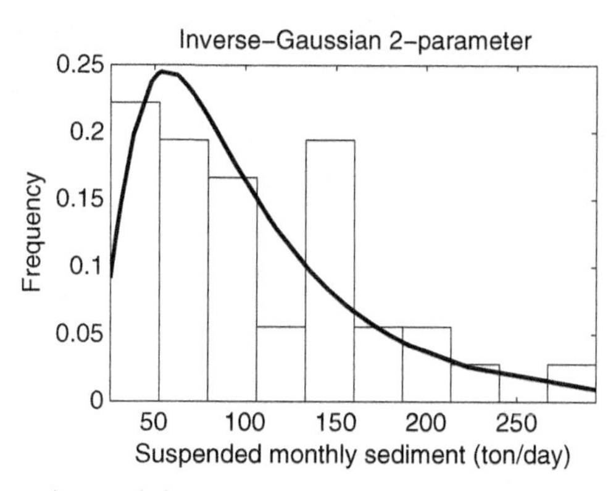

Figure 9.13 Comparison of fitted frequency distribution to empirical frequency distribution for monthly suspended sediment.

water resources engineering but may be useful. More importantly, these methods indicate a different way of deriving probability distributions. Following a similar line of thinking, it may be plausible to develop other methods. With the application of 2-parameter inverse Gaussian distribution, the 2-parameter inverse Gaussian distribution may be applied to study the frequency for the positively skewed random variables. Even though other distributions based on the Bessel functions discussed in this chapter may also properly model the frequency of the random variables, however the complexity of the parameter estimation may limit their applications.

References

Appell, P. (1925). Sur les fonctions hypergéometriques de plusieurs variables. In *Memoir. Sci. Math.* Paris: Gauthier-Villars.

Barndorff-Nielsen, O. E. (1997). Normal inverse Gaussian distributions and stochastic volatility modelling. *Scandinavian Journal of Statistics* 24, pp. 1–13.

Blinnikov, S., and Moessner, R. (1998). Expansions for nearly Gaussian distributions. *Astronomy and Astrophysics Supplement Series* 13, pp. 193–205.

Bowers, N. L., Jr. (1966). Expansion of probability density functions as a sum of gamma densities with applications in risk theory. *Transactions of Society of Actuaries* 18, no. 52, pp. 125–147.

Cavalcante, C. C., Mota, J. C. M., and Romano, J. M. T. (2004) Polynomial expansion of the probability density function about Gaussian mixtures. In Machine Learning and Signal Processing, Proceedings of the 2004 14th IEEE Signal Processing Society Workshop. doi: 10.1109/MLSP.2004.1422970.

Johnson, N. L., Kotz, S., and Balakrishnan, N. (1994). *Continuous Univariate Distributions*, Vol 1., 2nd ed. New York: John Wiley and Sons.

Kendall, S. M., and Stuart, A. (1977). *The Advanced Theory of Statistics, Volume 1: Distribution Theory*, 4th ed. New York: Macmillan Publishing Co, Inc.

Kolassa, J. E. (2006). *Series Approximation Methods in Statistics*, 3rd ed. New York: Springer Science+Business Media, Inc.

Mckay, A. T. (1932). A Bessel function distribution. *Biometrika* 24, no. ½, pp. 39–44.

Nadarajah, S. (2008). Some product Bessel density distributions. *Taiwanese Journal of Mathematics* 12, no. 1, pp. 191–211.

Suetin, P. K. (2001). Jacobi Polynomials. In *Encylopedia of Mathematics*, edited by M. Hazewinkel. Springer Science+Business Media B.V./Kluwer Academic Publishers. www.encyclopediaofmath.org/index.php?title=Jacobi_polynomials&oldid=18958.

10

Frequency Distributions by Entropy Maximization

10.1 Introduction

A wide spectrum of frequency distributions is used in hydrologic, hydraulic, environmental, and water resources engineering. Examples of such distributions include normal, lognormal, gamma, Pearson and log-Pearson type III, extreme value type I, generalized extreme value, Burr XII, and Pareto distributions. These distributions satisfy different constraints or have different sufficient statistics. This means that there is a unique correspondence between constraints and a distribution. It is then hypothesized that entropy maximization yields the least biased distribution for given constraints. The objective of this chapter therefore is to present the general framework for deriving probability distributions based on (1) specification of constraints and (2) Shannon entropy maximization subject to the specified constraints, and derive a number of distributions using the framework that are commonly used in environmental and water engineering.

10.2 Entropy Maximization

A large number of distributions can be derived through entropy maximization with given certain constraints. Although different types of entropies can be maximized to derive distributions, the easiest is the maximization of Shannon entropy (Shannon, 1948). Singh (1997) has derived a large number of distributions using this approach. The Shannon entropy, $H(X)$, for a continuous random variable X whose PDF is $f(x)$ can be defined as

$$H(X) = -\int f(x) \ln f(x) dx. \tag{10.1}$$

By maximizing entropy, subject to given information, a general form of $f(x)$ can be expressed. Depending the given information, specific forms of $f(x)$ can then be derived. The information on X can expressed as

$$C_i = \int g_i(x) f(x) dx = E[g_i(x)] = \overline{g_i(x)}; i = 0, 1, 2, \ldots; g_0(x) = 1 \tag{10.2}$$

where $g_i(x)$ is an arbitrary function of x. If $g_0(x)$=1, then C_0=1, which is the area under the probability density function (PDF). If $g_1(x) = x$, then Equation (10.2) gives $C_1 = \mu$, which is the central tendency or the mean of the distribution. For sample data, it is computed as sample mean: $\mu = \bar{x}$. Likewise, if $g_2(x) = x^2$ then Equation (10.2) defines $C_2 = \sigma^2 + \mu^2$, which is the second moment about the origin (or second order noncentral moment) of the distribution. In a similar manner, higher moments of the distribution can be defined.

Entropy maximization can be applied in two ways. First, any distribution can be derived by maximizing entropy subject to appropriate constraints. Second, if a distribution is known then its parameters can be determined by applying the principle of maximum entropy. In the first case, there is a unique correspondence between constraints and distribution. This means that it is the constraints that characterize the distribution and defining the constraints is the key. The second case is much simpler. For any distribution, constraints can be defined by invoking the definition of entropy, and then by invoking the principle of maximum entropy the correspondence between constraints and distribution parameters can be determined. Singh and Rajagopal (1986) developed a new method for parameter estimation for hydrologic frequency analysis. Singh (1997) and Singh et al. (1986) have derived a number of distributions or estimated their parameters using the regular entropy method and parameter-space expansion method. For illustrative purposes, the second case is illustrated here. This case entails the following steps: (1) define constraints; (2) construct the Lagrangian function; (4) apply the principle of maximum entropy; (5) determine the PDF in terms of Lagrange multipliers; (6) derive the relations between constraints and Lagrange parameters; and (7) express the distribution in terms of constraints.

Equation (10.1) can be maximized subject to Equation (10.2) using the method of Lagrange multipliers. Hence, the Lagrangian function L can be constructed as:

$$L = -\int f(x) \ln f(x) dx - (\lambda_0 - 1)\left[\int f(x) dx - 1\right] - \sum_{i=1}^{m} \lambda_i \left[\int g_i(x) f(x) dx - C_i\right] \tag{10.3a}$$

where λ_i, $i = 0, 1, 2, \ldots, m$. are the unknown Lagrange multiplies. Differentiating Equation (10.3a) with respect to function $f(x)$ and equating the derivative to zero, we obtain

$$\frac{\partial L}{\partial f(x)} = -\ln f(x) - 1 - (\lambda_0 - 1) - \sum_{i=1}^{m} \lambda_i g_i(x) = 0 \tag{10.3b}$$

Equation (10.3b) yields

$$f(x) = \exp\left[-\sum_{i=1}^{m} \lambda_i g_i(x)\right] \tag{10.4}$$

The Lagrange multipliers λ_i, $i = 0, 1, 2, \ldots, m$, can be determined in terms of constraints expressed by Equation (10.2) as:

$$\frac{\partial \lambda_0}{\partial \lambda_i} = -C_i \tag{10.5a}$$

It can be shown that

$$\frac{\partial^2 \lambda_0}{\partial \lambda_i^2} = var[g_i(x)]; \quad \frac{\partial^2 \lambda_0}{\partial \lambda_i \partial \lambda_j} = cov\left(g_i(x), g_j(x)\right]; \; i \neq j \tag{10.5b}$$

Equations (10.5a) and (10.5b) constitute the fundamental equations for establishing relations between Lagrange multipliers and constraints and in turn between distribution parameters and constraints.

Furthermore, λ_0 is a strictly convex function of $(\lambda_1, \ldots \lambda_m)$ and the objective function may be written as:

$$Z(\lambda_1, \ldots, \lambda_m) = \lambda_0 + \sum_{i=1}^{m} a_i \lambda_i = \ln\left(\int \exp\left[-\sum_{i=1}^{m} \lambda_i g_i(x)\right]\right) + \sum_{i=1}^{m} a_i \lambda_i \tag{10.6}$$

In Equation (10.6), a_i stands for the sample statistical moment of the i-th constraint.

The entropy maximization procedure can be used to derive any distribution whose PDF can be expressed in closed form. Table 10.1 lists the PDFs derived using this procedure. In what follows, we will use two examples to illustrate the procedure.

Example 10.1 Derive the PDF with constraint given as

$$C_0 = \int_0^{\infty} f(x)dx = 1 \tag{10.7}$$

$$C_1 = \int_0^{\infty} xf(x)dx = \bar{x} \tag{10.8}$$

Solution: The objective is to derive the PDF, subject to Equations (10.7) and (10.8). To that end, we construct the Lagrange function as

$$L = -\int_0^{\infty} f(x)\ln f(x)dx - (\lambda_0 - 1)\left[\int_0^{\infty} f(x)dx - 1\right] - \lambda_1\left[\int_0^{\infty} xf(x)dx - \bar{x}\right] \tag{10.9}$$

Differentiating Equation (10.9) with respect to $f(x)$ and equating the derivative to zero, the result is

$$\frac{\partial L}{\partial f(x)} = -\ln f(x) - 1 - (\lambda_0 - 1) - \lambda_1 x = 0 \tag{10.10}$$

Equation (10.10) leads to

$$f(x) = \exp(-\lambda_0 - \lambda_1 x) \tag{10.11}$$

The next step is to determine the Lagrange multipliers a_0 and a_1 in terms constraints given by Equations (10.7) and (10.8). Inserting Equation (10.11) in Equation (10.7), the result is:

$$\exp(\lambda_0) = \int_0^{\infty} \exp(-\lambda_1 x)\, dx \, or \, a_0 = \ln \int_0^{\infty} \exp(-\lambda_1 x) dx \tag{10.12}$$

which can be simplified for $\lambda_1 > 0$ as

$$\exp(\lambda_0) = \frac{1}{\lambda_1} \; or \; \lambda_0 = -\ln \lambda_1 \tag{10.13}$$

Equations (10.12)–(10.13) show that λ_0 is a function λ_1. Differentiating Equation (10.13) with respect to λ_1, we immediately obtain:

$$\frac{\partial \lambda_0}{\partial \lambda_1} = -\frac{1}{\lambda_1} \tag{10.14}$$

In addition, differentiating Equation (10.12) with respect to λ_1 and with the application of the Equations (10.7)–(10.8), we can write

$$\frac{\partial \lambda_0}{\partial \lambda_1} = -\frac{\int_0^{\infty} x \exp(-\lambda_1 x) dx}{\int_0^{\infty} \exp(-\lambda_1 x) dx} = -\frac{\int_0^{\infty} x \exp(-\lambda_0 - \lambda_1 x) dx}{\int_0^{\infty} \exp(-\lambda_0 - \lambda_1 x) dx} = -\bar{x} \tag{10.15}$$

Equating Equation (10.14) to Equation (10.15), we have:

$$\lambda_1 = \frac{1}{\bar{x}} \tag{10.16a}$$

Substituting Equations (10.13) and (10.16a) in Equation (10.11), the result yields

$$f(x) = \lambda_1 \exp(-\lambda_1 x) \tag{10.16b}$$

or

$$f(x) = \frac{1}{\bar{x}} \exp\left(-\frac{x}{\bar{x}}\right) \tag{10.17}$$

Equation (10.17) is the PDF of the exponential distribution with the parameter estimated from the mean of the random variable.

Example 10.2 Derive the PDF with the constraints given as

$$C_0 = \int_0^\infty f(x) = 1 \tag{10.18}$$

$$C_1 = \int_0^x xf(x) = \bar{x} \tag{10.19}$$

$$C_2 = \int_0^\infty \ln xf(x)dx = \overline{\ln x} \tag{10.20}$$

Solution: Same to Example 10.1, the objective here is to derive the PDF, subjected to Equations (10.18)–(10.20). With these constraints, we can express the maximum entropy-based PDF as:

$$f(x) = \exp\left(-\lambda_0 - \lambda_1 x - \lambda_2 \ln x\right) \tag{10.21}$$

In Equation (10.21), λ_0, λ_1, λ_2 are the Lagrange multipliers for the given constraints. Moreover, Lagrange multiplier λ_0 is a function of λ_1, λ_2. Inserting Equation (10.21) into Equation (10.18), we have:

$$\exp\left(\lambda_0\right) = \int_0^\infty \exp\left(-\lambda_1 x - \lambda_2 \ln x\right)dx \text{ or } a_0 = \ln \int_0^\infty \exp\left(-\lambda_1 x - \lambda_2 \ln x\right)dx \tag{10.22}$$

With some simple algebra, Equation (10.22) may be rewritten as:

$$\lambda_0 = (\lambda_2 - 1)\ln \lambda_1 + \ln \Gamma(1 - \lambda_2) \tag{10.23}$$

Taking the partial derivative of λ_0 vs. λ_1, λ_2 for Equation (10.22) we have:

$$\frac{\partial \lambda_0}{\partial \lambda_1} = -\frac{\int_0^\infty x \exp\left(-\lambda_1 x - \lambda_2 \ln x\right)dx}{\int_0^\infty \exp\left(-\lambda_1 x - \lambda_2 \ln x\right)dx} = -\int_0^\infty x \exp\left(-\lambda_0 - \lambda_1 x - \lambda_2 \ln x\right)dx = -\bar{x} \tag{10.24a}$$

$$\frac{\partial \lambda_0}{\partial \lambda_2} = -\int_0^\infty \ln x \exp\left(-\lambda_0 - \lambda_1 x - \lambda_2 \ln x\right)dx = -\overline{\ln x} \tag{10.24b}$$

Taking the partial derivative of λ_0 vs. λ_1, λ_2 for Equation (10.23) we have

$$\frac{\partial \lambda_0}{\partial \lambda_1} = \frac{\lambda_2 - 1}{\lambda_1} \tag{10.25a}$$

$$\frac{\partial \lambda_0}{\partial \lambda_2} = \ln \lambda_1 - \Psi(1 - \lambda_2) \tag{10.25b}$$

In Equations (10.25a)–(10.25b), $\lambda_1 > 0$, $\lambda_2 < 1$. Combining Equations (10.24a) with (10.25a), and Equations (10.24b) with (10.25b) we have:

$$\begin{cases} \dfrac{\lambda_2 - 1}{\lambda_1} = -\bar{x} \\ \ln \lambda_1 - \Psi(1 - \lambda_2) = -\overline{\ln x} \end{cases} \tag{10.26}$$

Then λ_1, λ_2 can be obtained by solving Equation (10.26) numerically. And Equation (10.21) may be rewritten as

$$f(x) = \frac{\hat{\lambda}_1^{1-\hat{\lambda}_2}}{\Gamma\left(1 - \hat{\lambda}_2\right)} x^{-\hat{\lambda}_2} \exp\left(-\hat{\lambda}_1 x\right) = \frac{\hat{\lambda}_1^{1-\hat{\lambda}_2}}{\Gamma\left(1 - \hat{\lambda}_2\right)} x^{1-\hat{\lambda}_2} - 1 \exp\left(-\hat{\lambda}_1 x\right) \tag{10.27}$$

Compare Equation (10.27) with the gamma PDF with parameter α, β given as:

$$f(x) = \frac{\beta^{\alpha}}{\Gamma(\alpha)} x^{\alpha-1} \exp(-\beta x) \tag{10.28}$$

We know that the constraints of Equations (10.18)–(10.20) yield maximum entropy-based gamma distribution. And the relations between Lagrange multipliers and parameters of gamma distribution is given as:

$$\beta = \lambda_1, \ \alpha = 1 - \lambda_2$$

Example 10.3 Derive the PDF with constraint given as

$$C_0 = \int_{e^c}^{\infty} f(x)dx = 1 \tag{10.29}$$

$$C_1 = \int_{e^c}^{\infty} \ln x f(x)dx = \overline{\ln x} \tag{10.30}$$

$$C_2 = \int_{e^c}^{\infty} \ln(\ln x - c) f(x)dx = \overline{\ln(\ln x - c)} \tag{10.31}$$

Solution: Similar to the last example, the objective is to derive maximum entropy-based PDF, subjected to Equations (10.29)–(10.31). We may construct the Lagrange function as:

$$L = -\int_{e^c}^{\infty} f(x) \ln f(x)dx - (\lambda_0 - 1)\left(\int_{e^c}^{\infty} f(x)dx - 1\right) - \lambda_1\left(\int_{e^c}^{\infty} f(x) \ln x dx - \overline{\ln x}\right)$$
$$- \lambda_2\left(\int_{e^c}^{\infty} f(x) \ln(\ln x - c)dx - \overline{\ln(\ln x - c)}\right) \tag{10.32}$$

Differentiating Equation (10.32) with respect to $f(x)$ and equating the derivative to zero, we have

$$\frac{\partial L}{\partial f(X)} = -\ln f(x) - 1 - (\lambda_0 - 1) - \lambda_1 \ln x - \lambda_2 \ln(\ln x - c) = 0 \tag{10.33}$$

From Equation (10.33), we have:

$$\begin{aligned} f(x) &= \exp(-\lambda_0 - \lambda_1 \ln x - \lambda_2 \ln(\ln x - c)) \\ &= \exp(-\lambda_0) x^{-\lambda_1} (\ln x - c)^{-\lambda_2} \end{aligned} \tag{10.34}$$

then we will determine the Lagrange multipliers $\lambda_0, \lambda_1, \lambda_2$ in terms of constraints given by Equations (10.29)–(10.31). Inserting Equation (10.34) into Equation (10.29), we have:

$$\exp(\lambda_0) = \int_{e^c}^{\infty} \exp(-\lambda_1 \ln x - \lambda_2 \ln(\ln x - c)) dx \tag{10.35a}$$

With some algebra, Equation (10.35a) may be solved as:

$$\exp(\lambda_0) = \frac{\exp(-c(\lambda_1 - 1))}{(\lambda_1 - 1)^{1-\lambda_2}} \Gamma(1 - \lambda_2) \tag{10.35b}$$

$$\lambda_0 = -c(\lambda_1 - 1) - (1 - \lambda_2) \ln(\lambda_1 - 1) + \ln \Gamma(1 - \lambda_2) \tag{10.35c}$$

Equations (10.35a)–(10.35c) indicate that λ_0 is a function of λ_1, λ_2. Differentiating Equation (10.35a) with respect to λ_1, λ_2, we have:

$$\frac{\partial \lambda_0}{\partial \lambda_1} = -\frac{\int_{e^c}^{\infty} \ln x \exp(-\lambda_1 \ln x - \lambda_2 \ln(\ln x - c)) dx}{\int_{e^c}^{\infty} \exp(-\lambda_1 \ln x - \lambda_2 \ln(\ln x - c)) dx} = -E(\ln x) \tag{10.36a}$$

$$\begin{aligned} \frac{\partial^2 \lambda_0}{\partial \lambda_1^2} &= \frac{\int_{e^c}^{\infty} \ln^2 x \exp(-\lambda_1 \ln x - \lambda_2 \ln(\ln x - c)) dx - \left(\int_{e^c}^{\infty} \ln x \exp(-\lambda_1 \ln x - \lambda_2 \ln(\ln x - c)) dx\right)^2}{\left(\int_{e^c}^{\infty} \exp(-\lambda_1 \ln x - \lambda_2 \ln(\ln x - c)) dx\right)^2} \\ &= E(\ln^2 x) - [E(\ln x)]^2 = Var(\ln x) \end{aligned} \tag{10.36b}$$

$$\frac{\partial \lambda_0}{\partial \lambda_2} = -\frac{\int_{e^c}^{\infty} \ln(\ln x - c) \exp(-\lambda_1 \ln x - \lambda_2 \ln(\ln x - c)) dx}{\int_{e^c}^{\infty} \exp(-\lambda_1 \ln x - \lambda_2 \ln(\ln x - c)) dx} = -E(\ln(\ln x - c)) \tag{10.36c}$$

In addition, the differentiation of Equation (10.35b) yields

$$\frac{\partial \lambda_0}{\partial \lambda_1} = -c + \frac{\lambda_2 - 1}{\lambda_1 - 1} \tag{10.37a}$$

$$\frac{\partial^2 \lambda_0}{\partial \lambda_1^2} = \frac{1 - \lambda_2}{(\lambda_1 - 1)^2} \tag{10.37b}$$

$$\frac{\partial \lambda_0}{\partial \lambda_2} = \ln(\lambda_1 - 1) - \Psi(1 - \lambda_2); \Psi(1 - \lambda_2) = \frac{d}{d(1 - \lambda_2)} \ln \Gamma(1 - \lambda_2) \tag{10.37c}$$

Setting the population moments equal to their sample moments, we have:

$$\begin{aligned} \frac{\partial \lambda_0}{\partial \lambda_1} &= -c + \frac{\lambda_2 - 1}{\lambda_1 - 1} = -\overline{\ln x} \\ \frac{\partial^2 \lambda_0}{\partial \lambda_1^2} &= \frac{1 - \lambda_2}{(\lambda_1 - 1)^2} = \overline{\ln^2 x} - \left(\overline{\ln x}\right)^2 \\ \frac{\partial \lambda_0}{\partial \lambda_2} &= \ln(\lambda_1 - 1) - \Psi(1 - \lambda_2) = -\overline{\ln(\ln x - c)} \end{aligned} \tag{10.38}$$

Solving Equation (10.38) numerically, we have the value of $\hat{\lambda}_1, \hat{\lambda}_2,$ and $\hat{c}$. Then we can rewrite the PDF with the constraints of Equations (10.29)–(10.31) as

$$\begin{aligned} f(x) &= \exp\left(\hat{c}\left(\hat{\lambda}_1 - 1\right) + \left(1 - \hat{\lambda}_2\right) \ln\left(\hat{\lambda}_1 - 1\right) - \ln \Gamma\left(1 - \hat{\lambda}_2\right) \right. \\ &\quad \left. -\hat{\lambda}_1 \ln x - \hat{\lambda}_2 \ln(\ln x - \hat{c})\right) \\ &= \exp \frac{\left(\hat{c}\left(\hat{\lambda}_1 - 1\right)\right)\left(\hat{\lambda}_1 - 1\right)^{1-\hat{\lambda}_2}}{\Gamma\left(1 - \hat{\lambda}_2\right)} x^{-\hat{\lambda}_1} (\ln x - \hat{c})^{-\hat{\lambda}_2} \end{aligned} \tag{10.39}$$

With some simple algebra, Equation (10.39) can be rewritten as:

$$f(x) = \frac{\hat{\lambda}_1 - 1}{x\Gamma\left(1 - \hat{\lambda}_2\right)} \left[\left(\hat{\lambda}_1 - 1\right)(\ln x - \hat{c})\right]^{-\hat{\lambda}_2} \exp\left(-\left(\hat{\lambda}_1 - 1\right)(\ln x - \hat{c})\right) \tag{10.39a}$$

Comparing Equation (10.39a) with the PDF of log-Pearson type III distribution given as:

$$f(x) = \frac{1}{\alpha x \Gamma(\beta)} \left(\frac{\ln x - c}{\alpha}\right)^{\beta - 1} \exp\left(-\frac{\ln x - c}{\alpha}\right), \tag{10.39b}$$

we have the relation of Lagrange multipliers and distribution parameters as:

$$\alpha = \frac{1}{\lambda_1 - 1}, \; \beta = 1 - \lambda_2 \tag{10.40}$$

and c is the location parameter depending on the dataset.

In a similar manner, other distributions can be derived by specifying appropriate constraints, as shown in Table 10.1.

Previously, we have discussed the entropy maximization of frequency distributions. In what follows, we will discuss a more general approach to derive the maximum entropy-based frequency distributions. The difference is in the constraint assignment. In the general case, the statistics governing the shape of density function are applied as the constraints, i.e., the first four moments. Additionally, for the positive random variables, the first moment in the logarithm domain may also be considered as a constraint. With this in mind, the constraints may be expressed as:

$$\text{Total probability}: \int f(x)dx = 1 \tag{10.41}$$

$$\text{First moment in logarithm domain}: \int_{x>0} \ln x f(x)dx = \overline{\ln x} \tag{10.42}$$

$$\text{First moment}: \int xf(x)dx = \bar{x} \tag{10.43}$$

$$\text{Second noncentral moment}: \int x^2 f(x)dx = \overline{x^2} \tag{10.44}$$

$$\text{Third noncentral moment}: \int x^3 f(x)dx = \overline{x^3} \tag{10.45}$$

$$\text{Fourth noncentral moment}: \int x^4 f(x)dx = \overline{x^4} \tag{10.46}$$

The constraint of fourth moment is necessary if and only if the excess kurtosis is significantly different from zero, which is evaluated as follow:

Let γ_2', G_2 denote the excess kurtosis and sample excess kurtosis, respectively, as:

$$\gamma_2' = \frac{n\sum_{i=1}^{n}(x_i - \bar{x})^4}{\left(\sum_{i=1}^{n}(x_i - \bar{x})^2\right)^2} - 3 \tag{10.47a}$$

$$G_2 = \frac{n-1}{(n-2)(n-3)}\left((n+1)\gamma_2' + 6\right) \tag{10.47b}$$

and the test statistic (T) can be computed as:

$$T = \frac{G_2}{SEK} \tag{10.47c}$$

$$SEK = 2\sqrt{\frac{6n(n-1)^2}{(n-2)(n+5)(n^2-9)}} \tag{10.47d}$$

Table 10.1 *Frequency distributions obtained from entropy maximization*

Distr.	Constraints	PDF	Relation between Parameters and $g_i(x)$	Entropy (H)
Uniform	$g_0(x) = 1$	$f(x) = \frac{1}{b-a}, x \in [a, b]$	$\exp(a_0) = b - a$	$H(X) = \ln(b - a)$
Exponential	$g_0(x) = 1$ $g_1(x) = x$	$f(x) = \frac{1}{k}\exp\left(-\frac{x}{k}\right)$, $x > 0, k > 0$	$\exp(-a_0) = a_1 = \frac{1}{k} = \frac{1}{\bar{x}}$	$H(X) = \ln k + 1$
Normal	$g_0(x) = 1$ $g_1(x) = x$ $g_2(x) = x^2$	$f(x) = \frac{1}{b\sqrt{2\pi}}$ $\exp\left[-\frac{(x-a)^2}{2b^2}\right]$	$a_0 = \frac{\ln \pi}{2} - \frac{\ln a_2}{2} + \frac{a_1^2}{4a_2}$ $a_1 = -\frac{a}{b_2}; a_2 = \frac{1}{2b^2}$	$H(X) = \ln\left(b\sqrt{2\pi e}\right)$
Log-normal	$g_0(x) = 1$ $g_1(x) = \ln x$ $g_2(x) = (\ln x)^2$	$f(x) = \frac{1}{xb\sqrt{2\pi}}$ $\exp\left(-\frac{(\ln x - a)^2}{2b^2}\right)$ $y = \ln x, x > 0$	$a_0 = \frac{\ln \pi}{2} - \frac{\ln a_2}{2} + \frac{(a_1-1)^2}{4a_2}$ $a = E(y); b = S_y$ $a_1 = 1 - \frac{a}{b^2}; a_2 = \frac{1}{2b^2}$	$H(X) = \ln\left(b\sqrt{2\pi e}\right) + a$
3-parameter log-normal	$g_0(x) = 1$ $g_1(x) = \ln(x - a)$ $g_2(x) = [\ln(x - a)]^2$	$f(x) = \frac{1}{(x-a)c\sqrt{2\pi}}$ $\exp\left(-\frac{(\ln(x-a)-b)^2}{2c^2}\right)$ $y = \ln(x - a); x > a$	$a_0 = \frac{\ln \pi}{2} - \frac{\ln a_2}{2} + \frac{(a_1-1)^2}{4a_2}$ $b = E(\ln(x - a)) = E(y)$ $c = S_y$ $a_1 = 1 - \frac{b}{c^2}$ $a_2 = \frac{1}{2c^2}$	$H(X) = \ln\left(c\sqrt{2\pi e}\right) + b$
EV I	$g_0(x) = 1$ $g_1(x) = x$ $g_2(x) = \exp(-ax)$	$f(x) = a\exp\left(-a(x - b) - e^{-a(x-b)}\right)$ $a > 0, x > b$	$a_0 = -\ln a - \frac{a_1}{a}\ln a_2 + \ln\left(\Gamma\left(\frac{a_1}{a}\right)\right)$ $a_1 = a; a_2 = \exp(ab)$	$H(X) = -\ln a + a$ $- 0.5772$
Log-EV I	$g_0(x) = 1$ $g_1(x) = \ln x$ $g_2(x) = \exp(-a(\ln x - b))$	$f(x) = a\exp\left(-a(\ln x - b) - e^{-a(\ln x - b)}\right)$ $a > 0, x > \exp(b), y = \ln x$	$a_0 = -\ln a + b - ba_1 + \frac{1-a_1}{a}\ln a_2 +$ $\ln\Gamma\left(\frac{a_1-1}{a}\right)$ $a_1 = a + 1, a_2 = 1$	
EV III	$g_0(x) = 1$ $g_1(x) = \ln(x - c)$ $g_2(x) = (x - c)^a$	$f(x) = \frac{a}{b-c}\left(\frac{x-c}{b-c}\right)^{a-1}$ $\exp\left(-\left(\frac{x-c}{b-c}\right)^a\right)$ $a > 0; b > 0$	$a_0 = \ln\Gamma\left(\frac{1-a_1}{a}\right) - \ln a - \frac{1-a_1}{a}\ln a_2$ $a_1 = 1 - a; a_2 = \frac{1}{(b-c)^a}$	

GEV	$g_0(x) = 1$ $g_1(x) = \ln\left(1 - \frac{b}{a}(x-c)\right)$ $g_2(x) = \left(1 - \frac{b}{a}(x-b)\right)^{\frac{1}{b}}$	$f(x) = \frac{1}{a}\left(1 - \frac{b}{a}\left(\frac{x-c}{b-c}\right)\right)^{\frac{1-b}{b}}$ $\exp\left(-\left(1 - \frac{b}{a}\left(\frac{x-c}{b-c}\right)\right)^{\frac{1}{b}}\right)$ $a > 0$ $x \in \left(-\infty, c + \frac{a}{b}\right)$ if b>0 $x \in \left(c + \frac{a}{b}, \infty\right)$ if b<0	$a_0 = \ln a + b(a_1 + 1) \ln a_2 +$ $\ln \Gamma(b(a_1 + 1))$ $a_1 = \frac{1-b}{b}; a_2 = 1$
Weibull	$g_0(x) = 1$ $g_1(x) = \ln(x)$ $g_2(x) = x^a$	$f(x) = \frac{a}{b}\left(\frac{x}{b}\right)^{a-1}$ $\exp\left(-\left(\frac{x}{b}\right)^a\right)$ $a > 0, b > 0, x \in (0, \infty)$	$a_0 = -\ln a + \ln \Gamma\left(\frac{1-a_1}{a}\right) -$ $\left(\frac{1-a_1}{a}\right) \ln a_2$ $a_1 = 1 - a, a_2 = \frac{1}{b^a}$
GA	$g_0(x) = 1$ $g_1(x) = x$ $g_2(x) = \ln x$	$f(x) = \frac{1}{a\Gamma(b)}\left(\frac{x}{a}\right)^{b-1} \exp\left(-\left(\frac{x}{a}\right)\right)$ $a > 0, b > 0, x \in (0, \infty)$	$a_0 = (a_2 - 1) \ln a_1 + \ln \Gamma(1 - a_2)$ $a_1 = \frac{1}{a}, a_2 = 1 - b$
P III	$g_0(x) = 1$ $g_1(x) = x$ $g_2(x) = \ln(x - c)$	$f(x) = \frac{1}{a\Gamma(b)}\left(\frac{x-c}{a}\right)^{b-1}$ $\exp\left(-\left(\frac{x-c}{a}\right)\right)$ $a > 0, b > 0, x \in (c, \infty)$	$a_0 = -a_1 c + (a_2 - 1) \ln a_1 +$ $\ln \Gamma(1 - a_2)$ $a_1 = \frac{1}{a}, a_2 = 1 - b$
LP III	$g_0(x) = 1$ $g_1(x) = \ln x$ $g_2(x) = \ln(\ln x - c)$	$f(x) = \frac{1}{ax\Gamma(b)}\left(\frac{\ln x - c}{b}\right)^{b-1} \exp\left(-\frac{\ln x - c}{a}\right)$ $a > 0, b > 0, x \in (\exp(c), \infty)$	$a_0 = -c(a_1 - 1) + (a_2 - 1)$ $\ln(a_1 - 1) + \ln \Gamma(1 - a_2)$ $a_1 = 1 + \frac{1}{a}, a_2 = 1 - b$
Beta	$g_0(x) = 1$ $g_1(x) = \ln x$ $g_2(x) = \ln(1 - x)$	$f(x) = \frac{\Gamma(a+b)}{\Gamma(a)\Gamma(b)} x^{a-1}(1 - x)^{b-1}$ $a > 0, b > 0, x \in (0, 1)$	$a_0 = \ln \Gamma(1 - a_1) + \ln(1 - a_2) -$ $\ln(2 - a_1 - a_2)$ $a_1 = 1 - a, a_2 = 1 - b$
2-parameter LG	$g_0(x) = 1$ $g_1(x) = \ln x$ $g_2(x) = \ln\left(1 + \left(\frac{x}{a}\right)^b\right)$	$f(x) = \frac{b}{a}\frac{\left(\frac{x}{a}\right)^{b-1}}{\left(1+\left(\frac{x}{a}\right)^b\right)^2}$ $a > 0, b > 1, x \in (0, \infty)$	$a_0 = \ln b + \ln \Gamma\left(\frac{1-a_1}{b}\right) +$ $\ln \Gamma\left(a_2 - \frac{1-a_1}{b}\right) - \ln \Gamma(a_2)$ $a_1 = 1 - b, a_2 = 2$
3-parameter LG	$g_0(x) = 1$ $g_1(x) = \ln\left(\frac{x-c}{a}\right)$ $g_2(x) = \ln\left(1 + \left(\frac{x-c}{a}\right)^b\right)$	$f(x) = \frac{b}{a}\frac{\left(\frac{x-c}{a}\right)^{b-1}}{\left[1+\left(\frac{x-c}{a}\right)^b\right]^2}$ $a > 0, b > 0, x \in (c, \infty)$	$a_0 = \ln b + \ln \Gamma\left(\frac{(1-a_1)}{b}\right) +$ $\ln \Gamma\left(a_2 - \frac{1-a_1}{b}\right) - \ln \Gamma(a_2)$ $a_1 = 1 - b, a_2 = 2$

Table 10.1 (*cont.*)

Distr.	Constraints	PDF	Relation between Parameters and $g_i(x)$	Entropy (H)
2-parameter Pareto	$g_0(x) = 1$ $g_1(x) = \ln(x)$	$f(x) = ba^b x^{b-1}$ $a > 0, b > 0, x \in (a, \infty)$	$a_0 = -\ln(a_1 - 1) + (a_1 - 1)\ln a$ $a_1 = 1 + b$	
2-parameter GP	$g_0(x) = 1$ $g_1(x) = \ln\left(1 - a\left(\frac{x}{b}\right)\right)$	$f(x) = \begin{cases} \frac{1}{b}\left(1 - \frac{x}{b}\right)^{\frac{1}{a}}, a \neq 0 \\ 1 - \exp\left(-\frac{x}{b}\right), a = 0 \end{cases}$ $b > 0, x \in (a, \infty)$	$a_0 = \ln\left(\frac{b}{a}\right) - \ln(1 - a_1)$ $a_1 = 1 - \frac{1}{a}$	
3-parameter GP	$g_0(x) = 1$ $g_1(x) = \ln\left(1 - \frac{a(x-c)}{b}\right)$	$f(x) = \begin{cases} \frac{1}{b}\left(1 - \frac{x-c}{b}\right)^{\frac{1}{a}}, a \neq 0 \\ \frac{1}{b}\left(1 - \exp\left(-\frac{x-c}{b}\right)\right), a = 0 \end{cases}$ $b > 0, x \in (c, \infty)$	$a_0 = \ln\left(\frac{b}{a}\right) - \ln(1 - a_1)$ $a_1 = 1 - \frac{1}{a}$	

In Equations (10.47a)–(10.47d), γ_2' stands for the excess kurtosis, g_2 stands for sample excess kurtosis, *SEK* stands for the standard error of kurtosis, and n stands for the sample size. For test statistics T, $|T|>2$ indicates the excess kurtosis is significantly different from zero and the noncentral fourth moment needs to be applied as a constraint.

Based on the preceding discussion, the maximum entropy-based distribution may be expressed as:

$$f(x) = \exp\left(-\lambda_0 - \sum_{i=1}^{N} \lambda_i x^i\right), N = 3 \ or \ 4 \tag{10.48a}$$

or

$$f(x) = \exp\left(-\lambda_0 - \lambda_1 \ln x - \sum_{i=1}^{N} \lambda_{i+1} x^i\right), N = 3 \ or \ 4 \tag{10.48b}$$

It is necessary to note that the Lagrange multipliers for the N-th noncentral moment need to be greater than 0.

Substituting Equation (10.48b) to Equation (10.41), the partition function is obtained as:

$$\exp(\lambda_0) = \int \exp\left(-\sum_{i=1}^{N} \lambda_i x^i\right) dx, or \ \lambda_0 = \ln\left(\int \exp\left(-\sum_{i=1}^{N} \lambda_i x^i\right) dx\right); \quad N = 3 \ or \ 4 \tag{10.49a}$$

or

$$\begin{aligned} \exp(\lambda_0) &= \int_0^\infty \exp\left(-\lambda_1 \ln x - \sum_{i=1}^{N} \lambda_{i+1} x^i\right) dx \quad or \\ \lambda_0 &= \ln\left(\int_0^\infty \exp\left(-\lambda_1 \ln x - \sum_{i=1}^{N} \lambda_{i+1} x^i dx\right); N = 3 \ or \ 4 \end{aligned} \tag{10.49b}$$

Again, λ_0 is a strictly convex function of $\lambda_1, \lambda_2, \ldots$, and the objective function can be written as:

$$Z(\boldsymbol{\lambda}) = \lambda_0 + \sum_{i=1}^{N} a_i \lambda_i \ or \ Z(\boldsymbol{\lambda}) = \lambda_0 + \sum_{i=1}^{N+1} a_i \lambda_i \tag{10.50}$$

In Equation (10.50), a_i stands for the i-th sample moment of the constraint.

Given that λ_0 is a strictly convex function of $\lambda_1, \lambda_2, \ldots$, the objective function Z is also a convex function of λs. Thus, minimizing the objective function will result in the maximum entropy. Applying Newton's method, the Lagrange multipliers can be determined as follows:

Let

$$g_i(x) = x^i \ or \begin{cases} g_1(x) = \ln x \\ g_{i+1}(x) = x^i \end{cases}; i = 1, 2, \ldots, N \tag{10.51}$$

The objective function may be approximated with the second-order Taylor series for the Lagrange multipliers as:

$$Z(\boldsymbol{\lambda}) \cong Z(\boldsymbol{\lambda}^0) - \mathbf{G}(\boldsymbol{\lambda}^0)(\boldsymbol{\lambda} - \boldsymbol{\lambda}^0) + \frac{1}{2}(\boldsymbol{\lambda} - \boldsymbol{\lambda}^0)^T \mathbf{H}(\boldsymbol{\lambda}^0)(\boldsymbol{\lambda} - \boldsymbol{\lambda}^0) \tag{10.52}$$

In Equation (10.52), G_i, $H_{i,j}$ are the elements of gradient vector $\mathbf{G}$ and Hessian matrix $\mathbf{H}$ and can be expressed as:

$$G_i = \frac{\partial Z}{\partial \lambda_i} = a_i - E(g_i(x)); H_{i,j} = \frac{\partial^2 Z}{\partial \lambda_i \partial \lambda_j} = \operatorname{cov}(g_i(x), g_j(x)) \tag{10.53}$$

10.3 Application

In this section, we will use the real-world data to illustrate the application of entropy theory in frequency analysis. The peak flow data and monthly sediment yield for the month of March are applied for analysis.

10.3.1 Peak Flow

The constraints for peak flow are the same as that illustrated in Example 10.3. The Lagrange multipliers may be estimated by solving Equation (10.38). The estimated Lagrange multipliers are listed in Table 10.2. Figure 10.1 compares the fitted frequency distribution with the empirical frequency distribution as well as the fitted frequency of log-Pearson III distribution (discussed in Chapter 2). Figure 10.1 shows indiscernible difference between the fitted frequency distribution computed from the entropy-based distribution and that computed from the log-Pearson III distribution with the parameters estimated using the indirect method of moments. Additionally, Figure 10.1 proves the appropriateness to apply the entropy theory to frequency analysis.

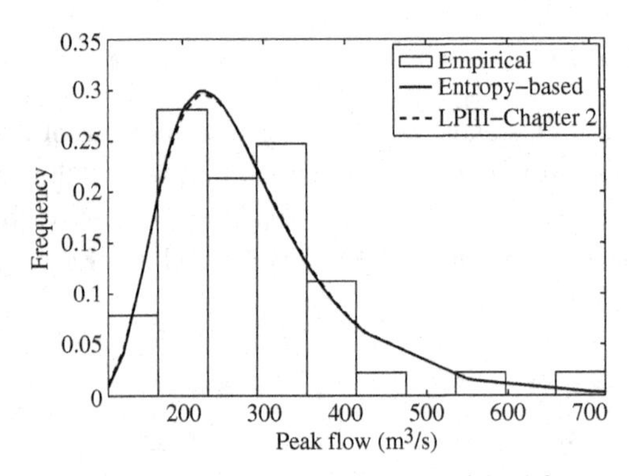

Figure 10.1 Comparison of fitted frequency and empirical frequency of peak flow.

10.3.2 Monthly Sediment Yield

The constraints for monthly sediment yield are defined as:

$$\begin{cases} \int xf(x)dx = E(x) = \bar{x} \\ \int x^2 f(x)dx = E(x^2) = \overline{x^2} \\ \int x^3 f(x)dx = E(x^3) = \overline{x^3} \end{cases} \quad (10.54)$$

To reduce the computation complexity, linear transformation was applied to the sediment variable as:

$$X^T = \frac{X - (1-d)\min(X)}{(1+d)\max(X) - (1-d)\min(X)}, d = 0.1 \quad (10.55)$$

Equation (10.55) indicates that the transformed variable X^T is in the range of (0,1).

Using the constraints defined in Equation (10.54), the entropy-based distribution can be written as:

$$f(x) = \exp(-\lambda_0 - \lambda_1 x - \lambda_2 x^2 - \lambda_3 x^3), \lambda_3 > 0 \quad (10.56)$$

Similarly, one can express λ_0 as a function of λ_1, λ_2, and λ_3:

$$\lambda_0 = \ln \int_0^1 \exp\left(-\lambda_1 x - \lambda_2 x^2 - \lambda_3 x^3\right) dx \quad (10.57)$$

Taking the derivative of λ_0 vs. λ_1, λ_2, and λ_3, we have:

$$\begin{cases} \frac{\partial \lambda_0}{\partial \lambda_1} = \frac{-\int x \exp(-\lambda_0 - \lambda_1 x - \lambda_2 x^2 - \lambda_3 x^3) dx}{\int_0^1 \exp(-\lambda_0 - \lambda_1 x - \lambda_2 x^2 - \lambda_3 x^3) dx} = -E(x) \approx -\bar{x} \\ \frac{\partial \lambda_0}{\partial \lambda_2} = \frac{-\int x^2 \exp(-\lambda_0 - \lambda_1 x - \lambda_2 x^2 - \lambda_3 x^3) dx}{\int_0^1 \exp(-\lambda_0 - \lambda_1 x - \lambda_2 x^2 - \lambda_3 x^3) dx} = -E(x^2) \approx -\overline{x^2} \\ \frac{\partial \lambda_0}{\partial \lambda_3} = \frac{-\int x^3 \exp(-\lambda_0 - \lambda_1 x - \lambda_2 x^2 - \lambda_3 x^3) dx}{\int_0^1 \exp(-\lambda_0 - \lambda_1 x - \lambda_2 x^2 - \lambda_3 x^3) dx} = -E(x^3) \approx -\overline{x^3} \end{cases} \quad (10.58)$$

Then, the Lagrange multipliers can be obtained by solving Equation (10.58). Table 10.2 lists the Lagrange multipliers estimated for monthly sediment yield. Based on the parameters estimated with $a_3 < 1.e - 10 \approx 0$, the entropy-based

Table 10.2 *Lagrange multipliers estimated for peak flow and monthly sediment yield*

Peak Flow	a_1	a_2	c
	17.95 a0 = –70.6	–33.7	3.54
Sediment	a_1	a_2	a_3
	–0.024 a0 = 0.83	4.14	< 1.e–10

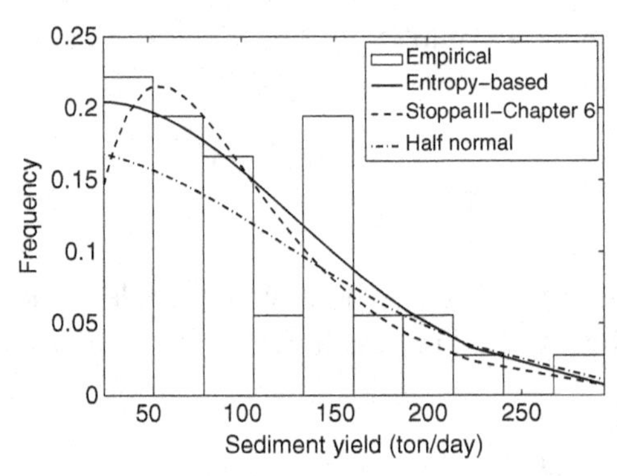

Figure 10.2 Comparison of fitted frequency and empirical frequency of sediment yield.

distribution derived converges to half normal (seminormal distribution). From Equation (4.36a) in Chapter 4, the parameter is computed as: $\sigma = 132.66$. Figure 10.2 compares the fitted frequencies of the entropy-based distribution to the empirical frequencies as well as the frequencies computed using the Stoppa III distribution (in Chapter 6) and half normal distribution (also called semi-normal distribution in Chapter 4). Figure 10.2 exhibits similar performances between entropy-based distribution and half-normal distribution. Though Stoppa III distribution may be applied to evaluate the frequency distribution of monthly sediment data, the entropy-based and half-normal distributions outperform the Stoppa III distribution.

10.4 Conclusion

This chapter presented maximum entropy-based probability distributions. We presented the Shannon entropy and explained that any given PDF with a closed form may be derived by maximizing the Shannon entropy. For the derived maximum entropy-based distribution, the parameters (i.e., Lagrange multipliers) are obtained with the use of the Lagrange function by equating the certain constraints to the population (or sample) moments. Additionally, there is a one-to-one relation between the Lagrange multipliers for the maximum entropy-based distribution and the parameters of the given PDF with closed form.

References

Shannon, E. E. (1948). A mathematical theory of communication. *Bell System Technical Journal* 27(3). doi: 10.1002/j.1538-7305.1948.tb01338.x.

Singh, V. P. (1997). *Entropy-Based Parameter Estimation in Hydrology*. Dordrecht: Springer.

Singh, V. P., and Rajagopal, A. K. (1986). A new method of parameter estimation for hydrologic frequency analysis. *Hydrological Science and Technology* 2, no. 3, pp. 33–40.

Singh, V. P., Rajagopal, A. K., and Singh, K. (1985). Derivation of some frequency distributions using principle of maximum entropy (POME). *Advances in Water Resources* 9, pp. 91–106.

11

Transformations for Frequency Distributions

11.1 Introduction

There is a multitude of frequency distributions used in environmental and water engineering that the basic distributions, i.e., the number of parameters of the transformed distribution = the number of parameters of the basic distribution + the number of parameters of the transformation. This suggests that a distribution with more parameters can be derived with the use of transformation. This is one way to derive generalized distributions that then lead to a number of special or limiting distributions. In other words, the result is a treelike structure.

The transformations widely used in environmental and water engineering are linear, logarithmic, power, and exponential. It may be noted that data can also be transformed so that they are approximately normally distributed. Further, not all transformations are applicable to each distribution, meaning certain transformations are distribution specific.

The objective of this chapter therefore is to discuss the transformation approach, basic distributions, transformations, and transformed distributions along with their special cases.

11.2 Transformation to Normal Distribution

Because of the popularity of the normal distribution, it is intuitively appealing to develop transformations that would convert nonnormally distributed observed data to the normally distributed data. The most popular examples of such transformations are power (Chander et al. 1978), logarithmic and SMEMAX and its modified version (Venugopal, 1980; Rasheed et al. 1982). Jain and Singh (1986) compared different methods of transformation used in flood frequency analysis and concluded that power transformation was adequate for a majority of flood data sets.

Box and Cox (1964) proposed the following system of scaled down transformations for normality:

$$y_j = \frac{\left(x_j^{\lambda} - 1\right)}{\lambda}, j = 1, 2, \ldots, \lambda \neq 0 \tag{11.1}$$

$$y_j = \ln x_j, j = 1, 2, \ldots, \lambda = 0 \tag{11.2}$$

where x_j is the variate of a given series of data, y_j is the transformed variate, and λ is the constant of transformation. This system is a more general transformation that can be shown to include logarithmic ($\lambda = 0.0$), reciprocal ($\lambda = 1$), and square root ($\lambda = 0.5$) transformations as special cases. It may be noted that constant λ appears nonlinearly in Equation (11.1) and cannot therefore be determined in closed form. However, it can be determined by trial and error such that it makes the coefficient of skewness of the transformed series nearly zero. It has been reported that its value generally lies between –1 and +1, and an increase in its value leads to an increase in the coefficient of skewness and vice versa.

Investigating bias in log-transformed frequency distributions, Wilson et al. (1980) noted that log transformation led to bias in moments and quantiles in the arithmetic space. This bias resulted from the process of inverse nonlinear transformation and the variance of log-space parameter estimates and depended on the sample size and population parameters. The implication is that a distribution fitted in log space may not fit the arithmetic series well. The bias can, however, be corrected by employing the maintenance-of-variance-extension (MOVE) developed by Hirsch (1982).

It can be shown that the inverse of Equation (11.1) may become imaginary for certain values y_j (Moog et al. 1999) and therefore this transformation is rejected for synthetically generated values (Hirsch, 1979). However, this problem can be overcome by imposing certain constraints. Combining MOVE and Box-Cox (BC) transformation improves order statistics of flow rates, particularly nonextremes.

The power-transformed data may not necessarily be normally distributed because their kurtosis may not be equal to 3.0. The transformation can be improved by accounting for the departure from 3.0 in the coefficient of kurtosis of the transformed series. Box and Tiao (1973) and Tiao and Lund (1970) related a factor β to kurtosis C_K as

$$C_K = \frac{\Gamma\left(\frac{5}{2}(1+\beta)\right)\Gamma\left(\frac{1}{2}(1+\beta)\right)}{\left[\Gamma\left(\frac{3}{2}(1+\beta)\right)\right]^2} - 3.0 \tag{11.3}$$

If $\beta = 0$, the transformed series is normal. The transformed series can be adjusted for kurtosis with the use of standard deviate corresponding to the computed value of β.

Gupta et al. (1989) proposed a two-step power transformation that seems to produce normally distributed data. The first step is to use Equation (11.1). For the data that needs correction for kurtosis, the following modulus transformation is used:

$$z_j = \text{sign}(y_j - \bar{y})|y_j - \bar{y}|^{\gamma} \tag{11.4}$$

in which z_j is the transformed data, and γ is positive. For a suitable value of γ, the transformed series will be normally distributed. In Equation (11.4), if γ tends to zero, the kurtosis of the transformed z-series would tend to one. If γ tends to infinity, the kurtosis of the transformed z-series would tend to infinity. If γ tends to one, the kurtosis of the transformed z-series would be the same as that of the original y-series. Hence, if the kurtosis of the y-series is greater than 3.0, then γ would be 1 and 0, and if kurtosis is less than 3.0, then γ would be greater than 1.

11.3 Transformation of Normal Distribution: The Johnson family

The idea of transformation originated with Edgeworth (1898), who called it the method of translation, and was pursued by Kapteyn (1903), van Uven (1917), and Frechet (1939). Recognizing that most observations in nature do not follow a normal distribution, the method of translation transforms the normal distribution into a skewed distribution. Edgeworth (1898) derived the lognormal distribution by applying a transformation to the normal distribution (also called the first law of Laplace) and called his method as a method of translation. Following this method, Rietz (1922) derived frequency distributions using certain transformations of normally distributed variates. This method was generalized by Johnson (1949) and D'Addario (1936), and the most popular example of this method is the Johnson family of frequency curves comprising the SL, SB, and SU systems. Snyder et al. (1978) obtained lognormal, log-lognormal, exnormal, and ex-exnormal distributions by applying different transforms to the normal distribution.

Let X and Y be two random variables and let Y follow a standard normal distribution:

$$f(y) = \frac{1}{\sqrt{2\pi}} \exp\left(-\frac{y^2}{2}\right) \tag{11.5}$$

Random variables X and Y are assumed to be related as

$$y = a + bg(x); a \in \mathbb{R}, b > 0 \tag{11.6}$$

In Equation (11.6) $g(x)$ is a nondecreasing monotonic function. Johnson (1949) proposed three forms of $f(x)$ or transformations of the normal distribution:

$$y = a + b \ln (x - c);\ a, c \in \mathbb{R},\ b > 0,\ x > c \tag{11.7}$$

$$y = a + b \ln \left(\frac{x - c}{c + d - x} \right); a \in \mathbb{R}, b > 0, x \in (c, c + d) \tag{11.8}$$

$$y = a + b \sinh^{-1} \left(\frac{x - c}{d} \right); a, c \in \mathbb{R}, b, d > 0, x \in (-\infty, \infty) \tag{11.9}$$

In Equations (11.6)–(11.9), a, b, c, and d are parameters. Figure 11.1 plots the transformations to unit normal variables with one set of parameters as an example. It is seen that the transformation functions [i.e., Equations (11.7)–(11.9)] are one-to-one monotone increasing transformation.

From the monotonic nondecreasing transformation [i.e., Equation (11.6)], the probability density function (PDF) of random variable X may be written as:

$$f_X(x) = f_Y(y(x)) \left| \frac{dy}{dx} \right| \tag{11.10}$$

The transformation of Equation (11.7) yields random variable X as log-normal distributed (i.e., Johnson's SL distribution):

$$f_Y(y(x)) = \frac{1}{\sqrt{2\pi}} \exp \left(-\frac{(b \ln (x - c) + a)^2}{2} \right), \quad \left| \frac{dy}{dx} \right| = \frac{b}{x - c} \tag{11.11}$$

Substitution of Equation (11.11) into Equation (11.10) yields

$$f_X(x) = \frac{1}{\sqrt{2\pi}(x - c)\left(\frac{1}{b}\right)} \exp \left(-\frac{\left(\ln (x - c) + \frac{a}{b} \right)^2}{2\left(\frac{1}{b}\right)^2} \right); x > c, b > 0 \tag{11.12}$$

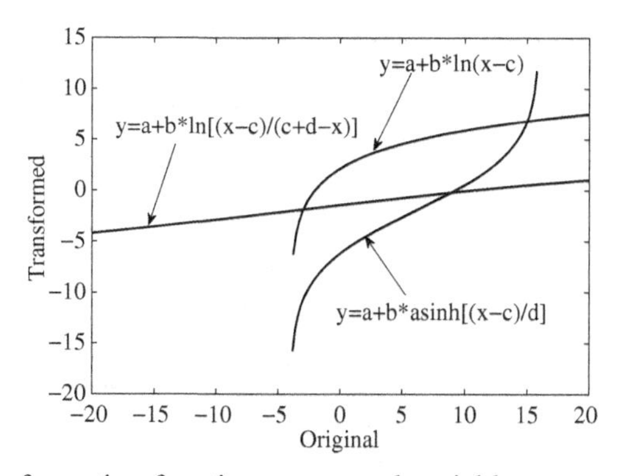

Figure 11.1 Transformation functions to normal variables.

Now, Equation (11.12) is expressed as a 3-parameter log-normal distribution with location parameter c, mean $-a/b$, and standard deviation $1/b$. If $c = 0$, we obtain the 2-parameter log-normal distribution.

The transformation expressed by Equation (11.8) yields Johnson's SB distribution. From Equation (11.8) we have:

$$\left|\frac{dy}{dx}\right| = \frac{bd}{(x-c)(c+d-x)} \tag{11.13}$$

Substitution of Equations (11.8) and (11.13) into Equation (11.10) yields the SB system as:

$$f_X(x) = \frac{bd}{\sqrt{2\pi}(x-c)(c+d-x)} \exp\left(-\frac{\left(a+b\ln\left(\frac{x-c}{c+d-x}\right)\right)^2}{2}\right); \tag{11.14}$$

$$x \in (c, c+d), a \in \mathbb{R}, b > 0$$

Figure 11.2 plots the Johnson SB PDF.

The transformation given by Equation (11.9) yields Johnson's SU distribution. From Equation (11.9) we have:

$$\sinh^{-1}\left(\frac{x-c}{d}\right) = \ln\left(\frac{x-c}{d} + \sqrt{\left(\frac{x-c}{d}\right)^2 + 1}\right) \tag{11.15a}$$

$$\left|\frac{dy}{dx}\right| = \frac{b}{\sqrt{(x-c)^2 + d^2}} \tag{11.15b}$$

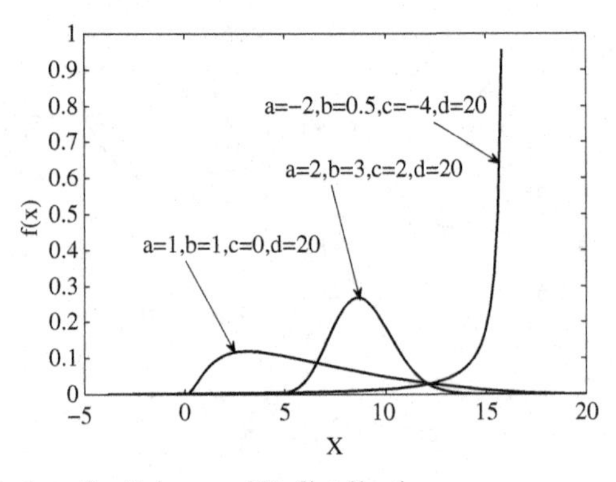

Figure 11.2 PDF plots for Johnson SB distribution.

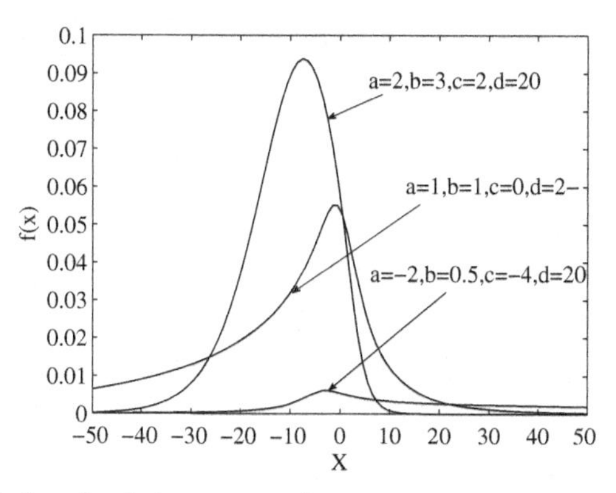

Figure 11.3 PDF plots for Johnson SU distribution.

Substitution of Equations (11.9) and (11.15b) into Equation (11.10) yields the SU system as:

$$f_X(x) = \frac{b\exp\left(-\frac{\left(a + b\sinh^{-1}\left(\frac{x-c}{d}\right)\right)^2}{2}\right)}{\sqrt{2\pi}\sqrt{(x-c)^2 + d}}; x \in (-\infty, \infty), d > 0, b > 0 \quad (11.16)$$

Figure 11.3 plots the Johnson SU PDF.

Johnson's SL, SB, and SU distributions cover the plane of moment ratios (β_1, β_2), i.e., the ratio of skewness to kurtosis, uniquely. The moment ratios of the distribution of X may be determined in terms of parameters a and b.

11.4 Transformation Based on the First Law of Laplace

Frechet (1939) used the Laplace distribution as the basic distribution for transformation. Applying certain transformations to the Laplace distribution, Johnson (1954) derived distributions called SL', SB', and SU' distributions that are discussed here. The standard Laplace distribution (also called double exponential distribution) can be expressed as:

$$f(y) = \frac{1}{2}\exp(-|y|); y \in (-\infty, \infty) \quad (11.17)$$

Equation (11.17) represents the standard Laplace distribution, adding the scale (β) and location (c) parameters, Equation (11.17) may be rewritten as:

$$f(y; c, \beta) = \frac{1}{2\beta}\exp\left(-\frac{|y-c|}{2\beta}\right); c \in \mathbb{R}, \beta > 0 \quad (11.17a)$$

Johnson (1954) proposed three transformations in for SL', SB', and SU':

$$\mathrm{SL}' : y = a + b\ln(x); x > 0, b > 0; \left|\frac{dy}{dx}\right| = \frac{b}{x} \tag{11.18}$$

$$\mathrm{SB}' : y = a + b\ln\left(\frac{x}{1-x}\right); x \in (0,1), b > 0; \left|\frac{dy}{dx}\right| = \frac{b}{x(1-x)} \tag{11.19}$$

$$\mathrm{SU}' : y = a + b\sinh^{-1}(x), b > 0; \left|\frac{dy}{dx}\right| = \frac{b}{\sqrt{1+x^2}} \tag{11.20}$$

Substitution of Equations (11.17)–(11.18) into Equation (11.10) yields Johnson's SL' distribution as:

$$f(x) = \begin{cases} \frac{b}{2}\exp(a)x^{-1+b}; x \in \left[0, \exp\left(-\frac{a}{b}\right)\right] \\ \frac{b}{2}\exp(-a)x^{-1-b}; x \in \left[\exp\left(-\frac{a}{b}\right), \infty\right) \end{cases} \tag{11.21}$$

The Johnson's SL' distribution is unimodal and has cusped mode at $x = \exp\left(-\frac{a}{b}\right)$. Figure (11.4) plots the PDF for Johnson's SL' distribution.

Substitution of Equations (11.17) and (11.19) into Equation (11.10) yields Johnson's SB' distribution as:

$$f(x) = \begin{cases} \frac{b}{2}\exp(a)x^{-1+b}(1-x)^{-1-b}; x \in \left(0, \left(1+\exp\left(\frac{a}{b}\right)\right)^{-1}\right] \\ \frac{b}{2}\exp(-a)x^{-1-b}(1-x)^{-1+b}; x \in \left[\left(1+\exp\left(\frac{a}{b}\right)\right)^{-1}, 1\right) \end{cases} \tag{11.22}$$

Johnson's SB' distribution is unimodal and has cusped mode at $x = \left(1+\exp\left(\frac{a}{b}\right)\right)^{-1}$ for $b > 1$ and multimodal if $b < 1$. The boundary between

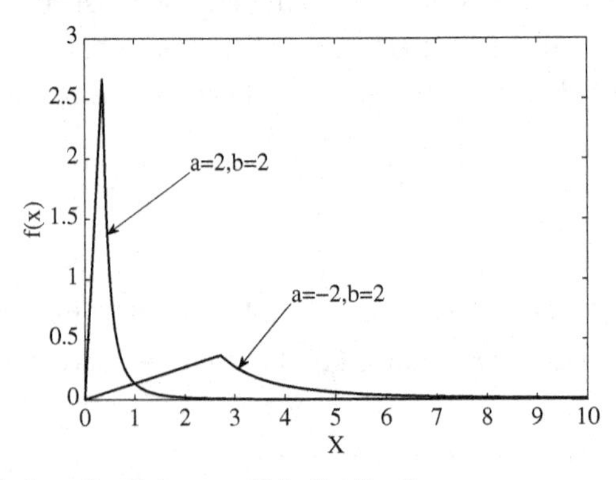

Figure 11.4 PDF plots for Johnson SL' distribution.

unimodal and multimodal occurs at $b = 1$. Figure 11.5 plots Johnson's SB' distribution under different mode conditions.

Substitution of Equations (11.17) and (11.20) into Equation (11.10) yields Johnson's SU' distribution as:

$$f(x) = \begin{cases} \frac{b}{2} \exp(a)\left(1+x^2\right)^{-\frac{1}{2}}\left(x+\sqrt{1+x^2}\right)^{b}; x \in \left(-\infty, -\sinh\left(\frac{a}{b}\right)\right] \\ \frac{b}{2} \exp(-a)\left(1+x^2\right)^{-\frac{1}{2}}\left(x+\sqrt{1+x^2}\right)^{-b}; x \in \left[-\sinh\left(\frac{a}{b}\right), \infty\right) \end{cases} \tag{11.23}$$

Johnson's SU' distribution is unimodal and has cusped mode at $x = -\sinh\left(\frac{a}{b}\right)$. Figure 11.6 plots the density function of Johnson SU' distribution.

The curves SL', SB', and SU' cover the plane (β_1, β_2). The curves SL' can be fitted using the first three moments and curves SB' and SU' using the first four moments.

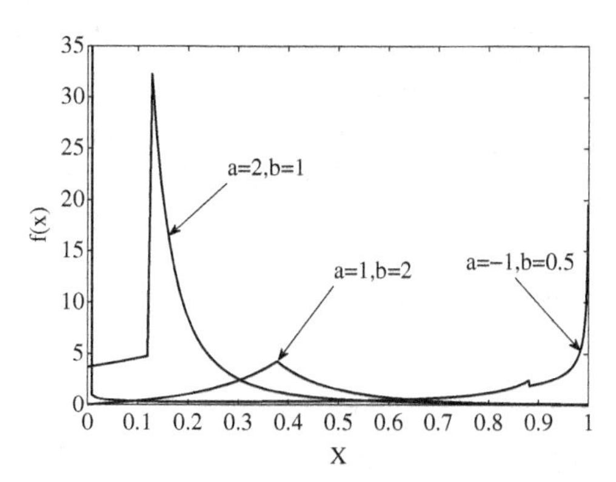

Figure 11.5 PDF plots for Johnson SB' distribution.

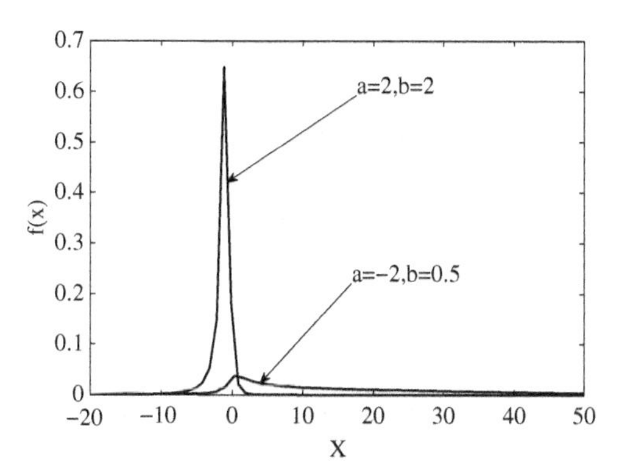

Figure 11.6 PDF plot for Johnson SU' distribution.

11.5 Transformation of Logistic Distribution

Applying three transformations to the standard logistic distribution, Tadikamalla and Johnson (1982) and Johnson and Tadikamalla (1992) derived distributions designated as L_L, L_B, and L_U. These transformations are given by Equations (11.18)–(11.20). The logistic distribution for random variable *Y* can be given as

$$f(y) = \frac{\exp(y)}{(1+\exp(y))^2}, \quad F(y) = \frac{1}{1+\exp(-y)} \tag{11.24}$$

Application of Equations (11.18) and (11.24) to Equation (11.10) leads to the log-logistic (i.e., L_L) distribution:

$$f(x) = b\exp(-a)x^{b-1}(x^b + \exp(-a))^{-2}, \; x \geq 0, \; b > 0 \tag{11.25}$$

The distribution function of the L_L distribution is given as:

$$F(x) = (1 + \exp(-a)x^{-b})^{-1} \tag{11.26}$$

The L_L distribution is unimodal with the mode occurring at $x = 0$ for $b \leq 1$ and results in a reverse J-shaped curve, and occurs at $x^b = \frac{(b-1)\exp(a)}{b+1}$, *for* $b > 1$. Figure 11.7 plots the density function for L_L distribution.

Application of Equations (11.18) and (11.20) to Equation (11.10) leads to the L_B distributions as:

$$f(x) = b\exp(a)x^{b-1}(1-x)^{b-1}((1-x)^b + x^b\exp(a))^{-2}; \quad x \in (0,1) \tag{11.27}$$

The L_B distribution reduces to the standard uniform distribution if $a = 0$, $b = 1 \Rightarrow f(x) = 1$. The L_B distribution is a power function distribution if $b = 1 \Rightarrow f(x) = \exp(a)(1 - x + x\exp(a))^{-2}$.

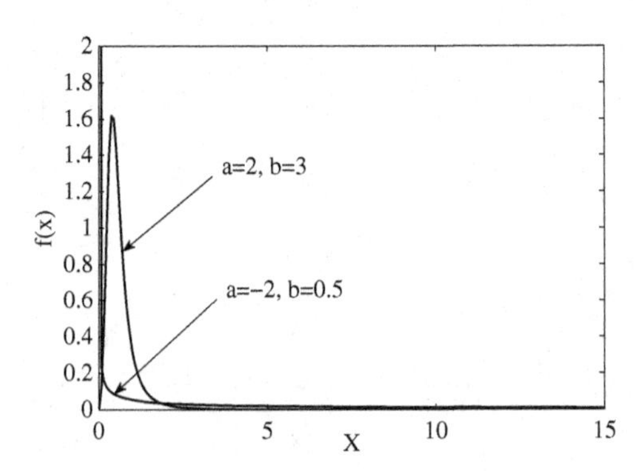

Figure 11.7 PDF plot for L_L distribution.

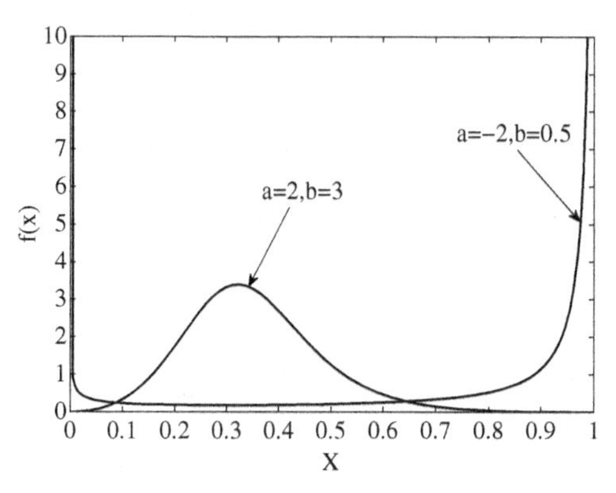

Figure 11.8 PDF plot for L_B distribution.

It is unimodal if $b > 1$ at a value of x satisfying

$$\exp(a) = \left(\frac{1-x}{x}\right)^b \left(\frac{b-1+2x}{b+1-2x}\right) \tag{11.28}$$

Furthermore, if $b < 1$, the L_B distribution is U-shaped and the antimode occurs between $(1 - b)/2$ and $(1 + b)/2$. Figure 11.8 plots the density function for L_B distribution.

Application of Equations (11.20) and (11.24) to Equation (11.10) leads to the L_U distribution:

$$f(x) = \frac{b\exp(a)}{\sqrt{1+x^2}} \frac{\left(x+\sqrt{1+x^2}\right)^b}{\left(1+\exp(a)\left(x+\sqrt{1+x^2}\right)^b\right)^2}, b > 0, x \in \mathbb{R} \tag{11.29}$$

The L_U distribution is unimodal with x satisfying

$$b\left(1 - \exp(a)\left(x+\sqrt{1+x^2}\right)\right) = \frac{x}{\sqrt{1+x^2}} \tag{11.30}$$

Figure 11.9 plots the PDF for L_U distribution.

11.6 Transformation of Beta Distribution

Majumder and Chakravarty (1990) and McDonald (1984) employed the beta distribution as the basic distribution for transformation. Let Y be a beta distributed random variable with the PDF as

$$f(y) = \frac{1}{B(p,q)} y^{p-1}(1-y)^{q-1}; y \in (0,1), p > 0, q > 0 \tag{11.31}$$

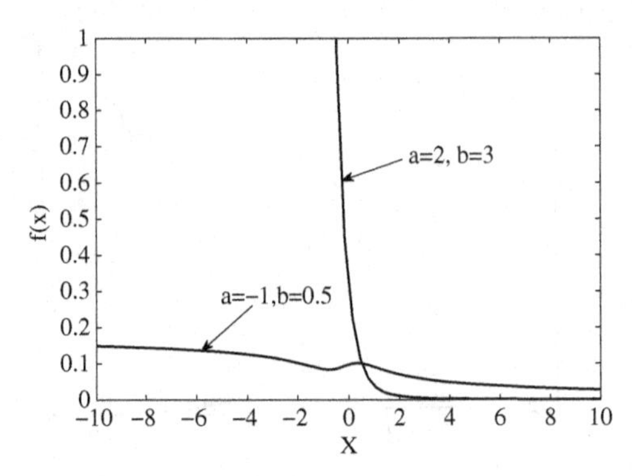

Figure 11.9 PDF plots for L_U distribution.

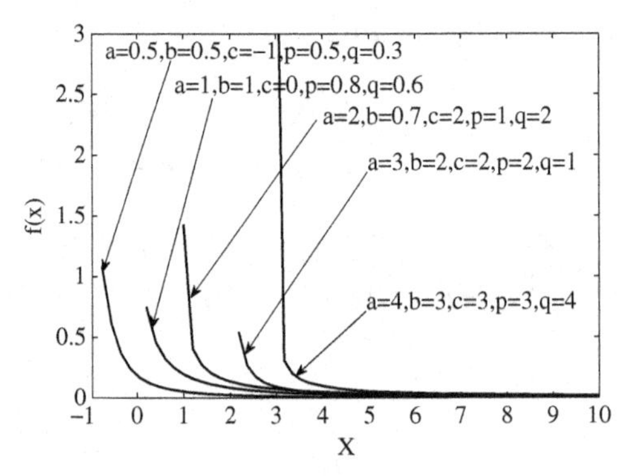

Figure 11.10 PDF plots for 5-parameter FP distribution.

where $B(.)$ is the beta function, and p and q are the shape parameters. Now a Pareto-type transformation is defined as:

$$y = \left(1 + \left(\frac{x-c}{b}\right)^{\frac{1}{a}}\right)^{-1}, \quad \left|\frac{dy}{dx}\right| = \frac{1}{ab}\left(\frac{x-c}{b}\right)^{\frac{1}{a}-1}\left(1 + \left(\frac{x-c}{b}\right)^{\frac{1}{a}}\right)^{-2} ; x \geq c, a, b > 0 \tag{11.32}$$

where c is the location parameter, b is the scale parameter, and a is the shape parameter. Substitution of Equations (11.31)–(11.32) into Equation (11.10) yields:

$$f(x; a, b, c, p, q) = \frac{1}{abB(p,q)}\left(\frac{x-c}{b}\right)^{\frac{q}{a}-1}\left(1 + \left(\frac{x-c}{b}\right)^{\frac{1}{a}}\right)^{-(p+q)} ; x \geq c; a, b, p, q > 0 \tag{11.33}$$

Equation (11.33) is the 5-parameter Feller-Pareto (FP) distribution (Tahmasebi and Behboodian, 2010). Figure 11.10 shows that the FP distribution is L-shaped for different values of parameters.

The FP distribution may be also expressed in a slightly different form (Mahmoud and Abd El Ghafour, 2013). The transformation function is given as:

$$y=\left(1+\left(\frac{x}{b}\right)^{a}\right)^{-c};\left|\frac{dy}{dx}\right|=\frac{ac}{b}\left(\frac{x}{b}\right)^{a-1}\left(1+\left(\frac{x}{b}\right)^{a}\right)^{-(c+1)};x,a,b,c>0 \quad (11.34)$$

Substitution of Equations (11.31) and (11.34) into Equation (11.10) yields:

$$f(x;a,b,c,p,q)=\frac{acx^{a-1}}{b^{a}B(p,q)}\left(1+\left(\frac{x}{b}\right)^{a}\right)^{-cq-1}\left(1-\left(1+\left(\frac{x}{b}\right)^{a}\right)^{-c}\right)^{p-1} \quad (11.35)$$

Other generalized distributions that can be obtained from the above Generalized Feller-Pareto (GFP) distribution [i.e., Equation (11.35)] are:

(a) Generalized Burr XII distribution may be obtained if $p = 1$:

$$f(x;a,b,c,1,q)=\frac{acqx^{a-1}}{b^{a}}\left(1+\left(\frac{x}{b}\right)^{a}\right)^{-cq-1};x\geq 0 \quad (11.36)$$

(b) Beta Lomax distribution with $c = 1$:

$$f(x;a,b,1,p,q)=\frac{ax^{a-1}}{bB(p,q)}\left(1+\frac{x}{b}\right)^{-q-1}\left(1-\left(1+\frac{x}{b}\right)^{-1}\right)^{p-1};x\geq 0 \quad (11.37)$$

Figure 11.11 plots the density functions for generalized Burr XII and Beta Lomax distributions.

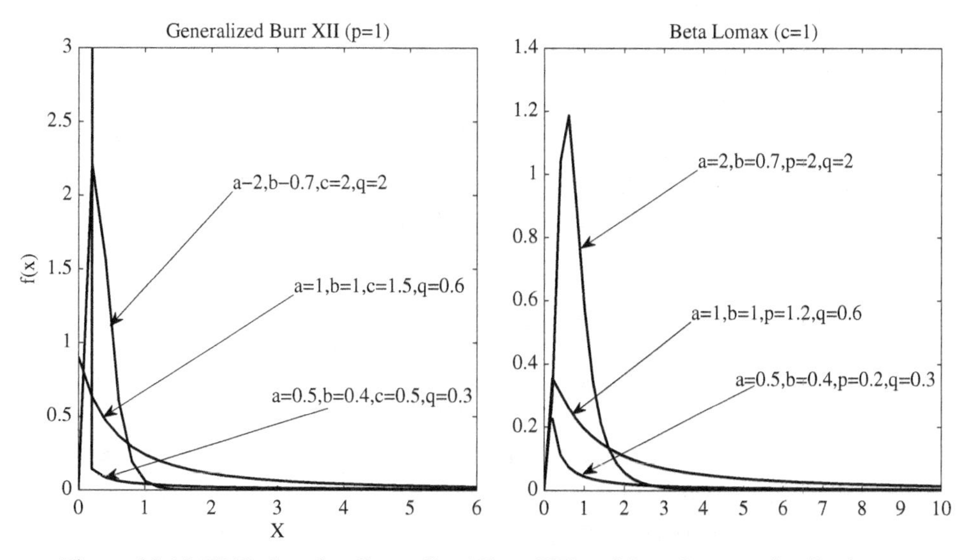

Figure 11.11 PDF plots for Generalized Burr XII and Beta Lomax distributions.

The FP distribution, FP(a, b, c, p, q), leads to four families of distributions which, in turn, lead to several other distributions. Thus, a treelike structure of frequency distributions is formed. This is briefly presented here.

(1) Pareto IV distribution that has four parameters (a, b, c, p). This is equivalent to FP (a, b, c, p, 1) as:

$$f(x; a, b, c, p, 1) = \frac{p}{ab}\left(\frac{x-c}{b}\right)^{\frac{1}{a}-1}\left(1+\left(\frac{x-c}{b}\right)^{\frac{1}{a}}\right)^{-(p+1)} ; x \geq c; a, b, p > 0 \quad (11.38)$$

The Pareto IV distribution then yields the following distributions as its special cases:

(a) Pareto I distribution has two parameters (b, p), which is equivalent to FP (*1*, b, b, p, 1):

$$f(x; 1, b, b, p, 1) = \frac{p}{b}\left(\frac{x}{b}\right)^{-(p+1)} ; x \geq 0, b, p > 0 \quad (11.38a)$$

(b) Pareto II distribution has three parameters (b, c, p), which is equivalent to FP (*1*, b, c, p, 1):

$$f(x; 1, b, c, p) = \frac{p}{b}\left(1+\left(\frac{x-c}{b}\right)\right)^{-(p+1)} ; x \geq c, c \in \mathbb{R}, b, p > 0 \quad (11.38b)$$

(c) Pareto III distribution has three parameters (a, b, c), which is equivalent to FP (a, b, c, 1, 1):

$$f(x; a, b, c, 1, 1) = \frac{1}{ab}\left(\frac{x-c}{b}\right)^{\frac{1}{a}-1}\left(1+\left(\frac{x-c}{b}\right)^{\frac{1}{a}}\right)^{-2} ; x \geq c, c \in \mathbb{R}, a, b > 0 \quad (11.38c)$$

(2) Transformed beta (TB) distribution with four parameters TB (a, b, p, q). This is equivalent to FP (1/a, b, 0, p, q) given as:

$$f\left(x; \frac{1}{a}, b, 0, p, q\right) = \frac{a}{bB(p,q)}\left(\frac{x}{b}\right)^{aq-1}\left(1+\left(\frac{x}{b}\right)^{a}\right)^{-(p+q)} \quad (11.39)$$

Figure 11.12 plots the PDF for transformed beta distribution with four parameters.

The TB distribution yields the following distributions as special cases:

(a) Burr XII distribution, or simply Burr distribution, contains three parameters: Burr (a, b, p). It is equivalent to TB (a, b, p, *1*) or FP (*1/a*, b, *0*, p, *1*):

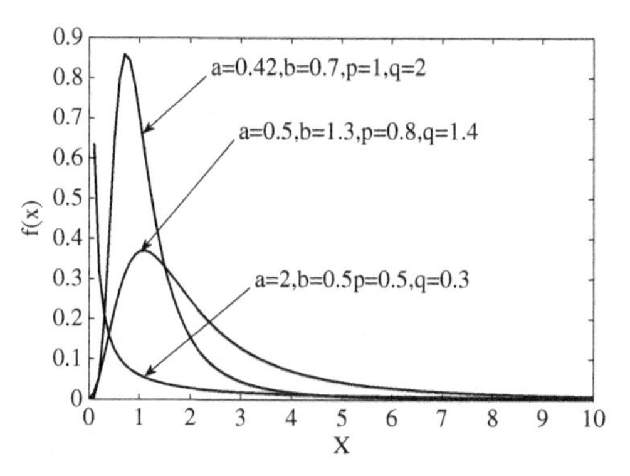

Figure 11.12 PDF plots for transformed beta distribution.

$$f\left(x;\frac{1}{a},b,0,p,1\right)=\frac{ap}{b}\left(\frac{x}{b}\right)^{a-1}\left(1+\left(\frac{x}{b}\right)^{a}\right)^{-(p+1)} \tag{11.39a}$$

(a) Two-parameter log-logistic distribution is equivalent to TB (a, b, *1*, 1), FP (*1/a*, b, *0*, 1, 1), or Burr (a, b, *1*):

$$f\left(x;\frac{1}{a},b,0,1,1\right)=\frac{a}{b}\left(\frac{x}{b}\right)^{a-1}\left(1+\left(\frac{x}{b}\right)^{a}\right)^{-2};a,b>0 \tag{11.39b}$$

(c) Paralogistic distribution contains two parameters (a, b), which is equivalent to TB (a, b, a, 1), FP (*1/a*, b, 0, a, 1), or Burr (a, b, a) as:

$$f\left(x;\frac{1}{a},b,0,a,1\right)=\frac{a^2}{b}\left(\frac{x}{b}\right)^{a-1}\left(1+\left(\frac{x}{b}\right)^{a}\right)^{-(a+1)};x\geq 0,a,b>0 \tag{11.39c}$$

(d) Generalized F-distribution contains two parameters (p, q), which is equivalent to TB $\left(1,\frac{p}{q},p,q\right)$, FP $\left(1,\frac{p}{q},0,p,q\right)$ as:

$$f\left(x;1,\frac{p}{q},0,p,q\right)=\frac{q}{pB(p,q)}\left(\frac{qx}{p}\right)^{q-1}\left(1+\frac{qx}{p}\right)^{-(p+q)} \tag{11.39d}$$

Equation (11.39d) represents the generalized F distribution with $d_1 = 2q$, $d_2 = 2p$; $2p,\ 2q \in \mathbb{N}^+$. Figure 11.13 plots the density functions for paralogistic and generalized F distributions.

(3) Generalized Pareto (GP) distribution that contains three parameters GP (μ, σ, ξ). The GP is equivalent to FP ($1, \frac{\sigma}{\xi}, \mu, \frac{1}{\xi}, 1$) given as:

$$f\left(x;1,\frac{\sigma}{\xi},\mu,\frac{1}{\xi},1\right)=\frac{1}{\sigma}\left(1+\frac{\xi(x-\mu)}{\sigma}\right)^{-\frac{1}{\xi}-1} \tag{11.40}$$

The GP distribution yields the following distributions as special cases:

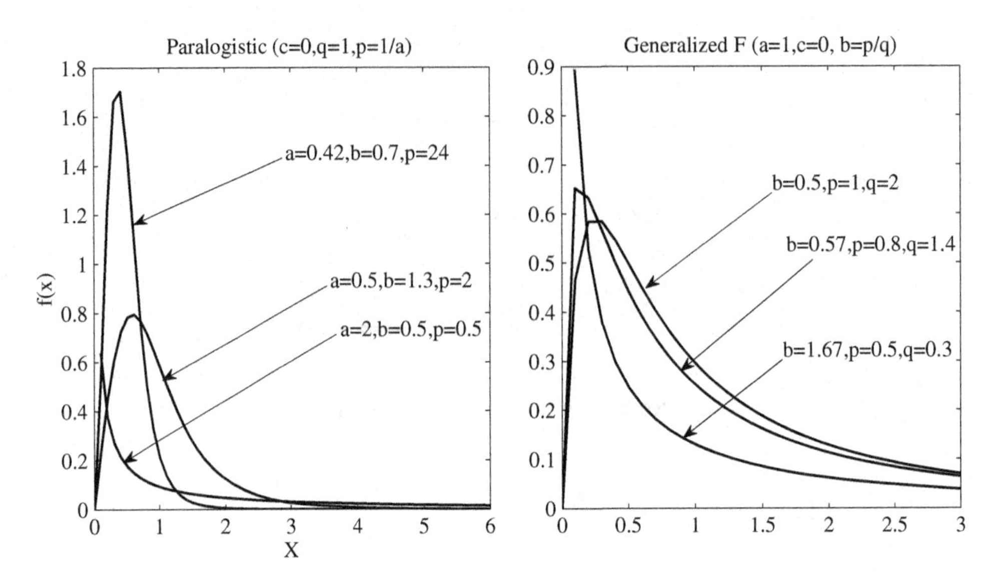

Figure 11.13 PDF plots for paralogistic and generalized F distributions.

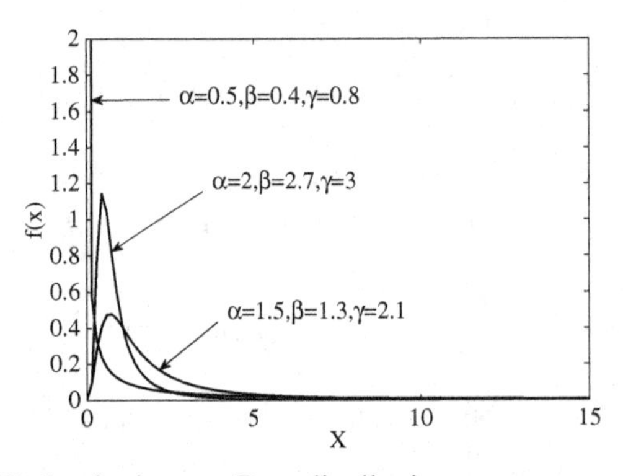

Figure 11.14 PDF plot for inverse Burr distribution.

(4) Inverse Burr distribution (IBurr) contains three parameters: IBurr (α, β, γ). This is equivalent to TB$(\alpha, \frac{1}{\beta}, \gamma, 1)$ and also to FP $(\alpha, 1/\beta, 0, p, 1)$ as:

$$
\begin{aligned}
IBurr(x; \alpha, \beta, \gamma) &= TB\left(x; \alpha, \frac{1}{\beta}, 1, \gamma\right) = FP\left(x; \alpha, \frac{1}{\beta}, 0, \gamma, 1\right) \\
&= \alpha\gamma\beta^{\alpha\gamma}x^{\alpha\gamma-1}\left(1 + (\beta x)^{\alpha}\right)^{-\gamma-1}
\end{aligned}
\tag{11.41}
$$

Figure 11.14 plots the PDF of inverse Burr distribution.

The inverse Burr distribution yields the following distributions as special cases:

(a) Log-logistic distribution is equivalent to inverse Burr distribution as IBurr $(\alpha, \beta, 1)$/TB$(\alpha, \frac{1}{\beta}, 1, 1)$/FP$(\alpha, \frac{1}{\beta}, 0, 1, 1)$:

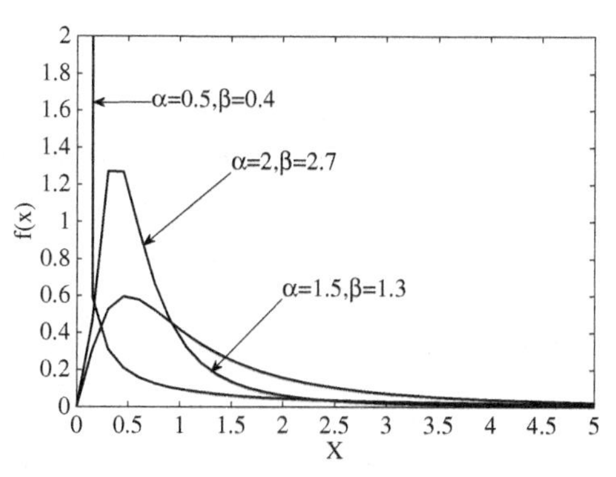

Figure 11.15 PDF plots for inverse para-logistic distribution.

$$f(x;\alpha,\beta) = \alpha\beta x^{\alpha-1}\left(1+\left(\frac{\beta}{x}\right)^{\alpha}\right)^{-2} \tag{11.41a}$$

(b) Inverse para-logistic distribution contains two parameters (α, β). The inverse para-logistic distribution is equivalent to IBurr (α, β, α)/TB $\left(\alpha,\frac{1}{\beta},1,\alpha\right)/FP\left(\alpha,\frac{1}{\beta},0,1,\alpha\right)$:

$$f(x;\alpha,\beta) = \alpha^2\beta^{\alpha^2}x^{\alpha^2-1}(1+(\beta x)^{\alpha})^{-\alpha-1} \tag{11.41b}$$

Figure 11.15 plots the PDF for inverse para-logistic distribution.

11.7 Transformation of Gamma Distribution

D'Addario (1936), Johnson and Kotz (1970), and McDonald (1984) applied transformations to the gamma distribution. Let random variable Y follow a gamma distribution given as:

$$f(y;\alpha,\beta) = \frac{\beta^{\alpha}}{\Gamma(\alpha)}y^{\alpha-1}\exp(-\beta y); y>0, \alpha,\beta>0 \tag{11.42}$$

There are three transformations that are commonly applied to the gamma distribution, which are discussed as follows:

(a) The transformation equation is a logarithm transformation as:

$$y = \ln(x); x>0 \Rightarrow \left|\frac{dy}{dx}\right| = \frac{1}{x} \tag{11.43}$$

Substitution of Equations (11.42)–(11.43) into Equation (11.10) yields the log-gamma distribution as:

$$f(x) = \frac{1}{x\Gamma(\alpha)}\beta^{\alpha} x^{-b} (\ln(x))^{a-1} \tag{11.44}$$

(b) The transformation equation is a linear function with location parameter as:

$$y = x - \mu; c > 0 \Rightarrow \left|\frac{dy}{dx}\right| = 1 \tag{11.45}$$

Substitution of Equations (11.42) and (11.45) into Equation (11.10) yields the Pearson III distribution as:

$$f(x) = \frac{\beta^{\alpha}}{\Gamma(\alpha)}(x - \mu)^{\alpha-1} \exp(-\beta(x - \mu)); x > \mu, \mu \in \mathbb{R} \tag{11.46}$$

(c) The transformation equation is a log-linear function with location parameter as:

$$y = \ln x - \mu \Rightarrow \left|\frac{dy}{dx}\right| = \frac{1}{x}, x \in \mathbb{R}^{+}, x > \exp(\mu) \tag{11.47}$$

Substitution Equations (11.42) and (11.47) into Equation (11.10) yields the log-Pearson III distribution as:

$$f(x) = \frac{\beta^{\alpha}}{x\Gamma(\alpha)}(\ln(x) - \mu)^{\alpha-1} \exp(-\beta(\ln(x) - \mu)) \tag{11.48}$$

The log-Pearson III distribution is the standard distribution for flood frequency analysis in the United States.

(d) The transformation equation is a power transformation as:

$$y = x^{p}; p > 0, x > 0 \Rightarrow \left|\frac{dy}{dx}\right| = px^{p-1} \tag{11.49}$$

Substitution of Equations (11.42) and (11.49) into Equation (11.10) yields the generalized gamma (GG) distribution GG($\alpha, \beta^{\frac{1}{p}}, p$) as:

$$f(x) = \frac{p\beta^{\alpha}}{\Gamma(\alpha)} x^{\alpha p-1} \exp(-\beta x^{p}) = \frac{p\left(\beta^{\frac{1}{p}}\right)^{\alpha p} x^{\alpha p-1}}{\Gamma(\alpha)} x^{\alpha p-1} \exp\left(\left(\beta^{\frac{1}{p}} x\right)^{p}\right) \tag{11.50}$$

Figure 11.16 plots the PDF for the GG distribution.

The chi-square distribution is obtained from the gamma distribution with $\beta = \frac{1}{2}$ or GG (α, $\frac{1}{2}$, 1). The exponential distribution is obtained from the gamma distribution with β = 1 or GG (α, 1, 1).

(e) The transformation equation is an inverse transformation as:

$$y = x^{-1}; x > 0 \Rightarrow \left|\frac{dy}{dx}\right| = x^{-2} \tag{11.51}$$

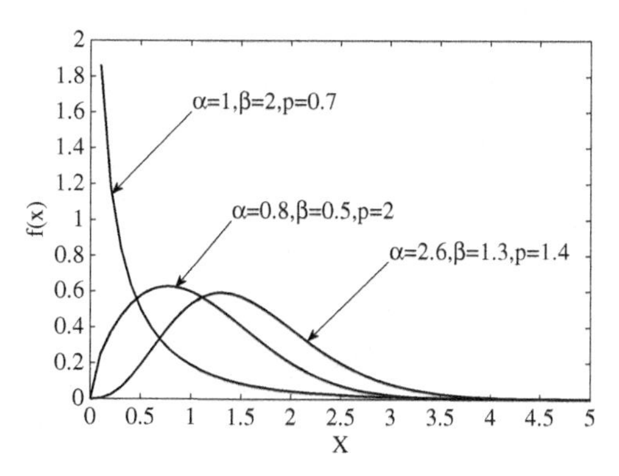

Figure 11.16 PDF plots for GG distribution.

Substitution of Equations (11.42) and (11.51) into Equation (11.10) yields the inverse gamma distribution as:

$$f(x;\alpha,\beta)=\frac{\beta^{\alpha}}{\Gamma(\alpha)}x^{-\alpha-1}\exp\left(-\frac{\beta}{x}\right);x>0,\alpha,\beta>0 \tag{11.52}$$

(f) The transformation is the inverse power transformation given as:

$$y=x^{-p};x>0\Rightarrow\left|\frac{dy}{dx}\right|=px^{-p-1} \tag{11.53}$$

Substitution of Equations (11.42) and (11.53) into Equation (11.10) yields the inverse GG distribution as:

$$f(x;\alpha,\beta,p)=\frac{p\beta^{\alpha}}{\Gamma(\alpha)}x^{-\alpha p-1}\exp\left(-\frac{\beta}{x^{p}}\right)=\frac{p}{x\Gamma(\alpha)}\left(\frac{\beta^{\frac{1}{p}}}{x}\right)^{\alpha p}\exp\left(-\left(\frac{\beta^{\frac{1}{p}}}{x}\right)^{p}\right) \tag{11.54}$$

Figure 11.17 plots the PDF of inverse gamma distribution.

11.8 Transformation of Student-*t* Distribution

Kloek and Dijk (1977, 1978) applied transformations to the Student-*t* distribution. Let a random variable Y follow the student-*t* distribution expressed as:

$$f(y;v)=\frac{\Gamma\left(\frac{v+1}{2}\right)}{\sqrt{v\pi}\Gamma\left(\frac{v}{2}\right)}\left(1+\frac{x^{2}}{v}\right)^{-\left(\frac{(v+1)}{2}\right)} \tag{11.55}$$

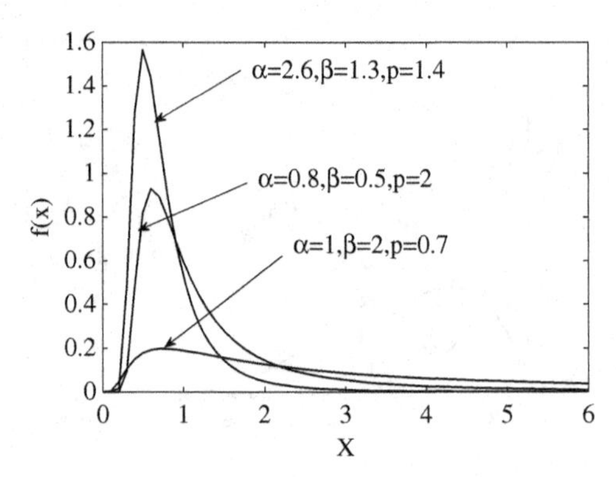

Figure 11.17 PDF plots for inverse gamma distribution.

We apply the linear transformation given as:

$$y = \frac{x}{b} - c; b > 0, c \in \mathbb{R} \tag{11.56}$$

The linear transformation [Equation (11.56)] yields a general distribution with location and scale parameters as:

$$f(x) = \frac{v^{\frac{v}{2}}}{bB\left(\frac{1}{2}, \frac{v}{2}\right)} \left(v + \frac{(x - bc)^2}{b^2}\right)^{-\left(\frac{v+1}{2}\right)} \tag{11.57a}$$

Let $\mu = bc$. Then, Equation (11.57a) may be rewritten as:

$$f(x) = \frac{v^{\frac{v}{2}}}{bB\left(\frac{1}{2}, \frac{v}{2}\right)} \left(v + \frac{(x - \mu)^2}{b^2}\right)^{-\frac{v+1}{2}} \tag{11.57b}$$

Applying the logarithm transformation equation,

$$y = \ln x \Rightarrow \left|\frac{dy}{dx}\right| = \frac{1}{x}, x > 0 \tag{11.58}$$

The logarithm transformation of the general Student-t distributed random variable yields the general log-student-t distribution as:

$$f(x) = \frac{v^{\frac{v}{2}}}{bxB\left(\frac{1}{2}, \frac{v}{2}\right)} \left(v + \frac{(\ln x - \mu)^2}{b^2}\right)^{-\frac{v+1}{2}} \tag{11.59}$$

Figure 11.18 plots the PDF for log-student-t distribution.

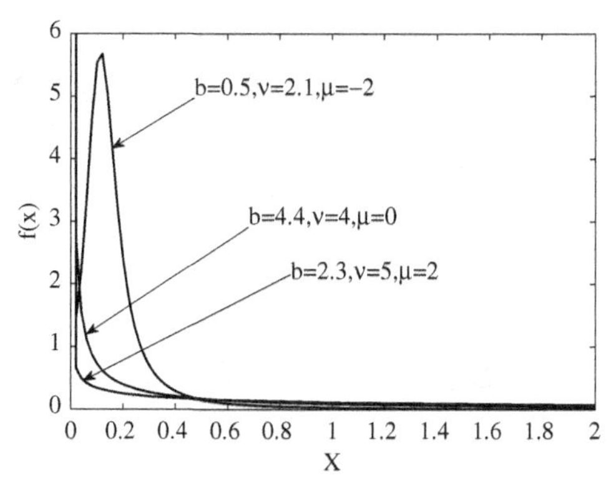

Figure 11.18 PDF plots for log-student-*t* distribution.

11.9 Application

In this section, the peak flow, monthly sediment, daily maximum precipitation, and annual rainfall are applied to evaluate the performance of the distributions through transformation. The Johnson S_L, log-logistic, para-logistic, and inverse gamma distributions are applied for frequency analysis of peak flow and maximum daily precipitation. Johnson S_L, log-logistic, and para-logistic distributions are applied for frequency analysis of monthly sediment and annual rainfall. The maximum likelihood estimation method is applied for parameter estimation. The log-likelihood functions of the preceding distribution candidates can be provided as:

Johnson S_L

$$\ln L = -n \ln\left(\frac{\sqrt{2\pi}}{b}\right) - n\sum_{i=1}^{n} \ln(x_i - c) - \sum_{i=1}^{n} \frac{b^2\left(\ln(x_i - c) + \frac{a}{b}\right)^2}{2} \tag{11.60}$$

Log-logistic

$$\ln L = n \ln b - an + (b-1)\sum_{i=1}^{n} \ln x_i - 2\sum_{i=1}^{n} \ln\left(x_i^b + \exp(-a)\right) \tag{11.61}$$

Para-logistic

$$\ln L = n \ln a - an \ln b + (a-1)\sum_{i=1}^{n} \ln x_i - (a+1)\sum_{i=1}^{n} \ln\left(1 + \left(\frac{x_i}{b}\right)^a\right) \tag{11.62}$$

Inverse Gamma

$$\ln L = \alpha n \ln \beta - n \ln \Gamma(\alpha) - (\alpha + 1) \sum_{i=1}^{n} \ln x_i - \sum_{i=1}^{n} \frac{\beta}{x_i} \tag{11.63}$$

Maximizing Equations (11.60)–(11.63), the parameters of each distribution candidate can then be estimated.

11.9.1 Peak Flow and Maximum Daily Precipitation

The peak flow and maximum daily precipitation are more right-skewed than the monthly sediment and annual rainfall. With the distribution candidates of Johnson S_L, log-logistic, para-logistic, and inverse gamma distributions, the maximum likelihood estimation method is applied to estimate the parameters with the use of genetic algorithm. Table 11.1 lists the parameters estimated for peak flow. Table 11.2 lists the parameters estimated for maximum daily precipitation. Figures 11.19–11.20 compare the fitted frequency distributions with the empirical frequency distribution for peak flow and maximum daily precipitation, respectively. Comparisons of fitted frequency distributions with empirical frequency distribution show that all the distribution candidates yield similar performances and can be applied to model peak flow and maximum daily precipitation.

Table 11.1 *Parameter estimated for peak flow*

	Parameters			
Distributions	a	b	c	p
Johnson S_L	–13.19	2.44	39.9	
Log-logistic	–28.58	5.12		
Paralogistic	3.5	417.21		
Inverse gamma	54.47	489.92		0.395

Table 11.2 *Parameters estimated for maximum daily precipitation*

	Parameters			
Distributions	a	b	c	p
Johnson S_L	–6.67	1.93	20.05	
Log-logistic	–21.53	5.42		
Paralogistic	3.82	83.55		
Inverse gamma	59.62	310.01		0.42

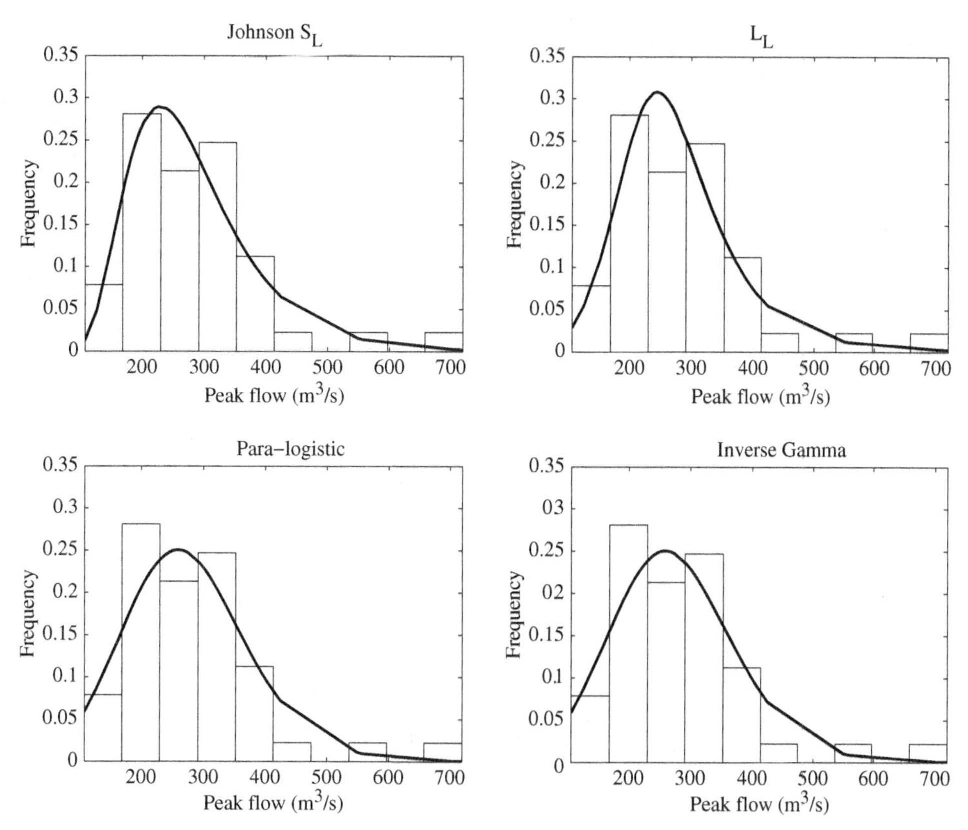

Figure 11.19 Comparison of fitted frequency distributions to empirical frequency distribution for peak flow.

11.9.2 Monthly Sediment and Annual Rainfall

For the less skewed monthly sediment and annual rainfall, Johnson S_L, log-logistic, and para-logistic distributions are applied for evaluation. The maximum likelihood method is applied for parameter estimation with the aid of genetic algorithm. Table 11.3 lists the parameters estimated for monthly sediment distribution. Table 11.4 lists the parameters estimated for annual rainfall distribution. Figures 11.21–11.22 compare the fitted frequency distribution distributions with the empirical frequency distribution. Comparisons again indicate that the distribution candidates yield similar performances and may be applied to model monthly sediment and annual rainfall.

11.10 Conclusions

There are two main transformation approaches. First, data can be normalized using a power or log or an appropriate transformation. Second, an appropriate

Table 11.3 *Parameters estimated for monthly sediment*

Distributions	Parameters a	b	c
Johnson S_L	−8.28	1.80	−7.97
Log-logistic	−12.47	2.77	
Paralogistic	2.30	148.53	

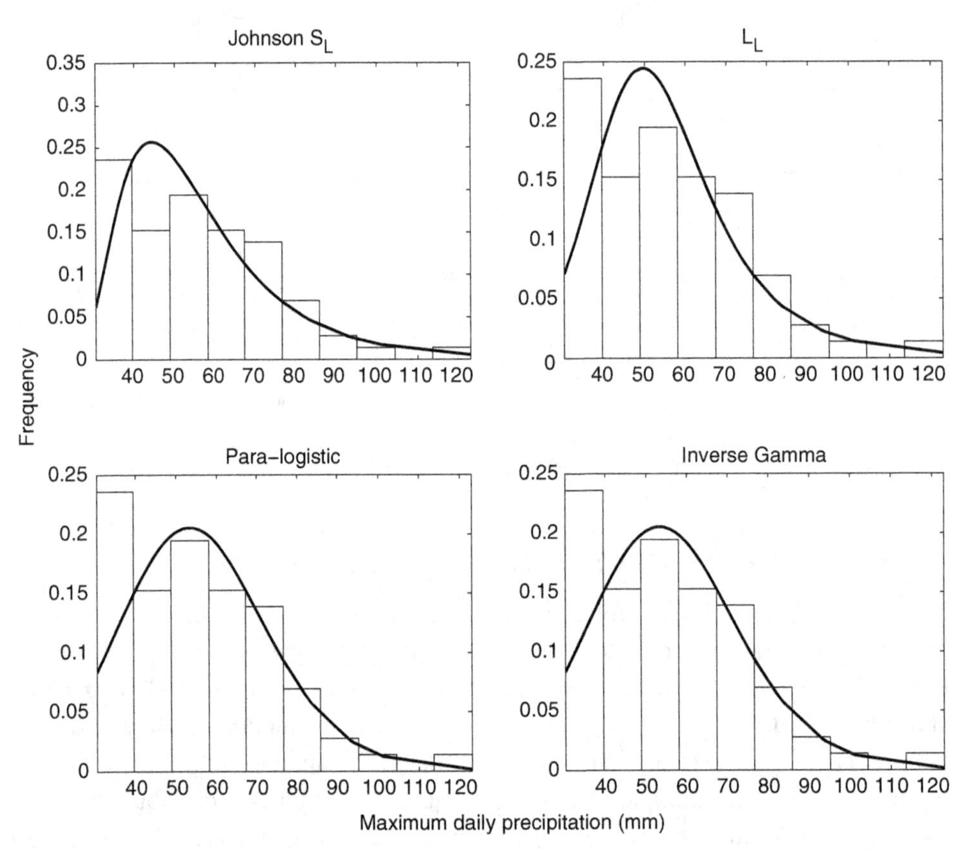

Figure 11.20 Comparison of fitted frequency distributions to empirical frequency distribution for maximum daily precipitation.

transformation can be applied to a basic probability distribution, which results in a transformed distribution. In this manner, a number of distributions have been derived. With the proper transformation, these distributions may be useful in hydrologic, hydraulic, environmental, and water resources engineering.

Table 11.4 *Parameters estimated for monthly sediment*

Distributions	Parameters a	b	c
Johnson S_L	–37.74	5.53	23.10
Log-logistic	–70.62	10.31	
Paralogistic	6.4	1.34×10^3	

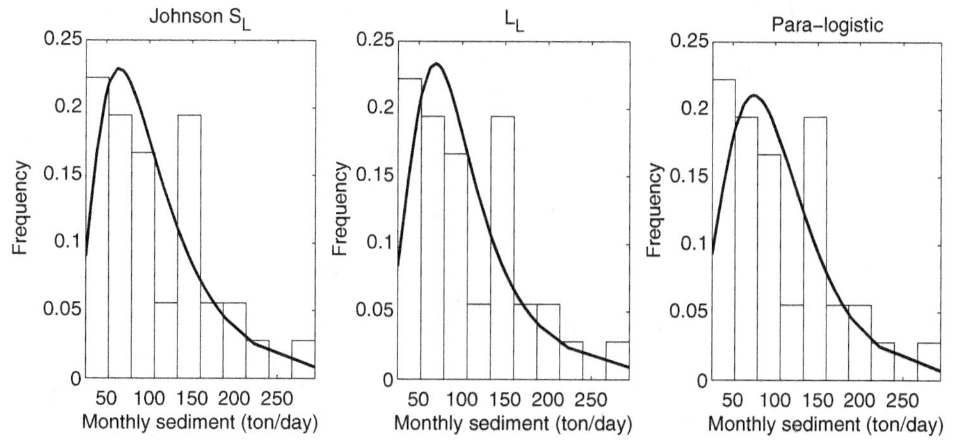

Figure 11.21 Comparison of fitted frequency distributions to empirical frequency distribution for monthly sediment.

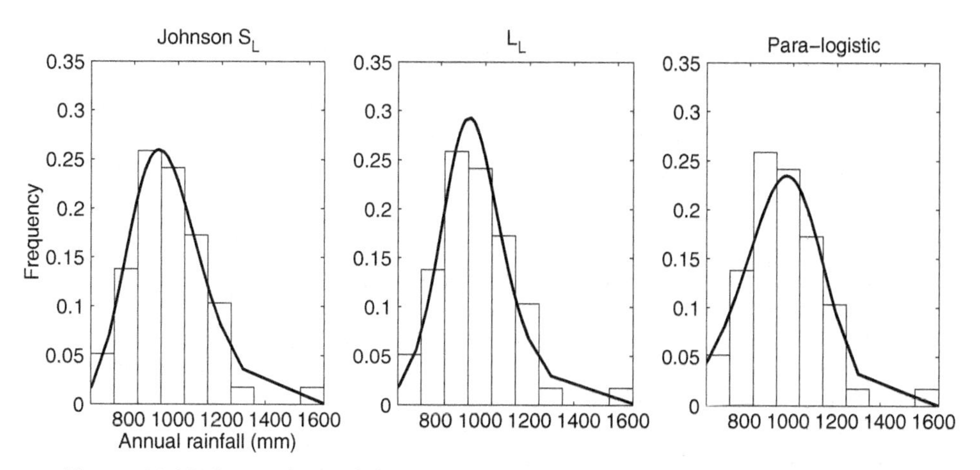

Figure 11.22 Comparison of fitted frequency distributions to empirical frequency distribution for annual rainfall.

References

Box, G. E. P., and Cox, G. P. (1964). An analysis of transformation. *Journal of the Royal Statistical Society* B26, pp. 211–252.

Box, G. E. P., and Tiao, G. C. (1973). *Bayesian Inference in Statistical Analysis.* New York: Addison Wesley Publishing Company, pp. 156–160.

Chander, S., Spolia, S. K., and Kumar, A. (1978). Flood frequency analysis by power transformation. *Journal of the Hydraulics Division, ASCE* 104, no. HY1, pp. 1495–1504.

D'Addario, R. (1936). Le transformate Euleriane. *Annali dell'Instituto di Statistica*, Vol. 8, p. 67. Bari, Italy: Universita di Bari.

Edgeworth, F. Y. (1898). On the representation of statistics by mathematical formulae. *Journal of the Royal Society* 1, pp. 670–700.

Frechet, M. (1939). Sur les formules de repartition des renenus. *Revue de l'Institut International de Statistique* 7, pp. 32–38.

Gupta, D. K., Asthana, B. N., and Bhargawa, A. N. (1989). Flood frequency analysis by two-step power transformation. *IE(I) Journal-CV* 70, pp. 127–131.

Hirsch, R. M. (1979). Synthetic hydrology and water supply reliability. *Water Resources Research* 15, pp. 1603–1615.

Hirsch, R. M. (1982). A comparison of four streamflow record extension techniques. *Water Resources Research* 18, pp. 1081–1088.

Jain, D., and Singh, V. P. (1986). A comparison of transformation methods for flood frequency analysis. *Water Resources Bulletin* 22, no. 6, pp. 903–912.

Johnson, N. L. (1949). Systems of frequency curves generated by methods of translation. *Biometrika* 36, pp. 147–176.

Johnson, N. L. (1954). Systems of frequency curves derived from the first law of Laplace. *Trabajos de Estadistica* 5, pp. 283–291.

Johnson, N. L., and Kotz, S. (1970). *Continuous Univariate Distributions (1).* New York: John Wiley & Sons.

Johnson, N. L., and Tadikamalla, P. R. (1992). Translated families of distributions. In *Handbook of the Logistic Distribution*, edited by N. Balakrishnan, pp. 189–208. New York: Marcel Dekker.

Kapteyn, J. C. (1903). *Skew Frequency Curves in Biology and Statistics.* Groningen: Noordhoff.

Kloek, T., and van Dijk, N. K. (1977). Efficient estimation of income distribution parameters. *Journal of Econometrics* 8, pp. 61–74.

Mahmoud, M. R., and Abe El Ghafour, A. S. (2013). Shannon entropy for the generalized Feller-Pareto (GFP) family and order statistics of GFP subfamilies. *Applied Mathematical Sciences* 7, no. 65, pp. 3247–3253.

Majumder, A., and Chakravarty, S. R. (1990). Distribution of personal income: Development of a new model and its application to U.S. data. *Journal of Applied Economics* 5, pp. 189–196.

McDonald, J. B. (1984). Some generalized functions for the size distribution of income. *Econometrics* 52, no. 3, pp. 647–663.

Moog, D. B., Whiting, P. J., and Thomas, R. B. (1999). Streamflow record extension using power transformation and application to sediment transport. *Water Resources Research* 35, no. 1, pp. 243–254.

Rasheed, H. R., Ramamoorthy, M. V., and Aldabbagh, A. S. (1982). Modified SMEMAX transformation for frequency analysis. *Water Resources Bulletin* 18, no. 3, pp. 509–512.

Rietz, H. L. (1922). Frequency distributions obtained by certain transformations of normally distributed variates. *Annals of Mathematics* 23, no. 4, pp. 292–300.
Snyder, W. M., Mills, W. C., and Knisel, W. G. (1978). A unifying set of probability transforms for stochastic hydrology. *Water Resources Bulletin* 14, no. 1, pp. 83–98.
Tadikamalla, P. R., and Johnson, N. L. (1982). Systems of frequency curves generated by transformations of logistic variables. *Biometrika* 69, pp. 461–465.
Tahmasebi, S., and Behboodian, J. (2010). Shannon entropy for the Feller-Pareto (FP) family and order statistics of FP subfamilies. *Applied Mathematical Sciences* 4, no. 10, pp. 495–504.
Tiao, G. C., and Lund, D. R. (1970). The use of OLUMY estimates in inference robustness studies of the location parameter of a class of symmetric distribution. *Journal of the American Statistical Association* 65, pp. 370–386.
van Uven, M. J. (1917). Logarithmic frequency distributions. *Koninklijke Akademie van Wetenschappen te Amsterdam Proceedings* 19, no. 4, pp. 670–694.
Venugopal, K. (1980). Flood analysis by SMEMAX transformation – a review. *Journal of the Hydraulics Division, ASCE* 106, no. HY2, pp. 338–340.
Wilson, B. G., Adams, B. J., and Karney, B. W. (1980). Bias in log-transformed frequency distributions. *Journal of Hydrology* 118, pp. 19–37.

12

Genetic Theory of Frequency

12.1 Basic Concept of Elementary Errors

Before we introduce the genetic theory of frequency, we will first briefly state its foundation, i.e., the hypothesis of elementary errors: the hypothesis of elementary errors vies an error E in a measurement process as the sum of a large number of component elementary errors, $e_1, e_2, \ldots, e_n$ as:

$$y = x_1 + x_2 + \cdots + x_n \tag{12.1}$$

in this equation x_i comes from different, independent sources (S_i) with its own probability characteristics, and it is negligible in comparison to the sum y. In addition, as shown in Wicksell (1917), the probability of x_i falling in $\left[x_i - \frac{1}{2}dx_i, x_i + \frac{1}{2}dx_i\right]$ is given as $f_i(x_i)dx_i$ in which $f_i(x_i)$ denotes the probability density of x_i of the independent source S_i.

12.2 General Discussion of Charlier Type A and B Curves

Based on the hypothesis of elementary errors, the genetic approach was first originated with Charlier (1905) and was extended by Wicksell (1917) with the density of y in Equation (12.1) given as:

$$f(y) = \int\int \cdots \int f_1(x_1).\, f_2(x_2) \ldots f_n(x_n) dx_1 dx_2 \ldots dx_n \tag{12.2}$$

The Charlier system is comprised of two systems; A-type and B-type. Charlier A- and B-type curves are the analytical expressions besides commonly known Gaussian curve of error and Pearson's generalized probability curves in mathematical statistics (Carver, 1921).

Charlier type-A curve is the solution expressed in terms of the normal function that represents the nearly symmetric variates very well. In general, the Charlier type A curve may be approximated as:

$$f(x) = A_0 g(x) + A_1 g'(x) + A_2 g''(x) + \cdots \tag{12.3}$$

In Equation (12.3), $g(x)$ is the auxiliary function fulfilling the assumptions as follows:

(1) $g(x)$ is continuous and differentiable;
(2) $g(x)$ and its derivatives approach to 0 for the lower and upper limit that x may take.

And A_0, A_1, A_2, ... are the constants that need to be estimated.

From Equation (12.2), the moments about origin for the random variable X may be given as:

$$\begin{aligned}
&\int_{-\infty}^{\infty} f(x)dx = A_0 \int_{-\infty}^{\infty} g(x)dx \\
&\int_{-\infty}^{\infty} xf(x)dx = A_0 \int_{-\infty}^{\infty} xg(x)dx + A_1 \int_{-\infty}^{\infty} xg'(x)dx \\
&\int_{-\infty}^{\infty} x^2 f(x)dx = A_0 \int_{-\infty}^{\infty} x^2 g(x)dx + A_1 \int_{-\infty}^{\infty} x^2 g'(x)dx + A_2 \int_{-\infty}^{\infty} x^2 g''(x)dx \\
&\qquad \cdots\cdots
\end{aligned} \tag{12.4}$$

In Equation (12.2), we may further evaluate the integral of $\int_{-\infty}^{\infty} x^m g^{(n)}(x)dx$ as:

$$\int_{-\infty}^{\infty} x^m g^{(n)}(x)dx = \begin{cases} (-1)^n \dfrac{m!}{(m-n)!} \displaystyle\int_{-\infty}^{\infty} x^{m-n} g(x)dx; & m \geq n \\ 0; & m < n \end{cases} \tag{12.4a}$$

12.3 Charlier Type A Curve

Following Charlier (1905) and Wicksell (1917) and set

$$X_i(\omega) = \int_{-\infty}^{\infty} f_i(x) e^{i\omega x} dx \tag{12.5}$$

We have the Charlier type A curve as:

$$f(y) = A_0 \varphi(y) + A_3 \varphi^{(3)}(y) + A_4 \varphi^{(4)}(y) + \cdots \tag{12.6}$$

where $\varphi(y) = \frac{1}{\sqrt{2\pi}\sigma} \exp\left(-\frac{(x-m)^2}{2\sigma^2}\right)$, $\varphi^{(i)}(y) = \frac{d^i}{y^i}\varphi(y)$. In other words, the auxiliary function for type A is normal density function.

Charlier (1905) listed the first several coefficients through the moments as:

$$\begin{cases} A_0 = \mu_0 = 1 \\ A_3 = -\dfrac{\mu_3}{3!} \\ A_4 = \dfrac{\mu_4 - 3\sigma^4}{4!} \\ A_5 = \dfrac{-\mu_5 + 10\sigma^2\mu_3}{5!} \\ A_6 = \dfrac{\mu_6 - 15\sigma^2\mu_4 + 15\sigma^6}{6!} \end{cases} \tag{12.7}$$

12.4 Charlier Type B Curve

As discussed in Wicksell (1917), Equation (12.5) is rewritten for type B curve as:

$$X_i(\omega) = b_1 + b_2 \exp\left(ia_1^{(i)}\omega\right) + b_3 \exp\left(ia_2^{(i)}\omega\right) \tag{12.8}$$

In Equation (12.8), $b_1 + b_2 + b_3 = 1$, b_2, b_3 are in the order of 1/n. And the density function of y is then solved as:

$$f(y) = \psi(y) + B_1\Delta_1\psi(y) + B_2\Delta_2\psi(y) + \cdots \tag{12.9}$$

$$\psi(y) = \frac{e^{-\lambda}}{\pi}\int_0^{\pi} e^{\lambda\omega}\cos(\alpha\sin\omega - y\omega)d\omega \tag{12.9a}$$

In Equations (12.9)–(12.9a), λ, α, B_1, B_2, ... are the parameters. Δ in Equation (12.9) denotes the backward difference. As examples:

$$\Delta_1\psi(y_i) = \psi(y_i) - \psi(y_{i-1}),\ \Delta_2\psi(y_i) = \psi(y_i) - 2\psi(y_{i-1}) + \psi(y_{i-2}) \tag{12.9b}$$

As stated in Wicksell (1917), the coefficients B is computed independently from the density of x_i and can be determined from the observations.

12.5 Extensions by Wicksell

Rather than assigning the variate as the sum of elementary errors, Wicksell (1917) proposed the generalized hypothesis of elementary errors with the basic assumptions as follows:

(i) Sources were taken in the order of time of action as: $S_1, S_2, \ldots, S_n$;
(ii) The elementary errors $X = \{x_1, x_2, \ldots, x_n\}$ are treated as impulses with their impacts proportional to the strengths, in addition, the elementary errors are very small in the order of $1/\sqrt{n}$;

(iii) The probability of $x_i \in \left[x - \frac{1}{2}dx, x + \frac{1}{2}dx\right]$ is given in the same fashion of the classic hypothesis as: $f_i(x)dx$.
(iv) Dependent on the size of time of action;
(v) When S_i occurred, the summation of the effects (y_i, i.e., impulse of elementary error and its impact) reaches the value y_{i-1} with the impact of S_i proportion to a function $\mathcal{G}(y_{i-1})$ that may be written as:

$$y_i = y_{i-1} + x_i\mathcal{G}(y_{i-1}). \tag{12.10}$$

Equation (12.10) can be rewritten as:

$$y = y_0 + x_1\mathcal{G}(x_0) + x_2\mathcal{G}(x_1) = x_3\mathcal{G}(y_2) + \cdots + x_s\mathcal{G}(y_{n-1}) \tag{12.10a}$$

Expressing the summation of error impulse as:

$$\mathcal{A}(y) = x_1 + x_2 + \cdots + x_n \tag{12.11}$$

From Equations (12.10)–(12.11), we obtain

$$\begin{cases} \mathcal{A}(y_i) - \mathcal{A}(y_{i-1}) = x_i \\ y_i - y_{i-1} = x_i\mathcal{G}(y_{i-1}) \\ \dfrac{\mathcal{A}(y_i) - \mathcal{A}(y_{i-1})}{y_i - y_{i-1}} = \dfrac{1}{\mathcal{G}(y_{i-1})} \end{cases} \tag{12.12}$$

In Equation (12.12) with the assumption of x_i being very small, the third equation is rewritten as:

$$\frac{d\mathcal{A}(y)}{dy} = \frac{1}{\mathcal{G}(y)} \tag{12.13}$$

Now, we can express the density function of $\mathcal{A}(y)$ with Type A and Type B curves as:

Type A curve:

$$f(\mathcal{A}(y)) = \varphi(\mathcal{A}(y)) + A_3\frac{d^3\varphi}{d\mathcal{A}(y)^3} + A_4\frac{d^4\varphi}{d\mathcal{A}(y)^4} + \cdots \tag{12.14}$$

where:

$$\varphi(\mathcal{A}(y)) = \frac{1}{\sqrt{2\pi}\sigma}\exp\left(-\frac{(\mathcal{A}(y) - m)^2}{2\sigma^2}\right) \tag{12.14a}$$

Type B curve:

$$f(\mathcal{A}(y)) = \psi(\mathcal{A}(y)) + B_1\Delta_1\psi(\mathcal{A}(y)) + B_2\Delta_2\psi(\mathcal{A}(y)) + \cdots \tag{12.15}$$

where:

$$\psi(\mathcal{A}(y)) = \frac{e^{-\lambda}}{\pi}\int_0^{\pi} e^{\lambda\omega}\cos\left(\alpha\sin\omega - \mathcal{A}(y)\omega\right)d\omega \tag{12.15a}$$

To this end, based on the probability transformation we obtain the density function of variable y as:

$$f(y) = f(\mathcal{A}(y))\frac{d\mathcal{A}(y)}{dy} = \frac{1}{\mathcal{G}(y)}f(\mathcal{A}(y)) \tag{12.16}$$

Substituting Equation (12.16) into Equations (12.14)–(12.15), we have:

Type A curve:

$$f(y) = \left|\frac{1}{\mathcal{G}(y)}\right|\left(\varphi(\mathcal{A}(y)) + A_3\frac{d^3\varphi}{d\mathcal{A}(y)^3} + A_4\frac{d^4\varphi}{d\mathcal{A}(y)^4} + \cdots\right) \tag{12.17}$$

Type B curve:

$$f(y) = \left|\frac{1}{\mathcal{G}(y)}\right|\left(\psi(\mathcal{A}(y)) + B_1\Delta_1\psi(\mathcal{A}(y)) + B_2\Delta_2\psi(\mathcal{A}(y)) + \cdots\right) \tag{12.18}$$

In Equation (12.17), given that $\mathcal{A}(y)$ being normally distributed, Equation (12.17) may then be simplified as:

$$f(y) = \frac{1}{\mathcal{G}(y)}\varphi(\mathcal{A}(y)) = \mathcal{A}(y)'\varphi(\mathcal{A}(y)) \tag{12.19}$$

Table 12.1 lists the $\mathcal{G}(y)$, $\mathcal{A}(y)$ functions proposed by Wicksell (1917) to obtain the Type A curve.

Example 12.1 Express the probability density function of Type A curve using No. 1 $\mathcal{G}(y)$ and $\mathcal{A}(y)$ listed in Table 12.1

Solution: for Type A curve proposed by Wicksell (1917), we have:

$$\varphi(\mathcal{A}(y)) = \frac{1}{\sqrt{2\pi}\sigma}\exp\left(-\frac{(\mathcal{A}(y) - m)^2}{2\sigma^2}\right) = \frac{1}{\sqrt{2\pi}\sigma}\left(-\frac{(\ln y - m)^2}{2\sigma^2}\right) \tag{12.20}$$

Table 12.1 *Lists of $\mathcal{G}(y), \mathcal{A}(y)$ functions*

No.	$\mathcal{G}(y)$	$\mathcal{A}(y)$
1	$\mathcal{G}(y) = y$	$\mathcal{A}(y) = \ln y$
2	$\mathcal{G}(y) = a(y + by^2)$	$\mathcal{A}(y) = \ln\left(\frac{y}{by+1}\right)$
3	$\mathcal{G}(y) = -y^2$	$\mathcal{A}(y) = \frac{1}{y}$
4	$\mathcal{G}(y) = \frac{1}{y}$	$\mathcal{A}(y) = y^2$
5	$\mathcal{G}(y) = ye^{-ay}$	$\mathcal{A}(y) = \int \frac{e^{az}}{z} dz$ $\mathcal{A}(y) = \ln y + \frac{a^2}{2!}y^2 + \cdots$, if a is small
6	$\mathcal{G}(y) = \frac{1}{y+b}$	$\mathcal{A}(y) = y^2 + 2by$
7	$\mathcal{G}(y) = e^{-ay}$	$\mathcal{A}(y) = e^{ay}$
8	$\mathcal{G}(y) = \sqrt{(p-y)(q+y)}$	$\mathcal{A}(y) = \sin^{-1}\left(\frac{2y-p+q}{p+q}\right)$
9	$\mathcal{G}(y) = e^{ay^2}$	$\mathcal{A}(y) = \int e^{-az^2} dz$

$$
\begin{aligned}
\frac{d^3\varphi}{d\mathcal{A}(y)^3} &= \varphi(\mathcal{A}(y))\left(\frac{3(\mathcal{A}(y) - m)}{\sigma^4} - \frac{(\mathcal{A}(y) - m)^3}{\sigma^6}\right) \\
&= \varphi(\mathcal{A}(y))\left(\frac{3(\ln y - m)}{\sigma^4} - \frac{(\ln y - m)^3}{\sigma^6}\right)
\end{aligned} \tag{12.21a}
$$

$$
\begin{aligned}
\frac{d^4\varphi}{d\mathcal{A}(y)^4} &= \varphi(\mathcal{A}(y))\left(\frac{3}{\sigma^4} - \frac{6(\mathcal{A}(y) - m)^2}{\sigma^6} + \frac{(\mathcal{A}(y) - m)^4}{\sigma^8}\right) \\
&= \varphi(\mathcal{A}(y))\left(\frac{\mathcal{A}(y) - m}{\sigma^4} - \frac{\sqrt{3}}{\sigma^2}\right)^2 = \varphi(\mathcal{A}(y))\left(\frac{(\ln y - m)}{\sigma^4} - \frac{\sqrt{3}}{\sigma^2}\right)^2
\end{aligned} \tag{12.21b}
$$

Substituting Equations (12.20)–(12.21b) into Equation (12.17), we have:

$$
\begin{aligned}
f(y) = \frac{1}{y\sqrt{2\pi}\sigma} \exp\left(-\frac{(\ln y - m)^2}{2\sigma^2}\right)&\left(1 - \frac{\mu_3}{3!}\left(\frac{3(\ln y - m)}{\sigma^4} - \frac{(\ln y - m)^3}{\sigma^6}\right)\right. \\
&\left. + \frac{\mu_4 - 3\sigma^4}{4!}\left(\frac{\ln y - m}{\sigma^4} - \frac{\sqrt{3}}{\sigma^2}\right)^2 + \ldots\right)
\end{aligned} \tag{12.22}
$$

In Equation (12.22), m, σ, μ_3, μ_4 represent the first four moments of variable $\ln y$. Furthermore, Equation (12.22) converge to log-normal distribution if the higher order terms may be ignored as:

$$f(y) = \frac{1}{y\sqrt{2\pi}\sigma} \exp\left(-\frac{(\ln y - m)^2}{2\sigma^2}\right) \tag{12.23}$$

Example 12.2 Express the probability density function of Type A curve using No. 4 $\mathcal{G}(y)$ and $\mathcal{A}(y)$ listed in Table 12.1

Solution: Applying No. 4 functions, we have:

$$\varphi(\mathcal{A}(y)) = \frac{1}{\sqrt{2\pi}\sigma} \exp\left(-\frac{(\mathcal{A}(y) - m)^2}{2\sigma^2}\right) = \frac{1}{\sqrt{2\pi}\sigma}\left(-\frac{(y^2 - m)^2}{2\sigma^2}\right) \tag{12.24}$$

$$\begin{aligned} \frac{d^3\varphi}{d\mathcal{A}(y)^3} &= \varphi(\mathcal{A}(y))\left(\frac{3(\mathcal{A}(y) - m)}{\sigma^4} - \frac{(\mathcal{A}(y) - m)^3}{\sigma^6}\right) \\ &= \varphi(\mathcal{A}(y))\left(\frac{3(y^2 - m)}{\sigma^4} - \frac{(y^2 - m)^3}{\sigma^6}\right) \end{aligned} \tag{12.25a}$$

$$\begin{aligned} \frac{d^4\varphi}{d\mathcal{A}(y)^4} &= \varphi(\mathcal{A}(y))\left(\frac{3}{\sigma^4} - \frac{6(\mathcal{A}(y) - m)^2}{\sigma^6} + \frac{(\mathcal{A}(y) - m)^4}{\sigma^8}\right) \\ &= \varphi(\mathcal{A}(y))\left(\frac{\mathcal{A}(y) - m}{\sigma^4} - \frac{\sqrt{3}}{\sigma^2}\right)^2 = \varphi(\mathcal{A}(y))\left(\frac{(y^2 - m)}{\sigma^4} - \frac{\sqrt{3}}{\sigma^2}\right)^2 \end{aligned} \tag{12.25b}$$

Substituting Equations (12.24)–(12.25b) into Equation (12.17), we have:

$$\begin{aligned} f(y) = \frac{|y|}{\sqrt{2\pi}\sigma} \exp\left(-\frac{(y^2 - m)^2}{2\sigma^2}\right)\left(1 - \frac{\mu_3}{3!}\left(\frac{3(y^2 - m)}{\sigma^4} - \frac{(y^2 - m)^3}{\sigma^6}\right)\right. \\ \left. + \frac{\mu_4 - 3\sigma^4}{4!}\left(\frac{y^2 - m}{\sigma^4} - \frac{\sqrt{3}}{\sigma^2}\right)^2 + \ldots\right) \end{aligned} \tag{12.26}$$

Ignoring the higher-order terms, Equation (12.26) may be simplified as:

$$f(y) = \frac{|y|}{\sqrt{2\pi}\sigma} \exp\left(-\frac{(y^2 - m)^2}{2\sigma^2}\right) \tag{12.27}$$

Example 12.3 Express the probability density function of Type A curve using No. 6 $\mathcal{G}(y)$ and $\mathcal{A}(y)$ listed in Table 12.1

Solution: Applying No. 6 functions, we have:

$$\varphi(\mathcal{A}(y)) = \frac{1}{\sqrt{2\pi}\sigma} \exp\left(-\frac{(\mathcal{A}(y) - m)^2}{2\sigma^2}\right) = \frac{1}{\sqrt{2\pi}\sigma}\left(-\frac{((y^2 + 2by) - m)^2}{2\sigma^2}\right) \tag{12.28}$$

$$\begin{aligned} \frac{d^3\varphi}{d\mathcal{A}(y)^3} &= \varphi(\mathcal{A}(y))\left(\frac{3(\mathcal{A}(y) - m)}{\sigma^4} - \frac{(\mathcal{A}(y) - m)^3}{\sigma^6}\right) \\ &= \varphi(\mathcal{A}(y))\left(\frac{3((y^2 + 2by) - m)}{\sigma^4} - \frac{((y^2 + 2by) - m)^3}{\sigma^6}\right) \end{aligned} \tag{12.29a}$$

$$\begin{aligned} \frac{d^4\varphi}{d\mathcal{A}(y)^4} &= \varphi(\mathcal{A}(y))\left(\frac{3}{\sigma^4} - \frac{6(\mathcal{A}(y) - m)^2}{\sigma^6} + \frac{(\mathcal{A}(y) - m)^4}{\sigma^8}\right) \\ &= \varphi(\mathcal{A}(y))\left(\frac{\mathcal{A}(y) - m}{\sigma^4} - \frac{\sqrt{3}}{\sigma^2}\right)^2 = \varphi(\mathcal{A}(y))\left(\frac{((y^2 + 2by) - m)}{\sigma^4} - \frac{\sqrt{3}}{\sigma^2}\right)^2 \end{aligned} \tag{12.29b}$$

Substituting Equations (12.27)–(12.29b) into Equation (12.17), we have:

$$\begin{aligned} f(y) = \frac{|y + b|}{\sqrt{2\pi}\sigma} &\exp\left(-\frac{((y^2 + 2by) - m)^2}{2\sigma^2}\right) \\ &\left(1 - \frac{\mu_3}{3!}\left(\frac{3((y^2 + 2by) - m)}{\sigma^4} - \frac{((y^2 + 2by) - m)^3}{\sigma^6}\right)\right. \\ &\left. + \frac{\mu_4 - 3\sigma^4}{4!}\left(\frac{(y^2 + 2by) - m}{\sigma^4} - \frac{\sqrt{3}}{\sigma^2}\right)^2 + \ldots\right) \end{aligned} \tag{12.30}$$

Ignoring the higher-order terms, Equation (12.30) may be simplified as:

$$f(y) = \frac{|y + b|}{\sqrt{2\pi}\sigma} \exp\left(-\frac{((y^2 + 2by) - m)^2}{2\sigma^2}\right) \tag{12.31}$$

References

Carver, H. C. (1921). The mathematical representation of frequency distributions. *American Statistical Association* 17, no. 134. pp. 720–731.

Charlier, C. V. L. (1905). Über die Darstellung willkürlicher Funktionen. Arkiv För Matematik, Astronomi Och Fysik. *Band* 2, no. 20, pp. 23–35.

Wicksell, S. D. (1917). On the genetic theory of frequency. Arkiv För Matematik, Astronomi Och Fysik. *Band* 12, no. 20, pp. 1–56.

Appendix

Datasets for Applications

Peak Flow Dataset (USGS 04208000)

Time	Flow (cfs)	Time	Flow (cfs)	Time	Flow (cfs)
4/15/22	6910	1/22/59	24800	5/26/89	14300
1/21/23	6810	12/13/59	7860	9/7/90	12200
12/1/27	10600	4/26/61	11800	12/30/90	15000
1/19/29	12300	2/27/62	5960	7/31/92	8780
1/12/30	7900	3/13/63	6830	12/31/92	10800
4/26/31	3810	3/5/64	12300	8/13/94	12900
3/22/32	5860	1/24/65	6830	1/16/95	10500
12/31/32	7900	2/11/66	7720	4/23/96	9680
4/4/34	6760	5/11/67	6370	6/1/97	13500
2/26/35	5990	7/17/68	8630	4/17/98	9450
2/27/36	8600	7/20/69	13600	1/24/99	6430
4/20/40	12200	1/29/70	6570	4/8/00	7730
12/29/40	6620	2/23/71	7160	12/17/00	7910
3/16/42	8970	4/15/72	8750	5/13/02	7180
12/30/42	8600	3/15/73	7180	7/22/03	14200
4/11/44	7750	1/19/74	8700	5/22/04	15000
5/18/45	9240	5/22/75	12200	1/12/05	11900
2/27/46	5430	2/17/76	14000	6/23/06	25400
6/2/47	12700	2/24/77	7970	1/6/07	12400
3/22/48	8700	12/14/77	12900	2/6/08	12300
1/28/49	5500	9/15/79	19700	3/9/09	11500
1/16/50	11000	12/25/79	9120	3/14/10	5350
12/8/50	7990	2/19/81	10600	2/28/11	19400
1/27/52	11300	12/23/81	6960	12/6/11	10600
1/18/53	4480	7/2/83	10400	10/30/12	13800
3/25/54	9360	3/21/84	9090	12/22/13	10500
10/16/54	14300	3/29/85	9080	6/27/15	11900
2/25/56	8890	11/5/85	8630	12/27/15	8420
5/20/57	9000	7/2/87	10500	1/12/17	10700
8/8/58	6730	3/26/88	6190		

Annual Rainfall Amount Dataset (U330058)

Year	Amount (mm)	Year	Amount (mm)
1953	753.618	1982	880.872
1954	954.278	1983	975.106
1955	796.29	1984	892.302
1956	1195.324	1985	980.694
1957	816.356	1986	800.354
1958	949.452	1987	769.874
1959	1149.096	1988	843.788
1960	687.832	1989	998.728
1961	872.49	1990	1668.78
1962	736.092	1991	611.632
1963	604.266	1992	1146.556
1964	1036.828	1993	1047.242
1965	852.17	1994	1028.954
1966	752.856	1995	905.51
1967	809.752	1996	1191.768
1968	900.938	1997	820.166
1969	837.184	1998	1023.112
1970	971.042	1999	909.574
1971	817.88	2000	1155.7
1972	1114.552	2001	834.644
1973	934.466	2002	1031.748
1974	1072.134	2003	1297.94
1975	1041.146	2004	1176.02
1976	905.002	2005	1045.718
1977	1014.984	2006	1115.822
1978	847.344	2007	1038.606
1979	998.474	2008	1066.8
1980	969.264	2009	904.494
1981	1020.318	2010	961.39

Monthly Discharge (USGS 09239500, for Month of June)

Year	Flow (cfs)	Year	Flow (cfs)	Year	Flow (cfs)
1910	996.1	1949	2296	1988	1577
1911	1699	1950	1715	1989	663.8
1912	2946	1951	2021	1990	1175
1913	890.4	1952	2841	1991	1815
1914	2347	1953	2167	1992	536.7
1915	1288	1954	323.1	1993	2072
1916	2014	1955	1065	1994	545.3
1917	3771	1956	1206	1995	2583
1918	2513	1957	3697	1996	2471
1919	773.1	1958	1708	1997	3222
1920	3115	1959	1748	1998	1580
1921	3582	1960	1546	1999	1893
1922	1484	1961	1108	2000	1241
1923	2570	1962	2028	2001	800
1924	1884	1963	689.4	2002	304.9
1925	1073	1964	1643	2003	1677
1926	1459	1965	2535	2004	702.7
1927	2240	1966	439.3	2005	1703
1928	2051	1967	1604	2006	1389
1929	3047	1968	2683	2007	583.2
1930	1549	1969	1008	2008	2655
1931	1092	1970	2509	2009	1620
1932	2155	1971	2715	2010	2036
1933	2709	1972	1732	2011	4005
1934	141.3	1973	2202	2012	180.5
1935	2050	1974	2282	2013	1070
1936	1421	1975	2557	2014	2404
1937	976.1	1976	1406	2015	1551
1938	2036	1977	374.1	2016	2096
1939	962	1978	3058	2017	1493
1940	1024	1979	1940	2018	674.7
1941	1171	1980	1971		
1942	1700	1981	860.4		
1943	1505	1982	2952		
1944	1634	1983	3550		
1945	1890	1984	3424		
1946	1320	1985	1856		
1947	1757	1986	1964		
1948	1268	1987	461		

TPN Sampling Data (Monthly)

Year	Month	TPN (mg/L)	Year	Month	TPN (mg/L)	Year	Month	TPN (mg/L)	Year	Month	TPN (mg/L)
1994	10	0.116	2000	1	0.271	2005	4	0.2	2010	7	0.078
1994	11	0.312	2000	2	0.481	2005	5	0.094	2010	8	0.083
1994	12	0.287	2000	3	0.195	2005	6	0.087	2010	9	0.109
1995	1	0.285	2000	4	0.141	2005	7	0.13	2010	10	0.17
1995	2	0.21	2000	5	0.146	2005	8	0.13	2010	11	0.162
1995	3	0.212	2000	6	0.061	2005	9	0.13	2010	12	0.267
1995	4	0.141	2000	7	0.049	2005	10	0.19	2011	1	0.217
1995	5	0.081	2000	8	0.082	2005	11	0.329	2011	2	0.161
1995	6	0.086	2000	9	0.082	2005	12	0.22	2011	3	0.224
1995	7	0.066	2000	10	0.148	2006	1	0.263	2011	4	0.154
1995	8	0.117	2000	11	0.149	2006	2	0.24	2011	5	0.095
1995	9	0.125	2000	12	0.229	2006	3	0.253	2011	6	0.042
1995	10	0.253	2001	1	0.229	2006	4	0.2	2011	7	0.04
1995	11	0.178	2001	2	0.224	2006	5	0.1	2011	8	0.05
1995	12	0.203	2001	3	0.257	2006	6	0.052	2011	9	0.149
1996	1	0.223	2001	4	0.187	2006	7	0.067	2011	10	0.188
1996	2	0.155	2001	5	0.11	2006	8	0.094	2011	11	0.233
1996	3	0.119	2001	6	0.074	2006	9	0.11	2011	12	0.255
1996	4	0.114	2001	7	0.089	2006	10	0.308	2012	1	0.356
1996	5	0.09	2001	8	0.135	2006	11	0.2645	2012	2	0.191
1996	6	0.043	2001	9	0.181	2006	12	0.23	2012	3	0.233
1996	7	0.045	2001	10	0.309	2007	1	0.24	2012	4	0.123
1996	8	0.107	2001	11	0.193	2007	2	0.14	2012	5	0.115
1996	9	0.194	2001	12	0.305	2007	3	0.12	2012	6	0.094
1996	10	0.331	2002	1	0.22	2007	4	0.099	2012	7	0.041
1996	11	0.237	2002	2	0.207	2007	5	0.061	2012	8	0.061
1996	12	0.286	2002	3	0.245	2007	6	0.073	2012	9	0.081
1997	1	0.201	2002	4	0.149	2007	7	0.1	2012	10	0.21
1997	2	0.232	2002	5	0.099	2007	8	0.098	2012	11	0.227
1997	3	0.299	2002	6	0.077	2007	9	0.12	2012	12	0.42
1997	4	0.218	2002	7	0.068	2007	10	0.25	2013	1	0.337

(*cont.*)

Year	Month	TPN (mg/L)	Year	Month	TPN (mg/L)	Year	Month	TPN (mg/L)	Year	Month	TPN (mg/L)
1997	5	0.077	2002	8	0.046	2007	11	0.2	2013	2	0.244
1997	6	0.076	2002	9	0.096	2007	12	0.24	2013	3	0.175
1997	7	0.058	2002	10	0.08	2008	1	0.28	2013	4	0.171
1997	8	0.055	2002	11	0.349	2008	2	0.2	2013	5	0.067
1997	9	0.204	2002	12	0.205	2008	3	0.2	2013	6	0.058
1997	10	0.163	2003	1	0.21	2008	4	0.215	2013	7	0.084
1997	11	0.187	2003	2	0.219	2008	5	0.13	2013	8	0.226
1997	12	0.282	2003	3	0.18	2008	6	0.098	2013	9	0.177
1998	1	0.387	2003	4	0.153	2008	7	0.06			
1998	2	0.258	2003	5	0.105	2008	8	0.055			
1998	3	0.217	2003	6	0.054	2008	9	0.12			
1998	4	0.165	2003	7	0.11	2008	10	0.18			
1998	5	0.14	2003	8	0.094	2008	11	0.21			
1998	6	0.065	2003	9	0.19	2008	12	0.216			
1998	7	0.077	2003	10	0.25	2009	1	0.212			
1998	8	0.075	2003	11	0.2745	2009	2	0.19			
1998	9	0.097	2003	12	0.23	2009	3	0.214			
1998	10	0.23	2004	1	0.273	2009	4	0.227			
1998	11	0.378	2004	2	0.193	2009	5	0.091			
1998	12	0.392	2004	3	0.15	2009	6	0.04			
1999	1	0.333	2004	4	0.12	2009	7	0.083			
1999	2	0.275	2004	5	0.08	2009	8	0.067			
1999	3	0.19	2004	6	0.073	2009	9	0.159			
1999	4	0.162	2004	7	0.076	2009	10	0.327			
1999	5	0.146	2004	8	0.12	2009	11	0.163			
1999	6	0.064	2004	9	0.15	2009	12	0.279			
1999	7	0.057	2004	10	0.18	2010	1	0.276			
1999	8	0.084	2004	11	0.2	2010	2	0.182			
1999	9	0.124	2004	12	0.24	2010	3	0.161			
1999	10	0.189	2005	1	0.15	2010	4	0.14			
1999	11	0.29	2005	2	0.19	2010	5	0.076			
1999	12	0.242	2005	3	0.251	2010	6	0.049			

Monthly Suspended Sediment (Month of May)

Year	Month	Sediment (Ton/Day)
1951	5	51.4
1952	5	38.5
1956	5	223.1
1961	5	77.5
1965	5	48.2
1967	5	146.4
1968	5	117
1969	5	294
1970	5	63.5
1971	5	95
1972	5	72.6
1973	5	189.3
1974	5	179.2
1977	5	25.9
1978	5	135.4
1979	5	84.1
1980	5	53.6
1981	5	69.4
1982	5	86.3
1983	5	145.5
1984	5	190.6
1988	5	74.8
1989	5	155.5
1990	5	92.5
1991	5	62.4
1992	5	51.8
1993	5	30.1
1994	5	31.8
1995	5	113.1
1996	5	158.8
1997	5	134.1
1998	5	88.9
1999	5	28.4
2000	5	144.4
2001	5	86.8
2002	5	170.7

Maximum Daily Precipitation

Year	Amount (mm)	Year	Amount (mm)	Year	Amount (mm)	Year	Amount (mm)
1948	61.2	1966	43.7	1984	34.3	2002	55.4
1949	36.3	1967	38.9	1985	54.4	2003	101.1
1950	44.7	1968	34.8	1986	30.7	2004	58.7
1951	38.6	1969	61	1987	50.5	2005	83.6
1952	49.3	1970	72.4	1988	94	2006	58.7
1953	62	1971	60.5	1989	36.6	2007	52.6
1954	65.3	1972	68.8	1990	83.1	2008	50.5
1955	41.7	1973	39.6	1991	33.3	2009	37.6
1956	67.3	1974	51.8	1992	56.1	2010	46.5
1957	68.8	1975	76.2	1993	42.7	2011	122.9
1958	73.9	1976	74.2	1994	93.2	2012	70.4
1959	71.9	1977	47.5	1995	51.8	2013	53.6
1960	38.1	1978	33	1996	47.5	2014	63.2
1961	81.5	1979	82	1997	50.3	2015	45.2
1962	39.1	1980	54.6	1998	51.3	2016	46.7
1963	30.2	1981	58.2	1999	58.7	2017	76.5
1964	81.5	1982	37.6	2000	56.1	2018	63.5
1965	38.1	1983	40.4	2001	36.3	2019	37.3

Total Flow Deficit for Drought

Start	End	Flow Deficit (m^3)	Start	End	Flow Deficit (m^3)	Start	End	Flow Deficit (m^3)
12/1/42	6/5/43	36226.13	8/31/66	8/18/67	115888.09	5/19/90	7/18/90	35438.85
6/20/43	5/2/44	89257.53	10/9/67	1/19/68	25722.76	8/20/90	4/7/91	62934.68
5/12/44	5/27/44	6701.03	4/5/68	5/11/68	4102.27	4/19/91	5/14/91	8349.82
6/18/44	8/27/44	30994.93	6/1/68	10/12/69	190354.55	5/23/91	9/26/91	66896.85
9/16/44	3/12/45	53108.59	2/21/70	5/28/70	14418.76	10/4/91	12/20/91	28196.06
5/4/45	10/4/45	74073.67	6/11/70	6/26/71	127947.56	6/29/92	7/26/92	11113.95
10/23/45	9/17/46	66394.09	7/17/71	8/4/71	5136.82	8/10/92	11/17/95	412908.64
10/3/46	10/8/46	1760.66	2/23/72	2/28/72	1473.41	11/26/95	11/5/96	111461.41
10/22/46	5/17/47	32013.34	3/18/72	5/3/72	7173.36	1/6/97	3/20/97	6034.55
6/2/47	6/25/47	13978.79	5/24/72	7/21/73	147355.72	4/22/97	6/1/97	12743.99
7/10/47	6/30/48	119224.51	7/30/73	9/26/73	21324.40	8/6/97	8/17/98	129847.64
7/22/48	2/27/49	65311.42	2/22/74	3/14/74	3274.17	10/7/98	10/22/98	5434.44
5/9/49	6/10/49	15791.62	4/2/74	8/9/74	52345.43	2/22/99	3/20/99	2907.87
6/24/49	8/9/49	15912.83	8/21/74	11/7/74	39676.84	4/11/99	6/22/99	26399.46
8/22/49	5/11/50	70633.02	2/22/75	5/24/75	16592.87	7/10/99	8/25/99	11896.53
6/15/50	9/26/50	49324.88	7/12/75	7/23/75	3917.82	9/9/99	8/30/01	237117.48
10/6/50	6/1/51	59098.22	8/14/75	7/6/76	99856.57	9/11/01	11/16/01	48019.08
6/15/51	9/12/51	40860.71	9/2/76	9/19/76	8162.74	2/21/02	6/30/02	39717.37
9/21/51	5/26/52	66690.13	10/14/76	10/28/76	6112.25	8/10/02	9/8/02	9627.14
6/14/52	8/25/53	160117.73	6/4/77	6/8/78	120429.68	9/27/02	10/24/02	19765.48
9/18/53	10/24/53	26830.18	6/18/78	4/18/79	93867.39	2/23/03	7/13/03	50025.38
11/4/53	6/25/54	39430.60	4/30/79	6/6/79	14706.09	7/28/03	10/13/03	35121.62
7/17/54	10/2/55	147403.80	6/26/79	6/2/80	109679.74	10/30/03	2/26/04	7140.93
10/13/55	3/20/57	159036.25	6/11/80	8/10/80	29990.44	5/27/04	6/29/04	11627.72
3/28/57	4/20/57	2096.38	8/24/80	4/24/81	71404.27	8/6/04	11/16/04	41500.80
5/22/57	5/26/57	1353.58	5/18/81	5/28/81	4363.87	4/21/05	1/26/07	244468.46
6/18/57	9/22/57	46507.67	7/15/81	9/4/81	9460.94	2/3/07	1/15/10	278694.26
10/3/57	10/28/57	15189.58	9/14/81	10/20/81	24879.73	2/22/10	4/15/10	5951.48

(*cont.*)

Start	End	Flow Deficit (m^3)	Start	End	Flow Deficit (m^3)	Start	End	Flow Deficit (m^3)
11/6/57	9/25/58	65462.62	1/25/82	5/18/82	12220.78	4/29/10	10/14/13	452796.91
2/22/59	9/27/59	66267.42	5/30/82	10/20/84	332193.45	11/2/13	3/21/15	149151.72
10/23/59	7/22/60	59777.88	10/29/84	1/2/85	8384.16	3/29/15	4/16/15	1603.02
8/2/60	8/13/60	2250.61	2/21/85	3/2/85	1972.60	4/25/15	5/14/15	6995.04
9/3/60	10/23/60	35702.18	5/9/85	5/18/85	1841.52	6/18/15	8/17/16	133961.51
4/11/61	6/22/61	33269.56	6/2/85	6/13/85	7700.65	9/7/16	9/27/16	16136.00
7/2/61	7/29/61	12792.72	7/14/85	6/5/86	79838.63	10/19/16	12/5/16	11817.72
8/12/61	6/17/63	229516.29	6/20/86	10/20/86	65201.49			
6/29/63	9/14/64	160702.50	11/1/86	11/12/86	1239.70			
10/15/64	5/16/65	39045.26	2/23/87	2/26/87	1224.30			
6/4/65	4/18/66	103610.34	4/17/87	5/25/87	8331.28			
6/5/66	8/22/66	35099.05	8/10/87	3/3/90	294351.9179			

Index

Printed in the United States
By Bookmasters